CARBOHYDRATE RECOGNITION IN CELLULAR FUNCTION

The Ciba Foundation is an international scientific and educational charity. It was established in 1947 by the Swiss chemical and pharmaceutical company of CIBA Limited—now CIBA-GEIGY Limited. The Foundation operates independently in London under English trust law.

The Ciba Foundation exists to promote international cooperation in biological, medical and chemical research. It organizes about eight international multidisciplinary symposia each year on topics that seem ready for discussion by a small group of research workers. The papers and discussions are published in the Ciba Foundation symposium series. The Foundation also holds many shorter meetings (not published), organized by the Foundation itself or by outside scientific organizations. The staff always welcome suggestions for future meetings.

The Foundation's house at 41 Portland Place, London W1N 4BN, provides facilities for meetings of all kinds. Its Media Resource Service supplies information to journalists on all scientific and technological topics. The library, open seven days a week to any graduate in science or medicine, also provides information on scientific meetings throughout the world and answers general enquiries on biomedical and chemical subjects. Scientists from any part of the world may stay in the house during working visits to London.

Ciba Foundation Symposium 145

CARBOHYDRATE RECOGNITION IN CELLULAR FUNCTION

A Wiley-Interscience Publication

1989

JOHN WILEY & SONS

Chichester · New York · Brisbane · Toronto · Singapore

Published in 1989 by John Wiley & Sons Ltd, Chichester, UK.

Suggested series entry for library catalogues:
Ciba Foundation Symposia

Ciba Foundation Symposium 145
x + 294 pages, 69 figures, 24 tables

Library of Congress Cataloging-in-Publication Data

Carbohydrate recognition in cellular function.
p. cm.—(Ciba Foundation symposium ; 145)
'Symposium on Carbohydrate Recognition in Cellular Function, held at the Ciba Foundation, London, 15–17 November 1988.'
'Editors: Gregory Bock (organizer) and Sara Harnett.'
'A Wiley–Interscience publication.'
Bibliography: p.
Includes index.
ISBN 0 471 92307 9
1. Cell receptors—Congresses. 2. Carbohydrates—Congresses. 3. Cellular recognition—Congresses. I. Bock, Gregory. II. Harnett, Sara. III. Symposium on Carbohydrate Recognition in Cellular Function (1988: Ciba Foundation) IV. Series.
QH603.C43C37 1989
574.87′6—dc20 89-9153
CIP

British Library Cataloguing in Publication Data

Carbohydrate recognition in cellular function.
1. Mammals. Carbohydrates. Metabolism
I. Series
599.01′924

ISBN 0 471 92307 9

Phototypeset by Dobbie Typesetting Limited, Devon.
Printed and bound in Great Britain by Bath Press, Bath, Avon.

Contents

Symposium on Carbohydrate Recognition in Cellular Function, held at the Ciba Foundation, London, 15–17 November 1988
The topic for this symposium was proposed by Dr T. Feizi

Editors: Gregory Bock (Organizer) and Sara Harnett

Participants

S. Barondes Department of Psychiatry, University of California, San Francisco, CA 94143, USA

J. Carver Department of Medical Genetics, University of Toronto, Toronto, Ontario, Canada M5S 1A8

M. P. Czech Department of Biochemistry, University of Massachusetts Medical Center, 55 Lake Avenue North, Worcester, MA 01605, USA

K. Drickamer Department of Biochemistry & Molecular Biophysics, Columbia University, 630 W 168th Street, New York, NY 10032, USA

B. Ernst CIBA–GEIGY AG, CH-4002 Basle, Switzerland

T. Feizi Section of Glycoconjugate Research, MRC Clinical Research Centre, Watford Road, Harrow, Middlesex HA1 3UJ, UK

J. Finne Department of Medical Biochemistry, University of Turku, Kiinamyllynkatu 10, SF-20520 Turku, Finland

M. Fukuda Carbohydrate Chemistry, La Jolla Cancer Research Foundation, Cancer Research Center, 10901 North Torrey Pines Road, La Jolla, CA 92037, USA

C. Gahmberg La Jolla Cancer Research Foundation, 10901 North Torrey Pines Road, La Jolla, CA 92037, USA

G. W. Hart Department of Biological Chemistry, The Johns Hopkins University School of Medicine, 725 North Wolfe Street, Baltimore, MD 21205, USA

O. Hindsgaul Department of Chemistry, University of Alberta, Edmonton, Alberta, Canada T6G 2G2

R. C. Hughes MRC National Institute for Medical Research, The Ridgeway, Mill Hill, London NW7 1AA, UK

T. M. Jessell Center for Neurobiology & Behavior, Howard Hughes Medical Institute, College of Physicians & Surgeons, Columbia University, 722 West 168th Street, PI Research Annex, 6th Floor, NY 10032, USA

A. Kobata Department of Biochemistry, Institute of Medical Science, University of Tokyo, 4-6-1 Shirokanedai, Minato-ku, Tokyo 108, Japan

Y. C. Lee Department of Biology & McCollum-Pratt Institute, The Johns Hopkins University, 34th & Charles Streets, Baltimore, MD 21218, USA

U. Lindahl Department of Veterinary Medical Chemistry, Swedish University of Agricultural Sciences, The Biomedical Center, Box 575, S-75123 Uppsala, Sweden

T. Muramatsu Department of Biochemistry, Kagoshima University School of Medicine, 1208-1 Usuki-cho, Kagoshima 890, Japan

Y. Nagai Department of Biochemistry, Faculty of Medicine, University of Tokyo, Hongo 7-3-1, Bunkyo-ku, Tokyo 113, Japan

T. W. Rademacher Oxford Glycobiology Unit, Department of Biochemistry, University of Oxford, South Parks Road, Oxford OX1 3QU, UK

E. Ruoslahti *(Chairman)* La Jolla Cancer Research Foundation, Cancer Research Center, 10901 North Torrey Pines Road, La Jolla, CA 92037, USA

M. Schachner Department of Neurobiology, University of Heidelberg, Im Neuenheimer Feld 364, D-6900 Heidelberg 1, Federal Republic of Germany

N. Sharon Department of Biophysics, The Weizmann Institute of Science, PO Box 26, 76100 Rehovot, Israel

P. Stanley Department of Cell Biology, Albert Einstein College of Medicine, 1300 Morris Park Avenue, Bronx, NY 10461, USA

J. E. Thomas-Oates *(Ciba Foundation Bursar)* Department of Biochemistry, Imperial College of Science & Technology, Exhibition Road, London SW7 2AZ, UK

J. Wang Department of Biochemistry, Michigan State University, East Lansing, MI 48824-1319, USA

P. M. Wassarman Department of Cell & Developmental Biology, Roche Institute of Molecular Biology, Nutley, NJ 07110, USA

W. M. Watkins Division of Immunochemical Genetics, MRC Clinical Research Centre, Watford Road, Harrow, Middlesex HA1 3UJ, UK

Introduction

E. Ruoslahti

La Jolla Cancer Research Foundation, Cancer Research Center, 10901 North Torrey Pines Road, La Jolla, CA 92037, USA

The role of carbohydrate recognition in cellular function is poorly understood. The functions of DNA and proteins (DNA, of course, being a substituted carbohydrate) are generally known. It is much less clear what carbohydrates do. A number of functions have been proposed and it is possible that they all are correct, but they have not been clearly defined. When we ask what carbohydrates do, some standard replies are that bacteria bind to them, that they are immunogenic, and that they glue clams to the hulls of boats. We need a more precise answer; carbohydrates clearly do not exist to help bacteria gain entrance to the body.

In this field the study of molecular structure has progressed more rapidly than the understanding of function. I am sure we should all like to see that corrected. Nevertheless, the structural studies have been very interesting, starting with the classic work of Dr Winifred Watkins, whom we are pleased to have with us at this meeting, and progressing to some recent surprises with fascinating functional implications, such as the nuclear and cytoplasmic glycosylation.

A persuasive argument used by those who defend the functionality of carbohydrates is that they must be important because there is a tremendous variation to their structure. However, we do not fully understand the basis of that variation. It is likely to be important, but in what way? I find it puzzling that there seems to be no species conservation for the expression of a given carbohydrate group in a given place. There are some very specific, developmentally regulated expression patterns for certain carbohydrate structures, but they do not seem to be conserved from species to species. It is possible that a particular carbohydrate structure may be important only in one or very few places. Then it may be simply coincidentally expressed at many other places. This is going to be difficult to sort out. In this sense, carbohydrate expression may be similar to the alternative splicing of mRNA; there may be only one protein where alternative splicing is important to a given cell, and many

1989 Carbohydrate recognition in cellular function. Wiley, Chichester (Ciba Foundation Symposium 145) p 1–5

other proteins may follow that splicing pattern without it changing their function.

It is important to discover what carbohydrates do, because this is in some ways the last frontier in macromolecular chemistry and function. Molecular biology and the production and use of synthetic carbohydrates will be the most useful tools for investigating the functions of carbohydrates. If a given carbohydrate is found to be expressed at a certain place, one can study the result of a genetic modification that changes that carbohydrate in that place.

I would like to contribute to the discussion on the functionality of carbohydrates by describing evidence that suggests that a glycosaminoglycan component of a proteoglycan plays an important role in its function. We don't understand completely how it does this, but the result is rather striking. We have been studying decorin, a proteoglycan that fibroblasts and other mesenchymal cells secrete into their culture medium. This proteoglycan is also known as PGII and by a number of other names, including PG40, which was the name we originally gave it when we cloned it and showed that the core protein has a molecular mass of 40 000 Da. Decorin is made by IMR 90 human fibroblasts, and can also be made from a cDNA construct by Chinese hamster ovary (CHO) cells. We can reproduce the synthesis of decorin fairly accurately in the CHO cells except that the sulphate label seems to be a little more dispersed than in the IMR 90 cells, probably meaning that the glycosaminoglycan chain is shorter in some of the recombinant molecules. The core protein looks the same in the expression system. The control-transfected cells do not seem to make substantial quantities of the Chinese hamster equivalent of decorin.

We can increase the expression of decorin in the CHO cells by using an amplified expression system. We hoped that the expression of high levels of the proteoglycan might give us some indication of its function. We observed an interesting morphological change as we went from the control-transfected cells to the highest expressors. The highest expressors had a very different appearance; they were flat and occupied three times as much space on the dish as the original cells. Also, the saturation density of the decorin-transfected cells decreased in proportion to the amount of decorin that they made. The highest expressors had a saturation density of only about 40% of that of the original cells (Yamaguchi & Ruoslahti 1988).

Because these cells showed differences in adhesion and morphology, we thought of fibronectin as a possible explanation for the changes. Indeed, there is a quantitative change in fibronectin. The original CHO cells are semi-transformed and have little fibronectin in their matrix, whereas the decorin-transfected cells have more fibronectin, which is consistent with their being flatter and more spread out.

Yu Yamaguchi then deleted the glycosaminoglycan chain in this proteoglycan by mutating the serine residue which carries it. The core protein alone was expressed, being changed at only one place—serine to threonine. The CHO cells that expressed the core protein had a different morphology again and the other features obtained with the expression of the intact proteoglycan were not seen.

We seem to have found two things. First, this proteoglycan appears to control what classically has been called the contact inhibition of cell proliferation. When cells make contact they are supposed to stop growing and become quiescent. Indeed, those transfected cells that express the intact proteoglycan seem to show that morphology. This may give us a clue to what contact inhibition is. Second, and this is the punch-line for this symposium, if the glycosaminoglycan chain is omitted, the functional properties of the molecule change. We think this might work by the decorin molecule being bifunctional; the core protein has one role, and the glycosaminoglycan has another, perhaps a binding function. Thus here we have a carbohydrate structure that seems to have a function. Although this is only one of many functions that we do not yet thoroughly understand, we intend to elucidate this particular aspect further.

Reference

Yamaguchi Y, Ruoslahti E 1988 Expression of human proteoglycan in Chinese hamster ovary cells inhibits cell proliferation. Nature (Lond) 336:244–246

DISCUSSION

Feizi: Does decorin have a domain homologous with carbohydrate-binding proteins?

Ruoslahti: No, it does not.

Lindahl: What is the structure of the polysaccharide chain?

Ruoslahti: It is a chondroitin sulphate, with some contribution from dermatan sulphate. The amount of dermatan sulphate varies from cell to cell.

Schachner: Are there other chondroitin sulphate or dermatan sulphate proteoglycans still present in the cells? If you added, say, chondroitin sulphate to the mutant core-protein-expressing cells, would they become phenotypically normal? I am wondering about the necessity of having a membrane-bound proteoglycan as auxiliary component to an adhesion molecule.

Ruoslahti: If we were to add the other half of this proteoglycan, the chondroitin sulphate, we should see the same defective function. We have not done that, so I am not sure what would happen.

Sharon: You referred to bacterial attachment to cell surface sugars. Speaking teleologically, it is not advantageous for the host to have these sugars on, say, the surface epithelial cells. This bacterial binding is most likely opportunistic, and not a natural function of the sugars. But at a later stage of infection, when bacteria encounter phagocytes, surface sugars on the phagocytes may serve as attachment sites for lectin-carrying bacteria, resulting in the killing of the bacteria. This seems to be a proven case of a biological function for sugars. It has been shown, mainly in our laboratory, that type 1 fimbriated, mannose-specific *Escherichia coli*, which bind preferentially to oligomannose and hybrid-type N-linked units, attach readily to human polymorphonuclear leucocytes and to human and mouse peritoneal macrophages. As a result of this attachment, there is activation of the phagocytic cells followed by killing of the bacteria, a mechanism named lectinophagocytosis (Ofek & Sharon 1988).

There is also evidence for the complementary phenomenon where lectins on the surface of phagocytic cells may attach to sugars on bacteria. We have demonstrated this for the interaction of *Klebsiella pneumoniae* with guinea pig alveolar macrophages or human monocytes (Athamna et al 1989). We believe that lectinophagocytosis may function *in vivo* in the preimmune state or in opsonin-deficient sites, such as the renal medulla or the peritoneum during peritoneal dialysis.

Finne: The normal bacteria of the colon bind to certain glycolipids on the surface of the colon (Strömberg et al 1988). That may be a physiological function.

Ruoslahti: The binding of symbiotic bacteria to a carbohydrate structure would also be beneficial to the host.

Sharon: This is controversial, because some facts do not fit into the theory that lectin–carbohydrate interactions are required for the initiation of nodulation, which is the first step in establishing the symbiotic relationship for nitrogen fixation.

Wang: The role of carbohydrate recognition in the establishment of a symbiosis between nitrogen-fixing bacteria and the root of certain leguminous plants (the lectin recognition hypothesis) has indeed been controversial. But, I think we should view the problem at two different levels: that of the experimental results and of their interpretation. This hypothesis was initially advanced by Bohlool & Schmidt (1974) and was based on correlative studies showing that soybean agglutinin bound to those strains of *Rhizobium* that normally infect soybean roots, but did not react with non-symbiotic strains. Dazzo et al (1976) similarly reported a corresponding lectin from white clover which binds specifically to its symbiotic nitrogen-fixing bacterium, *Rhizobium trifolii*. Thus, it was proposed that legume lectins control host specificity by interacting with saccharide components on the symbiotic bacteria. There have been many inconsistencies and experimental deficiencies in various studies that purport either to support or to refute the lectin recognition hypothesis and these have been succinctly reviewed (Bauer 1981, Pueppke 1983).

At the level of experimental results, at least in the soybean system, I think there is good evidence that specific saccharides, such as D-galactose, do inhibit the adhesion between *Rhizobium japonicum* and the root hairs (see, for example, Stacey et al 1980). What has not been so clear is the interpretation of such an observation: is the 'recognizer' (lectin) really on the soybean host and the 'recognizee' (carbohydrate) on the bacterial surface or *vice versa*? The reasons for raising these issues of interpretation include: (a) some soybean lines that lack detectable root lectin (corresponding to soybean agglutinin) are, nevertheless, infected and nodulated (Pueppke 1983); and (b) there is evidence that *Rhizobium japonicum*, which infects soybeans, can bind specifically to Sepharose or polyacrylamide beads covalently derivatized with galactose (S.-C. Ho, personal communication). Therefore, Nathan Sharon's statement that 'some facts do not fit the theory' should be restricted to those hypotheses requiring that the host lectin determine the specificity of bacterial infection. Nevertheless, I believe that there is evidence supporting a role for lectin–carbohydrate interactions in the nodulation process.

References

Athamna A, Ofek I, Keisari Y, Markovich S, Dutton GCS, Sharon N 1989 Lectinophagocytosis mediated by interaction between a macrophage surface lectin and capsular polysaccharides of *Klebsiella pneumoniae*. (Abstr) Annual Meeting Israel Society of Microbiology, Haifa, January

Bauer WD 1981 Infection of legumes by *Rhizobia*. Annu Rev Plant Physiol 32:407–449

Bohlool BB, Schmidt EL 1974 Lectins: a possible basis for specificity in the *Rhizobium*–legume root nodule symbiosis. Science (Wash DC) 185:269–271

Dazzo FB, Napoli CA, Hubbell DH 1976 Adsorption of bacteria to roots as related to host specificity in the *Rhizobium*–clover symbiosis. Appl Environ Microbiol 32:166–171

Ofek I, Sharon N 1988 Lectinophagocytosis: a molecular mechanism of recognition between cell surface sugars and lectins in the phagocytosis of bacteria. Infect Immun 56:539–547

Pueppke SB 1983 Soybean lectin: does it have an essential role in the *Rhizobium*–soybean symbiosis? In: Goldstein IJ, Etzler ME (eds) Chemical taxonomy, molecular biology, and function of plant lectins. Alan R Liss, New York, p 225–236

Stacey G, Paau AS, Brill WJ 1980 Host recognition in the *Rhizobium*–soybean symbiosis. Plant Physiol 66:609–614

Strömberg N, Ryd M, Lindberg AA, Karlsson K-A 1988 Studies on the binding of bacteria to glycolipids. Two species of *Propionobacterium* apparently recognize separate epitopes on lactose of lactosylceramide. FEBS (Fed Eur Biochem Soc) Lett 232:193–198

Oligosaccharide-protein interactions: a three-dimensional view

J. P. Carver*, S. W. Michnick*, Anne Imberty† and D. A. Cumming°

**Department of Medical Genetics, University of Toronto, Toronto, Ontario, Canada M5S-1A8, °Genetics Institute, Andover, Massachusetts, USA* and *†INRA, LPCM, 44072 Nantes, France*

Abstract. For carbohydrates to serve as recognition elements in cellular function, there must be 'receptors' which are capable of distinguishing between the multitude of oligosaccharide structures generated by a cell. Generally these receptors are assumed to be proteins, and the plant lectins have been used as model systems to examine the molecular basis for specificity in such interactions. Three aspects of the specificity of oligosaccharide–protein interactions will be discussed: (1) the conformational flexibility of oligosaccharides will be demonstrated through a quantitative analysis of nuclear magnetic resonance measurements; (2) a comparison of the measured and calculated values for the entropy barrier to oligosaccharide binding will be used to argue that the barrier arises from a loss of this conformational flexibility upon binding to the lectin (this conclusion is also supported by X-ray crystallographic studies); and (3) the thermodynamic model can be extended to the binding of glycoproteins to receptors and the high affinity of these interactions explained by either multivalency or fixation of the oligosaccharide in the 'correct' three-dimensional structure through interaction with the protein moiety.

1989 Carbohydrate recognition in cellular function. Wiley, Chichester (Ciba Foundation Symposium 145) p 6–26

A large number of cellular recognition events are now thought to involve the specific binding of particular classes of carbohydrate structure on one cell surface to 'receptor' proteins on the surface of another cell (for a review see Rademacher et al 1988). Therefore, there is considerable interest in seeking a molecular understanding of the basis for specificity and affinity in these interactions. Our work (Cumming & Carver 1987a,b, Carver & Cumming 1987) on the three-dimensional structures of oligosaccharides and glycopeptides in solution has led us to a series of conclusions which suggest some answers to these questions. We present these views here in the hope of stimulating further discussion.

The interpretation of the intramolecular NOEs of oligosaccharides

In previous publications we (Cumming & Carver 1987a,b, Carver & Cumming 1987) and others (Thøgersen et al 1982, Ha et al 1988, Tran et al 1989) have shown that the torsional potential energy surfaces for the glycosidic linkages of oligosaccharides are generally quite shallow. As a result, the torsional angles corresponding to the lowest energy or 'preferred' three-dimensional structure, although they are the most probable, define a conformation that occurs in less than *one percent of the ensemble*. Thus a representation of the molecular structure by this conformation ignores more than 99% of the ensemble.

Typically, transglycosidic nuclear Overhauser effect measurements (NOEs) are used to estimate glycosidic torsion angles. The magnitudes of these nuclear magnetic resonance (NMR) measurements are a function of the distances between the respective hydrogens in the molecule. Thus if enough measurements are made a picture of the possible three-dimensional structures can be constructed. However, because the NOEs build up over hundreds of milliseconds and because oligosaccharides are expected to sample all available conformations during this time period, the NOEs must be treated as ensemble averages. If the NOEs are interpreted as corresponding to a single three-dimensional structure, when in reality they arise from averages over many such structures, then the derived geometry is a 'virtual' structure and will be significantly in error. Instead, potential energy surfaces must be generated and used to calculate statistical weights for the different three-dimensional structures present in solution. From these weights, ensemble average NOEs can be calculated for comparison with experimental values.

Examples of potential energy surfaces for the Man(α1–3)Manβ linkage of the glycopeptide GnGn(+F) (see Table 1 for this nomenclature) are shown in Fig. 1. The potential energy function used to generate the surface shown in the lower panel of Fig. 1 is the HSEA potential of Thøgersen et al (1982). The HSEA potential includes terms for van der Waals interactions and a torsional angle potential for rotation about the C1 to O1 bond of the glycosidic linkages; the latter potential simulates the exo-anomeric effect. In the upper panel of Fig. 1, the surface has been calculated using the HSEA potential with an added hydrogen bond potential (Mackie et al 1986). We denote this new combined potential 'HEAH' (Cumming & Carver 1987a,b). Both these potential energy surfaces are two-dimensional subsets of the *n*-dimensional conformational hyperspace for the Man(α1–3) linkage. The two dimensions chosen in these cases are the torsional angles phi (M3-H1, M3-C1, M3-O1, Mβ-C3) and psi (M3-C1, M3-O1, Mβ-C3, Mβ-H3) and hence represent the situation where there are no other variations in geometry. For all the potential energy functions, the ensemble samples approximately 10% of torsional angle space, a fraction which corresponds to the sterically available area, whereas half the population is found

TABLE 1 Structures and nomenclature for the glycopeptides and oligosaccharides referred to in the text

Structure	*Short form notation*
α-Methyl-D-mannoside	M-OMe
Man(α1–2)Manα,β	M2M
Man(α1–2)Man(α1–2)Manα,β	M2M2M
Man(α1–3)Man(β1–O)iPr	M3Mβ
Man(α1–2)Man(α1–3)Man(β1–O)Me	M2M3M-OMe
Man(α1–6) \ Man(α1–O)Me Man(α1–3) /	M3[M6]M-OMe
Man(α1–6) \ Man(α1–6) Man(α1–3) / \ Man(β1–4)GlcNAc(β1–4)GlcNAc(β1–N)Asn Man(α1–3) /	M5Gn2-Asn
GlcNAc(β1–2)Man(α1–O)Me	Gn2M-OMe
Fuc(α1–6) GlcNAc(β1–2)Man(α1–6) \ | Man(β1–4)GlcNAc(β1–4)GlcNAc(β1–N)Asn GlcNAc(β1–2)Man(α1–3) /	GnGn(+F)

in about one percent of conformational space. In Table 2, calculated NOE values are compared to experiment. Calculated values are given for the minimum energy conformation on each potential energy surface and for averaging over each of the surfaces. It is clear that NOE values calculated for the conformers at the minima are totally inadequate whereas those calculated as ensemble averages over the HSEA and HEAH surfaces straddle the observed values and agree better with experiment. One possible explanation for the lack of better agreement is that the contribution of intramolecular hydrogen bonds is not negligible, but is overestimated for aqueous solutions by the hydrogen bond potential of Mackie et al (1986) used in HEAH. Thus this analysis can be used not only to demonstrate that oligosaccharides are flexible but also to discriminate between various proposed sets of potential energy functions.

The discrimination can be refined further if specifically deuterated compounds are used. For example, in a previous study (Dime et al 1987, Cumming et al 1986) we synthesized Man(α1–3)Man(β1–O)iPr with the Mβ-H3 and the M3-H2, -H3, -H4, -H6, and -H6′ replaced by deuterium. Because of the geometry of the Man(α1–3) linkage, variations in the phi and psi values will have a far greater effect on the distances between M3-H1 and Mβ-H2 (or Mβ-H4) than between M3-H1 and Mβ-H3. However, in the undeuterated disaccharide the NOE from M3-H1 to Mβ-H2 is very small because the positive direct NOE is almost cancelled by a three-spin effect via the Mβ-H3 which experiences a large direct NOE. In the deuterated disaccharide this direct NOE to Mβ-H3 is eliminated as is, therefore, the three-spin effect on Mβ-H2. As a result, the observable NOE to the latter is larger and directly observable in the deuterated compound (Cumming et al 1986). The data and comparisons with calculated values are listed in the right hand column of Table 2. Whereas the ratio of the NOE to Mβ-H2 for the HSEA and HEAH ensembles, respectively, was 2.7:1 for GnGn(+F), the corresponding ratio is 4.8:1 for the deuterated disaccharide. Thus the latter compound is more sensitive to the nature of the potential energy surface and represents an ideal model for the refinement of these functions. The observed NOE for the deuterated disaccharide is, however, still small compared to that for the undeuterated glycopeptide, GnGn(+F), because of the differences in correlation time for the two molecules. If the deuterated Man(α1–3) linkage could be incorporated into a glycopeptide, an NOE substantially larger than that for the natural glycopeptide should be observed.

The above discussion has focused on equilibrium one-dimensional NOEs. We have developed a programme to calculate the ensemble averaged values for cross-peak intensities in two-dimensional NOE experiments (S. W. Michnick, D. A. Cumming & J. P. Carver, unpublished results). The results with GnGn(+F) indicate the same problem as was found with the one-dimensional case. Neither the HSEA nor the HEAH surfaces completely reproduce the data, although averaging over the surfaces gives much closer agreement than any single three-dimensional structure. Because more NOEs are observable in a NOESY experiment than in a one-dimensional experiment, this approach promises to yield even more constraints with which to refine the potential energy surface.

For more branched oligosaccharide structures we have found more restricted distributions of torsional angles; as a result the ensemble averaging occurs over a much reduced fraction of torsional hyperspace. For some of these cases the NOE values calculated from a single conformation and those calculated as the ensemble average agree to within the experimental error of the measurements and therefore cannot be distinguished.

Calculation of thermodynamic properties

The thermodynamic properties associated with torsional rotations can be calculated from the same ensemble representation used to interpret the NOE measurements. This ensemble is derived from the calculated potential energy

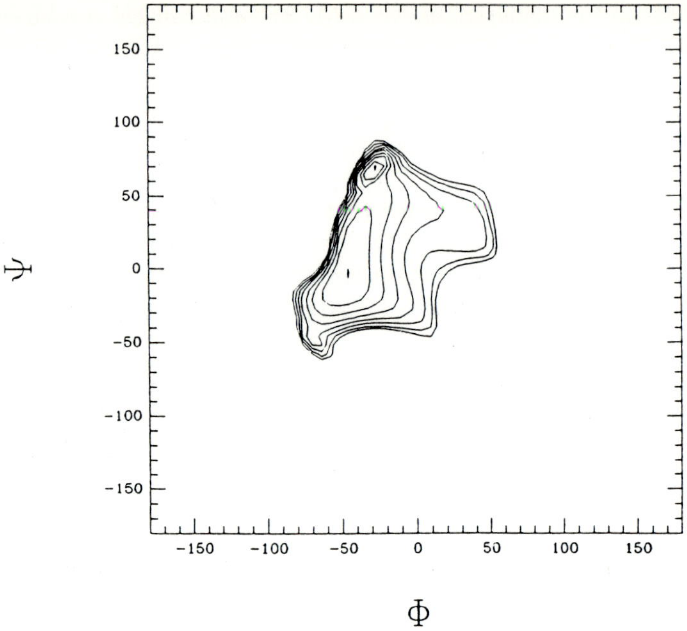
150
100
50
0
-50
-100
-150
Ψ
-150
-100
-50
0
50
100
150
Φ

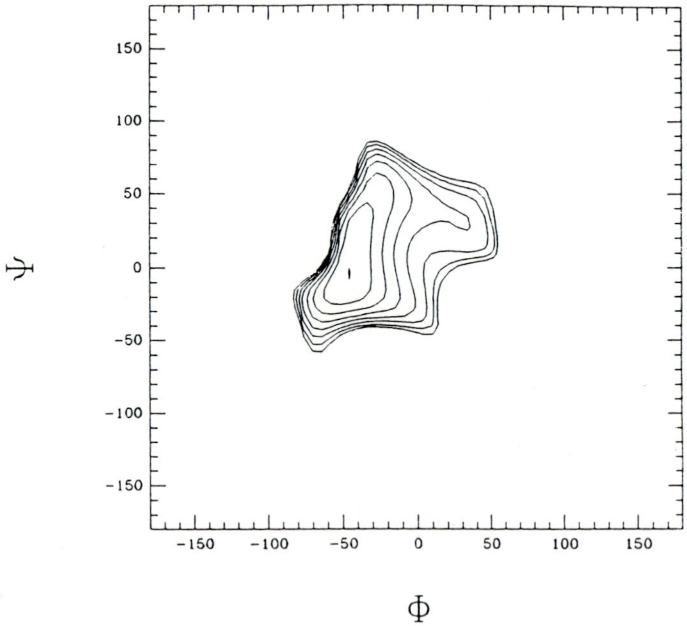
150
100
50
0
-50
-100
-150
Ψ
-150
-100
-50
0
50
100
150
Φ

TABLE 2 Man(α1–3) linkage: relative[a] NOE on Mβ-H2 for irradiation of M3-H1

Compound	*GnGn(+F)*	*M3Mβ 6-deuterio*
Calculated		
Minimum in HSEA	0.16	0.23
Minimum in HEAH	1.4	14
Averaged, HSEA	0.22	0.92
Averaged, HEAH	0.60	4.4
Observed		
mean ± sd	0.51 ± 0.03	1.8 ± 0.4

[a]The NOE on Mβ-H2 is expressed relative to that observed on M3-H2 (for GnGn(+F)) or M3-H5 (for M3Mβ).

surface by standard statistical mechanical methods. Statistical weights, p_i (of the ith three-dimensional structure), are derived from the calculated potential energies (E_i) for each three-dimensional structure by assuming a Boltzman distribution:

$$p_i = [\exp(-E_i/kT)]/Q \tag{1}$$

where Q is the partition function and is given by:

$$Q = \sum_i \exp(-E_i/kT) \tag{2}$$

The ensemble average energy, $\bar{E}$, is then given by:

$$\bar{E} = \sum_i p_i E_i \tag{3}$$

FIG. 1. Potential energy surfaces for rotation about the two torsional angles phi (M3-H1, M3-C1, M3-O1, Mβ-C3) and psi (M3-C1, M3-O1, Mβ-C3, Mβ-H3) for the glycosidic linkage Man(α1–3)Man in GnGn(+F). Each contour corresponds to an increase in energy of 4.2 kJ (1 kcal), relative to the minimum in the surface. Lower panel: Surface for the HSEA potential. The minimum is at (phi,psi) = (−48°, −2°) and has a statistical weight of 0.013. Upper panel: Surface for the HEAH potential. The global minimum is at (phi,psi) = (−30°,68°), has a statistical weight of 0.044 and is about 4 kJ lower in energy than the second minimum which is located at the same position as the minimum in the upper panel.

The ensemble entropy, S, can be calculated from:

$$S = -R\sum_{i} p_i \ln(p_i) \tag{4}$$

and the Helmholtz free energy as:

$$A = \bar{E} - TS \tag{5}$$

where T is the absolute temperature.

Examples of the values obtained for a variety of oligosaccharides are given in Table 3. For each oligosaccharide the potential energy algorithm used corresponds to that which generates ensemble average NOEs in best agreement with experimental values. The entropy value for the Man(α1–6)Manβ linkage of M5Gn2-Asn is significantly less than that for the disaccharides listed, despite the presence of a third torsional angle in the 1–6 linkage. This decrease reflects the more restricted ensemble which is characteristic of more branched compounds.

The thermodynamics of protein–oligosaccharide interactions

The phenomenon of entropy–enthalpy compensation has been observed with a wide variety of protein–ligand interactions (Eftink et al 1983, Hindsgaul et al 1985). In particular, when the enthalpy and entropy of binding of increasingly large ligands are compared, the compensation phenomenon is exhibited as a much smaller decrease in the free energy of binding than that found for the enthalpy. This arises because the entropy of binding becomes increasingly unfavourable in parallel with the increased stabilization represented by the enthalpy. The relationship is best shown by plotting $-\Delta H$ against $-T\Delta S$ (see Fig. 2). In situations where perfect compensation occurs, a line with a gradient of one and an y-intercept of ΔG is obtained (see dashed line in Fig. 2). Generally the

TABLE 3 Some examples of calculated ensemble thermodynamic properties[a]

Oligosaccharides[b]	*Linkage rotated*	*E*	*A*	*TS*	*Potential energy algorithm*
M2M3M-OMe	(α1–3)	−9.2	−26	17	HSEA
Gn2M-OMe	(β1–2)	−6.7	−23	16	HSEA
M5Gn2-Asn	(α1–6)	−31	−42	11	HSEA

[a]The thermodynamic parameters were calculated according to equations (1)–(4) sampling the potential energy surface on a 120×120 grid. For (1–6) linkages the three-dimensional potential energy surface was sampled with a 60×60×60 grid. E, A and TS are in kJ/mol at 300 K.
[b]See Table 1 for this nomenclature.

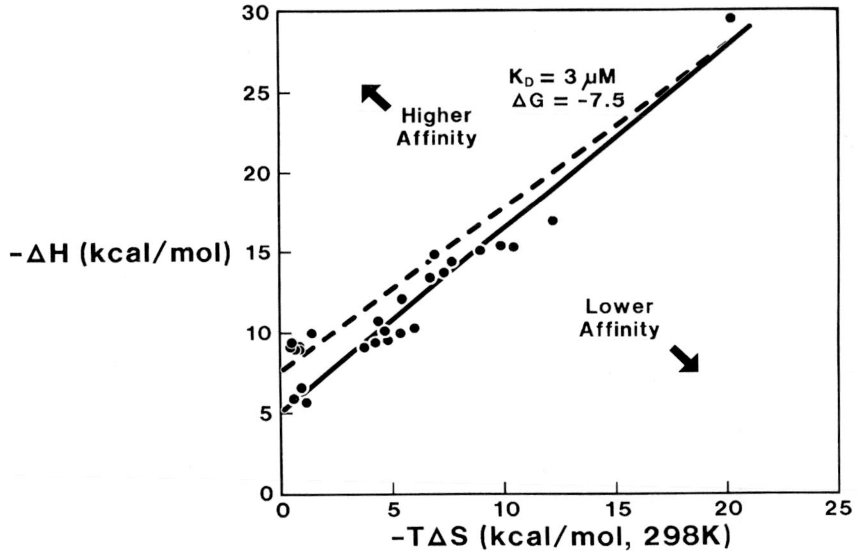

FIG. 2. Enthalpy–entropy compensation for oligosaccharides binding to lectins and antibodies. Each point corresponds to a different oligosaccharide–protein interaction; the coordinates are the values of ΔH and $-T\Delta S$ for that interaction. The values of ΔH and $-T\Delta S$ increase for larger ligands so that the points corresponding to reducing monosaccharides are at the extreme lower left and the point for a tetrasaccharide appears at the upper right. The dashed line shows the locus of points which would be generated if perfect compensation occurred so that the free energy change upon binding remained constant at -7.5 kcal (-30 kJ). The experimentally observed values are scattered about a somewhat steeper line, indicating that the free energy of binding becomes slightly more negative as the size of the ligand increases. This figure is modelled after Fig. 1 of Hindsgaul et al (1985). The data plotted originate from a number of other laboratories (references given in Hindsgaul et al 1985) as well as from this laboratory (Kronis & Carver 1985, MacKenzie 1985).

gradient is slightly greater than one, indicating that $-\Delta G$ increases gradually as the ligand size increases but not nearly as fast as it would if all the gain in enthalpy were available as free energy. In previous discussions of the origin of this phenomenon for protein–ligand interactions (Eftink et al 1983, Hindsgaul et al 1985), the usual model described a gradually increasing restriction, or 'freezing', of the distribution of binding site conformations on the protein as the size of the ligand increased. Although this may well be the correct model for many cases, we suggest here that the opposite is true for oligosaccharide–protein interactions, i.e. that it is the loss of conformational entropy in *the ligand* on binding that accounts for the entropy barrier to binding and that this loss increases as the extent of the oligosaccharide interaction with the protein surface increases because more linkages are restricted in their conformation.

There are essentially two lines of argument which support this idea: (1) calculated values of the conformational entropy for a number of ligands are at least as great as the values of ΔS observed on complex formation; and (2) in X-ray crystallographic studies of oligosaccharide–protein complexes, the protein structures appear unperturbed by the formation of the complex. These data are reviewed in the following sections.

Comparison of calculated and observed entropy values

Table 4 shows the observed and calculated values for the entropy of binding of several oligosaccharides to the plant lectin, concanavalin A (Con A). If the glycosidic torsional angles of the oligosaccharides are completely fixed upon binding to the lectin, then one might expect the ΔS value measured for the formation of the complex to have a similar value to the conformational entropy calculated from the ensemble of ligand conformations generated as described earlier. Such a simplistic model ignores several important contributions to the observed entropy, namely the losses of rotational and translational entropy of the whole ligand molecule and the effects of the removal of solvent from the binding site and ligand on complexation. However, the comparisons given in Table 4 do permit us to establish that the magnitudes of the calculated conformational entropies are similar to the observed entropy barriers to binding. Thus it is plausible to attribute the major part of the entropy barrier to the loss of conformational entropy of the ligand.

In a previous 'docking' study (Carver et al 1985), we have proposed a model for the interaction of the trimannoside shown in Table 4 with Con A. In this model, which is supported by extensive solution spectroscopic data, both the Man(α1–3)- and the Man(α1–6)-linked residues interact significantly with the protein binding site. Thus it is reasonable to expect that the torsional angles about these two glycosidic linkages will be considerably restricted in the complex. In the extreme case where the torsional angles become fixed, the resulting loss in entropy on binding will be equal to that calculated for the ensemble of

TABLE 4 Comparison of experimental and calculated thermodynamic quantities[a] for oligosaccharides binding to concanavalin A

Ligand	ΔG_{obsd}	ΔH_{obsd}	$-T\Delta S_{obsd}$	$-TS_{calc}$
Man(α1–3)[Man(α1–6)]Man(α1–O)Me	−33.6	−63.0	29.4	31.9
Man(α1–O)Me	−21.9	−38.2	16.8	—
Man(α1–2)Man	−23.1	−33.2	9.7	18.5
Manα(1–2)Man(α1–2)Man	−29.0	−37.0	9.2	36.5

[a]In kJ/mol at 300 K.

un-complexed trimannoside using equation (4), as listed in Table 4. The complete agreement between the observed and calculated values must be fortuitous, given the contributions which have been neglected; however, it would appear reasonable to conclude that the proposed model identifies a major contributing mechanism. For the linear Man(α1–2)-linked oligomannosides there is little agreement between the calculated and observed entropies. This is not unexpected because there is general agreement (Brewer et al 1975) that this class of oligosaccharides does not bind in a unique orientation. Instead it has been proposed that only one mannoside residue binds at any one time and that any one of the residues may do so. Thus the loss in entropy due to the restriction of torsional angle values might be expected to be confined at most to one linkage. In fact the observed value for Man(α1–2)Man is half that found for the simple methyl glycoside of mannose, suggesting that the possibility of both residues binding is expressed as a significant reduction of the entropy barrier below that for the simple glycoside. The observation that the entropy barrier for Man(α1–2)Man(α1–2)Man is unchanged relative to Man(α1–2)Man suggests that the further increase in the entropy barrier which might be expected by having three residues capable of binding is offset. The most likely cause is that only one out of the three residues experiences a reduction in the torsional angles (similar to the dimannoside) upon binding to Con A.

The crystal structure of a pea lectin–trimannoside complex

Recently, the 2.5 Å resolution structure of a complex between pea lectin and Man(α1–3)[Man(α1–6)]Man(α1–O)Me has been solved (Rini 1986). The primary interaction with the lectin is through one mannose residue only and consists of four hydrogen bonds involving the oxygens at positions 3, 4, 5 and 6 of the sugar and the backbone NH groups of Gly 99 and Ala 217 together with the carboxyl of the side chain of Asp 81. In the native pea lectin structure (Einspahr et al 1986), without bound sugar, the side chain of Asp 81 is a ligand of the bound Ca^{2+} ion and is therefore fixed within the molecule. The trimannoside complex is formed without any change in the position of this side chain, or in the positions of the backbone atoms which form the other hydrogen bonds. Thus, in the region of sugar binding, no change in protein conformation is observed between the native pea lectin structure and that of the complex with the trimannoside.

Origins of high affinity and specificity

If a conformational entropy barrier prevents oligosaccharide/proteins from achieving high affinities, how then are these generated in real biological interactions? There are three possible mechanisms—one for which there is extensive direct evidence, the others for which the evidence is indirect. The first is multivalency. This mechanism permits high affinity through the summation of a number of small favourable free energies of binding. However, because each oligosaccharide–protein interaction is weak, the degree of specificity need not be high. For virus binding to cell surfaces this is an ideal mechanism. A well characterized example is the binding of the influenza virus to sialylated cell surfaces (Weis et al 1988). The dissociation constant for sialyllactose binding to isolated haemagglutinin is in the millimolar range, yet the intact virus binds essentially irreversibly. Although there is linkage specificity in the binding, a number of modified sialic acids bind equally well.

The second mechanism is more subtle and is probably the basis for specificity in the recognition of substrates by glycosyltransferases. The K_ms for many glycosyltransferase reactions with isolated oligosaccharide substrates are generally in the 100 micromolar to millimolar range, which is consistent with values found for lectin binding. Yet the normal substrate is an oligosaccharide covalently linked to a glycoprotein or glycolipid. Transferases achieve exquisite specificity by having a unique binding site for each reaction. For example, there are at least three different GlcNAc transferases which form GlcNAc(β1–4) linkages, each enzyme forming the linkage to a hydroxyl in a different environment. Affinities for the oligosaccharide substrates covalently linked to glycoproteins might be expected to be higher because a significant amount of conformational entropy may already have been lost through the stabilization of a subset of three-dimensional structures at the protein surface. Because the degree of interaction with the protein surface would be expected to be different at different glycosylation sites, the affinities of a given glycosyltransferase for a given oligosaccharide structure would be expected to vary from site to site. It is frequently found that different glycosylation sites on the same glycoprotein are not processed in the same way as the glycoprotein passes through the biosynthetic pathway. We have previously proposed (Carver & Cumming 1987) that this effect be termed 'site-directed processing' to reflect the role of the protein environment at the glycosylation site in determining the oligosaccharide structure at that site. We suggest that the differential activity of a given glycosyltransferase towards the same oligosaccharide substrate at different protein sites arises from the differential losses in conformational entropy required to form the complex at each site. At some glycosylation sites, those where the oligosaccharide is a better substrate for the transferase, the conformational entropy barrier will have been considerably reduced through the binding of part of the oligosaccharide to its own protein surface. At other glycosylation sites, those where the oligosaccharide is a poorer substrate, the interaction with the protein surface may be negligible. Thus the conformational entropy barrier will be almost intact and the affinity of the interaction with the transferase much weaker. The prediction is that the K_ms should differ among glycosylation sites.

Finally, there is the case of the high affinity, highly specific receptor. Here one might expect that the conformational entropy of the covalently bound oligosaccharide would be considerably reduced by extensive interactions with the protein surface. The extreme case would be a binding site at which the stabilization of the oligosaccharide by the protein is so extensive that its mobility is essentially that of the protein. Here the element recognized may be partly oligosaccharide and partly protein—essentially a combined epitope. Under these circumstances almost all the enthalpy from the extensive hydrogen bond and van der Waals networks would be available as free energy and very high affinities and specificities would be possible.

Acknowledgments

This work was supported by a grant from the Medical Research Council of Canada (MT-3732) to JPC and a Terry Fox Special Initiatives Grant from the National Cancer Institute of Canada to JPC and Harry Schachter. We thank the members of the Carbohydrate Research Centre at the University of Toronto, Serge Perez in Nantes and Igor Tvaroska in Bratislava for ongoing, stimulating exchanges of ideas. JPC also thanks Karl Hardman for discussions in which the original concept of site-directed processing was developed.

References

Brewer CF, Sternlicht H, Marcus DM, Grollman AP 1975 ^{13}C NMR studies of the interaction of concanavalin A with saccharides. In: Chowdhury TK, Weiss AK (eds) Concanavalin A. Plenum Press, New York, p 55–70

Carver JP, Cumming DA 1987 Site-directed processing of N-linked oligosaccharides: the role of three-dimensional structure. Pure Appl Chem 59:1465–1476

Carver JP, MacKenzie AE, Hardman KD 1985 Docking of a trisaccharide to the concanavalin A sugar-binding site explains the glycopeptide specificity of this lectin. Biopolymers 24:49–63

Cumming DA, Carver JP 1987a Virtual and solution conformations of oligosaccharides. Biochemistry 26:6664–6676

Cumming DA, Carver JP 1987b Re-evaluation of rotamer populations for (1,6) linkages: reconciliation with potential energy calculations. Biochemistry 26:6676–6683

Cumming DA, Grey AA, Krepinsky JJ, Carver JP 1986 Specific deuteration of a trimannoside glycoside confirms the existence of a disputed inter-residue NOE. J Biol Chem 261:3208–3213

Dime DS, Rachaman E, Dime CE, Grey AA, Carver JP, Krepsinky JJ 1987 Syntheses of model oligosaccharides of biological significance. X. Synthesis of a specifically deuterated 2-propyl 3,6-di-O-[α-D-mannopyranosyl]-β-D-mannopyranoside. J Labelled Compd Radiopharm 24:725–739

Eftink MR, Anusiem AC, Biltonen RL 1983 Enthalpy–entropy compensation and heat capacity changes for protein–ligand interactions: general thermodynamic models and data for the binding of nucleotides to ribonuclease A. Biochemistry 22:3384–3896

Einspahr H, Parks EH, Suguna K, Subramanian E, Suddath FL 1986 The crystal structure of pea lectin at 3.0-Å resolution. J Biol Chem 261:16518–16527

Ha SN, Madsen LJ, Brady JW 1988 Conformational analysis and molecular dynamics simulations of maltose. Biopolymers 27:1927–1952

Hindsgaul O, Khare DP, Bach M, Lemieux RU 1985 Molecular recognition. III. The binding of the H-type 2 human blood group determinant by the lectin I of *Ulex europaeus*. Can J Chem 63:2653–2658

Kronis KA, Carver JP 1985 Thermodynamics of wheat germ agglutinin–sialyl-oligosaccharide interactions by proton nuclear magnetic resonance. Biochemistry 24:834–840

MacKenzie AE 1985 A study of the binding of oligosaccharides and glycopeptides to concanavalin A. PhD thesis, University of Toronto

Mackie W, Sheldrick B, Akrigg D, Perez S 1986 Crystal and molecular structure of mannotriose and its relationship to the conformations and packing of mannan and glucomannan chains and mannobiose. Int J Biol Macromol 8:43–51

Rademacher TW, Parekh RB, Dwek RA 1988 Glycobiology. Annu Rev Biochem 57:785–838

Rini JM 1986 Structural analysis of carbohydrate recognition by pea lectin: an X-ray crystallographic study. PhD thesis, Department of Medical Genetics, University of Toronto, Canada

Tran V, Buleon A, Imberty A, Perez S 1989 Relaxed potential energy surfaces of maltose. Biopolymers 28:679

Thøgersen H, Lemieux RU, Bock K, Meyer B 1982 Further justification for the exo-anomeric effect. Conformational analysis based on nuclear magnetic resonance spectroscopy of oligosaccharides. Can J Chem 60:44–57

Weis W, Brown JH, Cusack S, Paulson C, Skehel JJ, Wiley DC 1988 Structure of the influenza virus haemagglutinin complexed with its receptor, sialic acid. Nature (Lond) 333:426–432

DISCUSSION

Barondes: Have interactions similar to the ones you describe been analysed in real glycoprotein systems?

Carver: We are working on that. This is a model in search of some data! Ole Hindsgaul has tried to synthesize a sterically restricted substrate for transferases, in an effort to demonstrate that some three-dimensional structures have higher affinity than others. I tried to persuade Harry Schachter to use glycoproteins as substrates in transferase assays to test this, but he raised the obvious question of where you get a homogeneous glycoprotein substrate in a cell.

Sharon: I and others previously held the naive view that carbohydrates have almost fixed three-dimensional structures, such as the Y, T or 'broken wing' conformations described by Jean Montreuil (1984). Your data show that oligosaccharides may exist in many conformations. A carbohydrate-binding protein may have a site which is complementary only to one of the rare conformations. When the carbohydrate binds, the equilibrium will be pulled over. What is the thermodynamic cost of that alteration?

Carver: There is an entropy cost if a single conformation is bound, because all the conformational entropy is lost, regardless of which conformation it is. The rare conformations are of higher energy, which suggests that if one of these binds there will be an enthalpy cost. Presumably, for spontaneous binding to occur, this cost is more than compensated by the interactions with the protein surface, making the overall free energy change negative. This compensation is possible because the potential energy surfaces are shallow—it only costs a few kJ at every linkage to make substantial rotations.

Sharon: Are any proteins known to bind to a rare, higher energy conformation?

Carver: I don't know of any published examples.

Fukuda: Using NMR spectroscopy, can you first study the oligosaccharide conformation and then add the protein to study its impact on that conformation?

Carver: That is possible in principle. We can use transient NOE experiments. For example, if the conformational distribution is affected severely by the presence of a protein, the NMR experiment in the presence of the protein will give different NOE results, because the distribution of conformations is influenced by the exchange between the binding site and the free solution. I don't know if anyone else has done that sort of analysis with carbohydrates. It has been done with other enzyme (Clore et al 1986) or cofactor systems, such as NADP-binding proteins (Birdsall et al 1984).

Fukuda: Can two-dimensional NOE experiments be used to study this?

Carver: In such experiments the spectrum is spread in two dimensions and cross-peaks correspond to NOEs between hydrogens which would not be observed in the one-dimensional experiments. Those cross-peaks can be quantified by integration. The theory for the analysis is slightly different, but it works, and you can use the same ensemble-averaging to predict the intensity of the cross-peaks. More constraints are available, because more NOEs are observed. Using that method in the presence of the protein, we should be able to get more information, but it has not been done yet.

Czech: Ligands for hormones and growth factors range from quite small peptides, some of which are presumably very flexible, to quite rigid, large peptides or proteins. Does your analysis apply to the binding of peptides and, if so, can you make predictions about affinities that have already been accurately measured?

Carver: The concept of enthalpy–entropy compensation came originally from analysis of oligopeptide interactions with proteins (Lumry & Rajender 1970), but the explanation given for the entropy loss was that the protein gradually froze. I have not seen an analysis based on the conformational flexibility of the peptides themselves, but I would predict that the same phenomenon occurs.

Czech: Wouldn't your analysis predict that small peptide ligands (or neurotransmitter molecules, which are even smaller) would exhibit lower affinity for their receptors than more rigid peptides, such as insulin?

Carver: The rigid structure has to be the correct rigid structure for the binding interaction. For all the enthalpy to be available as free energy, the hydroxyls must all be oriented correctly—otherwise the interaction will be of low affinity.

Czech: Some rigid structures, such as insulin, have high affinities for their receptor. However, some smaller peptide ligands also show quite high affinities.

Ruoslahti: Some of those small peptides are quite rigid. The binding site is usually in an apparently stable turn of the peptide. It is not necessarily correct that the smaller the compound the less rigid it is.

Carver: If there are intramolecular interactions within the peptide that restrict its number of conformations, it is already pre-set.

Ruoslahti: When fucose is added to the GnGn structure it appears to become more rigid, because the range of possible conformations become smaller. Is it correct to say that the larger a carbohydrate structure, the more rigid it is? Your idea that the carrier protein would stabilize the structure of an oligosaccharide is very innovative. However, couldn't you get strong binding just by making an oligosaccharide larger, especially by adding branches to it, and then having the correct receptor for it? In some of the model systems, you may be losing affinity because the binding protein isn't quite right; you are not looking at the exact intended interaction.

Carver: It is possible that in using sugars derived from eukaryotic cell surfaces to bind to plant lectins we are investigating an inappropriate system. Kornfeld et al (1981) showed that pea lectin has an enhanced affinity for oligosaccharides if fucose is present on the chitobiose core. This is because pea lectin interacts primarily with a mannose in the trimannosyl core and with this fucose. The crystal structure of Man(α1–3)[Man(α1–6)]Man(α1–O)Me in the binding site of pea lectin (Rini 1986) shows that only one of the mannoses is interacting, whereas both mannoses interact with concanavalin (Con A) (Carver et al 1985). Thus plant lectins enhance affinity by having a primary monosaccharide-binding site, where most of the free energy is gained, and a secondary site that is different for different lectins. That's how they achieve their specificity, but this secondary interaction only adds a few kJ to the overall free energy.

Your first point is valid. If the structure can be made more rigid, in addition to having appropriate three-dimensional structure, by adding more branches (as happens in the oligomannose series), binding is enhanced. But for GnGn(+F), use of the extra interaction with fucose necessitates loss of conformational entropy in many intermediate linkages. Therefore, for larger substrates, the overall entropy change increases because entropy is lost per linkage and, although you gain additional enthalpic interactions, you always lose, unless the intermediate part can be made rigid.

Hindsgaul: That is true for linear structures, but the NMR data for structures like Lewis-a and the X trisaccharide indicate that they are as rigid as rocks.

Are there any thermodynamic parameters for antibodies against those structures? They would prove whether your entropy idea is correct, at least in those systems.

Carver: Not for antibodies, but the data for the Lewis-a trisaccharide fall on the line in Fig. 2.

Hindsgaul: That's for binding to a lectin, but we may be looking at a beautiful artefact in all these binding studies. Do you agree that structures such as Lewis-a and the X trisaccharides are as rigid as possible for oligosaccharides?

Carver: I consider them to be much more rigid than the structures that I have been talking about, but not 'as rigid as rocks'.

Feizi: We have studied monoclonal antibodies directed at structures related to the Le^a and Le^b antigens (Gooi et al 1985a,b). They seem to have much higher affinities than antibodies to unsubstituted backbone structures such as anti-I, anti-i and Fc10.2 (Feizi 1981, Gooi et al 1983).

Carver: Do you have thermodynamic data?

Feizi: No, but the anti-Le^a and -Le^b antibodies bind strongly to their target antigens at 20 °C whereas the others require a cold environment, 0–4 °C.

Stanley: Can you extend the NMR analysis to large glycoproteins with many carbohydrates or even a glycoprotein with just one class of oligosaccharides? If the protein is organizing interactions close to its surface, how far along a carbohydrate chain would that extend?

Carver: We can only speculate. We are doing two-dimensional NMR analysis of ribonuclease B to obtain some parameters for the three-dimensional structure of the oligosaccharides. Most of the sugar chains have five or six mannose residues and the structure is fairly homogeneous. The experiments aren't complete yet, but we have preliminary evidence for oligosaccharide–protein interactions even in solution. There are some very strange chemical shifts for the sugar moiety, indicating an interaction with the protein surface, and some unusual NOEs. So it appears that we are able to look at a sugar on a glycoprotein, but that's about as far as we can go at the moment. It is difficult to study large glycoproteins or those with large sugars or more than one sugar-binding site unless we can incorporate stable isotope labels and look at specific signals for particular sugars.

Rademacher: Are you now saying that the Man(α1–3) linkage is flexible? This leads to the question of what NMR experiments can tell us. For rigid bonds, the NOEs should approach a maximum. Under this condition, a molecule is effectively fixed, and only small dynamic movements are possible. Between the extremes of no detectable NOE and a maximum NOE it is difficult to analyse the conformation of sugar linkages. The shallowness of an NOE data curve depends on the type of energy calculation used. In my research group we use molecular orbital calculations and find deep potential energy wells, but Dr Carver uses the HSEA method and finds shallow wells. This is purely an academic exercise. We still consider some linkages to be, within the limits of experimentation, fairly rigid.

Carver: I am saying that the Man(α1–3) linkage is flexible. I am not sure what you mean when you say that only maximum NOEs can be interpreted. In some of our early work, where we felt that we were seeing the maximum possible NOE, we concluded that there was no averaging going on. But the more compounds one looks at, the more one finds that the observed NOEs are intermediate in value. Intermediate situations might correspond either to a single fixed structure that has an intermediate distance or to averaging. That's when you have to start looking at all possible constraints.

Rademacher: I was confused on the entropy issue. The work on the binding of carbohydrates to antibodies by Dr E. A. Kabat is relevant here. For a long time it was thought that carbohydrates bind to antibodies via a linear array of subsites. If one subsite binds mannose at 10^{-5} and another binds fucose at 10^{-5}, the effective binding constant may approach 10^{-10} for a disaccharide. How does this fit into your theory of restriction of entropic motion? That leads to another question about whether we really want protein–oligosaccharide interactions to be high affinity; perhaps the purpose of much of the heterogeneity in glycoproteins is to create low affinity interactions.

Carver: In the interaction of fucose and mannose with pea lectin, the fucose interacts only weakly; most of the free energy, about 20 kJ, comes at the monosaccharide site through the mannose interaction—less than 10 kJ comes from the fucose site. Fucose doesn't bind detectably to pea lectin by itself. Even if the fucose and the mannose interactions were both 10^{-5}, for the combination to be a high affinity interaction, those two have to be appropriately oriented in space. Only a few of the total range of possible conformations will have that orientation. Unless you have a rigid connection between them which gives the appropriate orientation, you won't get the 10^{-10} augmented affinity. That's why the entropy consideration is important.

Rademacher: We have compared the solution conformation of the 90% Man_9 terminal sugar on IgM in the presence and absence of the peptide. Again, within experimental error, we see no effect of the peptide on the solution conformation. So here is a case where the protein is *not* influencing the conformation of the oligosaccharide.

Carver: I was trying to make the point that there would be a continuum of possibilities between sugars that don't interact with the protein, and therefore act as flexible ligands, and those that do interact. I am glad to hear of an extreme example where there isn't an interaction. This is relevant to the issue of how far we can push NMR analysis. When an oligosaccharide is not interacting with the protein, the oligosaccharide will be moving more rapidly than the protein and can therefore be visualized by straightforward NMR experiments because it has a different relaxation time. We would like to gain experimental evidence for cases where the sugar is moving on the same time-scale as the protein. Then the oligosaccharide and protein protons are indistinguishable and we need to use different strategies.

Gahmberg: On the subject of the contribution of both the carbohydrate and the polypeptide portions to the binding of external ligands, we have studied the binding of an *E. coli* strain to glycophorin A. This strain is blood group M-specific and the specificity is determined by two amino acids, the first and the fifth. The specificity is also dependent on the carbohydrate, because removal of the sialic acids from the O-linked chains abolishes binding. We have synthesized the M-specific polypeptide and it does not bind to the bacterium. The oligosaccharide doesn't bind either. Only with the combination of both the oligosaccharide and the M-specific polypeptide do you get strong binding.

Carver: How big a peptide do you need?

Gahmberg: The peptide I used has seven amino acids.

Sharon: Is this also true for the interaction with lectins, e.g. that of *Vicia graminea*?

Gahmberg: Yes.

Ernst: Have you measured temperature-dependent NOEs?

Carver: Within the accessible range in aqueous solution, from a few degrees above zero to about 80 °C, we see very little change in NOE. That is consistent with the ensemble model; the Boltzman distribution is so broad that as you change the temperature over that range, the population is hardly changed.

Sharon: How much additional information can you obtain from ^{13}C NMR using ^{13}C-labelled galactose-specific sites?

Carver: ^{13} NMR data are useful in so far as the carbon allows you to visualize the protons. The carbon chemical shifts and attempts at NOE heteronuclear measurements are not particularly useful because they are less sensitive to three-dimensional structure.

Hart: I'd like you to comment on the mobility of protein in your model. How does site-specific oligosaccharide variability fit into that? Perhaps site-specific variability arises because of different favoured energy states in the protein.

Carver: A weak point in the use of two-dimensional NMR spectroscopy for determining protein three-dimensional structure is the uniform assumption that there is only one three-dimensional structure; that is to say, flexibility is ignored. I agree that the protein has some flexibility. If you use distances derived from NOE measurements to constrain some model, you will get the right answer only if that distance is fixed in the real molecule. It appears from crystal structures that proteins are, in general, less flexible than oligosaccharides. There is concern about how representative crystal structures are, but they usually contain 60% water and residues away from the points of contact within the lattice demonstrate similar temperature factors, which measure mobility, as residues around the contacts. Surface residues can be highly mobile; those involved in secondary structure are less mobile. It is conceivable that at a glycosylation site mobile surface amino acids will become less mobile by virtue of their interaction with the sugar, and the sugar will become less mobile by virtue of its interaction with the amino acids. In other words, the global free energy will represent a less flexible structure. But you can't generalize. It is necessary to look site-by-site, and as yet we don't have the tools to do that, because of the very heterogeneity you are talking about. Naturally isolated materials present a very difficult problem. Our approach at the moment is to try to chemically glycosylate fragments of glycoprotein with asparagine-linked structures, and then to try to reconstruct the situation in the intact glycoprotein by piecing together the data from a series of such fragments.

Hart: You could do that with the transferases that are becoming available.

Carver: You still have to get a core on them.

Sharon: Can you use soybean agglutinin, which doesn't show microheterogeneity?

Carver: Yes, we are doing that.

Hughes: Once you know the exact characteristics of the oligosaccharide-binding site of pea lectin, and the conformation of the oligosaccharide within it, will it be possible to modify the binding site by protein engineering to make an enzyme rather than a lectin, rather like what is being done with antibodies?

Carver: With antibodies, you are selecting for activity rather than designing it. We need to understand more about the function of enzymes before we can design a catalytic site, but in principle it is possible.

Hughes: And the product might be more specific than the glycosidases we have now.

Carver: Absolutely!

Ruoslahti: You could look for a lectin that binds the transition-state intermediate. That might give you an enzyme.

Carver: One needs to know the three-dimensional structure of the transition state for that glycosidase, and then model it.

Lindahl: You referred to different models for carbohydrate–protein interactions, based on either specific, usually high affinity, single-site binding, or low affinity, cooperative interactions. A single biological activity, namely the inactivation of thrombin by the protease inhibitor antithrombin, can be promoted by either type of interaction. The blood anticoagulant action of the glycosaminoglycan heparin is a typical example of the first model. A specific pentasaccharide sequence in this polysaccharide binds antithrombin with high affinity, and the antithrombin is thus 'activated' to complex (and thereby inactivate) thrombin, which is also bound to the same polysaccharide chain. The structure of the pentasaccharide sequence, with a unique disposition of sugar residues and sulphate substituents, is essential for the process (see Danielsson et al 1986).

The second model of interaction is demonstrated by thrombomodulin, an endothelial cell surface protein which binds thrombin (and thereby converts this protease into an activator for protein C, another modulator of blood coagulation). We found that (rabbit) thrombomodulin is a proteoglycan which carries a single polysaccharide (chondroitin/dermatan sulphate) chain. The intact molecule promoted the reaction between antithrombin and thrombin, and thus showed blood anticoaglulant activity similar to (although weaker than) that of heparin. This activity (but not protein C activation) is dependent on the presence of the polysaccharide chain; it disappeared when the polysaccharide was removed by digestion with bacterial chondroitinase (Bourin et al 1988). On the other hand, the free polysaccharide chain, released from the protein by alkaline β-elimination, showed only minimal activity; expression of heparin-like activity required that the target thrombin molecule was bound to the protein C-activating site of thrombomodulin. So while the biological activity is specific and is due to the polysaccharide chain, it is based on low affinity, non-covalent carbohydrate–protein interactions (between the polysaccharide chain and the target thrombin, and presumably also the antithrombin) and requires that the polysaccharide chain is covalently bound to the protein. The biological activity depends on collaboration between the protein and polysaccharide components of the thrombomodulin.

Wassarman: Can you recover activity by adding the polysaccharide to polysaccharide-free thrombomodulin?

Lindahl: Mixing the polysaccharide chain and the polysaccharide-free form of thrombomodulin gives no activity; they have to be covalently joined.

Hughes: Does the glycosaminoglycan chain on thrombomodulin have the pentasaccharide sequence that in heparin binds antithrombin?

Lindahl: It is most unlikely, because the polysaccharide chain is susceptible to degradation by chondroitinase.

Hughes: But it could still have the pentasaccharide as a very small proportion of the total structure.

Lindahl: I think not, because the radiolabelled polysaccharide chain shows only weak binding to antithrombin immobilized on Sepharose.

Kobata: The examples Dr Lindahl described fit with what is known about the glycohormones. Deglycosylated human chorionic gonadotropin (hCG) is known to bind to the target cells, but it cannot express any hormonal activity. Calvo & Ryan (1985) reported that the glycopeptide obtained from hCG by pronase digestion haptenically inhibits the action of the hCG on the target cells. These results indicated that both the polypeptide and the carbohydrate moieties are important in the expression of the hormonal activity of hCG. That might explain why we can't easily determine the function of the carbohydrate moiety in many glycoproteins.

References

Birdsall B, Bevan AW, Pascual C, Roberts GCK, Feeney J, Gronenborn AM, Clore GM 1984 Multinuclear NMR characterization of two coexisting conformational states of the *lactobacillus casei* dihydrofolate reductase trimethoprim $NADP^+$ complex. Biochemistry 23:4733–4742

Bourin M-C, Öhlin A-K, Lane DA, Stenflo J, Lindahl U 1988 Relationship between anticoagulant activities and polyanionic properties of rabbit thrombomodulin. J Biol Chem 263:8044–8052

Calvo FO, Ryan RJ 1985 Inhibition of adenylyl cyclase activity in rat corpora luteal tissue by glycopeptides of human chorionic-gonadotropin and the alpha-subunit of human chorionic-gonadotropin. Biochemistry 24:1953–1959

Carver JP, Mackenzie AE, Hardman KD 1985 Docking of a trisaccharide to the concanavalin A sugar-binding site explains the glycopeptide specificity of this lectin. Biopolymers 24:49–63

Clore GM, Gronenborn AM, Carlson G, Meyer EF 1986 Stereochemistry of binding of the tetrapeptide acetyl-pro-ala-pro-tyr-NH_2 to porcine pancreatic elastase. Combined use of two dimensional transferred nuclear Overhauser enhancement measurements, restrained molecular dynamics, X-ray crystallography and molecular modeling. J Mol Biol 190:259–267

Danielsson Å, Raub E, Lindahl U, Björk I 1986 Role of ternary complexes, in which heparin binds both antithrombin and thrombin or Factor X_a. J Biol Chem 261:15467–15473

Feizi T 1981 The blood group Ii system: a carbohydrate antigen system defined by naturally monoclonal or oligoclonal autoantibodies of man. Immunol Commun 10:127–156

Gooi HC, Williams LK, Uemura K, Hounsell EF, McIlhinney RAJ, Feizi T 1983 A marker of human foetal endoderm defined by a monoclonal antibody involves type 1 blood group chains. Mol Immunol 20:607–613

Gooi HC, Jones NJ, Hounsell EF, Scudder P, Hilkens J, Hilgers J, Feizi T 1985a Novel antigenic specificity involving the blood group antigen Le^a in combination with onco-developmental antigen SSEA-1, recognised by two monoclonal antibodies to human milk-fat globule membranes. Biochem Biophys Res Commun 131:543–550

Gooi HC, Jones NJ, Hilkens J, Hilgers J, Feizi T 1985b Lewis blood group-related specificities of monoclonal antibodies designated MAM-3a, b and c against human milk-fat globule membranes. Glycoconjugate J 2:409–420

Kornfeld K, Reitman ML, Kornfeld R 1981 The carbohydrate-binding specificity of pea and lentil lectins. Fucose is an important determinant. J Biol Chem 256:6633–6640

Lumry R, Rajender S 1970 Enthalpy–entropy compensation phenomena in water solutions of proteins and small molecules. A ubiquitous property of water. Biopolymers 9:1125–1222

Montreuil J 1984 Spatial conformation of glycans and glycoproteins. Biol Cell 51:115–132

Rini JM 1986 Structural analysis of carbohydrate recognition by pea lectin: an X-ray crystallographic study. PhD thesis, Department of Medical Genetics, University of Toronto, Canada

Multifunctional glycoprotein receptors for insulin and the insulin-like growth factors

Michael P. Czech, Robert E. Lewis and Silvia Corvera*

University of Massachusetts Medical Center, Department of Biochemistry, 55 Lake Avenue North, Worcester, MA 01655, USA and **Department of Pathology and Laboratory Medicine, University of Pennsylvania Medical School, Philadelphia, PA 19104, USA*

Abstract. Insulin and the insulin-like growth factors (IGF) I and II are structurally related peptides that elicit a large number of similar biological effects in target cells. Three well-characterized receptor complexes bind one or more of these peptides with high affinity. Two of these receptors, denoted as type I, are ligand-activated tyrosine kinases with similar heterotetrameric $\alpha_2\beta_2$ subunit structures which bind insulin or IGF-I, respectively, with highest affinity. Ligand-stimulated tyrosine autophosphorylation of these receptors further activates their intrinsic tyrosine kinase activities both *in vitro* and in intact cells. Rapid signal transduction follows such receptor autophosphorylation and tyrosine kinase activation, leading to increased serine phosphorylation of many cellular proteins and decreased serine phosphorylation of several others. Experiments in our laboratory have identified three distinct insulin-activated serine kinase activities in cell-free extracts that appear to account for the insulin-stimulated serine phosphorylation of the insulin receptor itself, ATP citrate lyase, and acetyl CoA carboxylase, respectively. A third receptor in this group binds IGF-I and II, lacks kinase activity and is denoted as type II IGF receptor. Amino acid sequences of this receptor deduced from isolated rat cDNA clones show a high degree of homology with those of the bovine cation-independent mannose 6-phosphate (Man-6-P) receptor. We demonstrated that these receptors are indeed identical. The IGF-II/Man-6-P receptor rapidly recycles between the cell surface membrane and intracellular membrane compartments, providing for the rapid uptake of both IGF-II and mannose 6-phosphate-linked lysosomal enzymes. Insulin action markedly increases the proportion of receptors in the plasma membrane and the uptake of bound ligands. We also observe that large amounts of the extracellular domain of the IGF-II/Man-6-P receptor are released into the serum of fetal, neonatal and adult rats. The biological role of this receptor in IGF-II function is yet to be determined.

1989 Carbohydrate recognition in cellular function. Wiley, Chichester (Ciba Foundation Symposium 145) p 27–44

The insulin-like growth factors (IGF) I and II exhibit substantial sequence homology to the hormone insulin, which regulates important metabolic effects in humans and other animals (Czech 1977, Massague & Czech 1982). The IGF peptides mimic insulin action on other cellular transport and metabolic pathways and, in many cell types, enhance cell proliferation more potently than insulin. Three major cell surface receptors bind these peptides with the overlapping specificities depicted in Fig. 1. Insulin and IGF-I receptors, denoted as type I, bind their respective ligands with high affinity (dissociation constant, $K_d = 1$ nM), but also bind both heterologous peptides with lower affinity. IGF-II binds to the IGF-I receptor with relatively high affinity (almost that of IGF-I), but insulin binds this receptor with low affinity. All three peptide ligands can

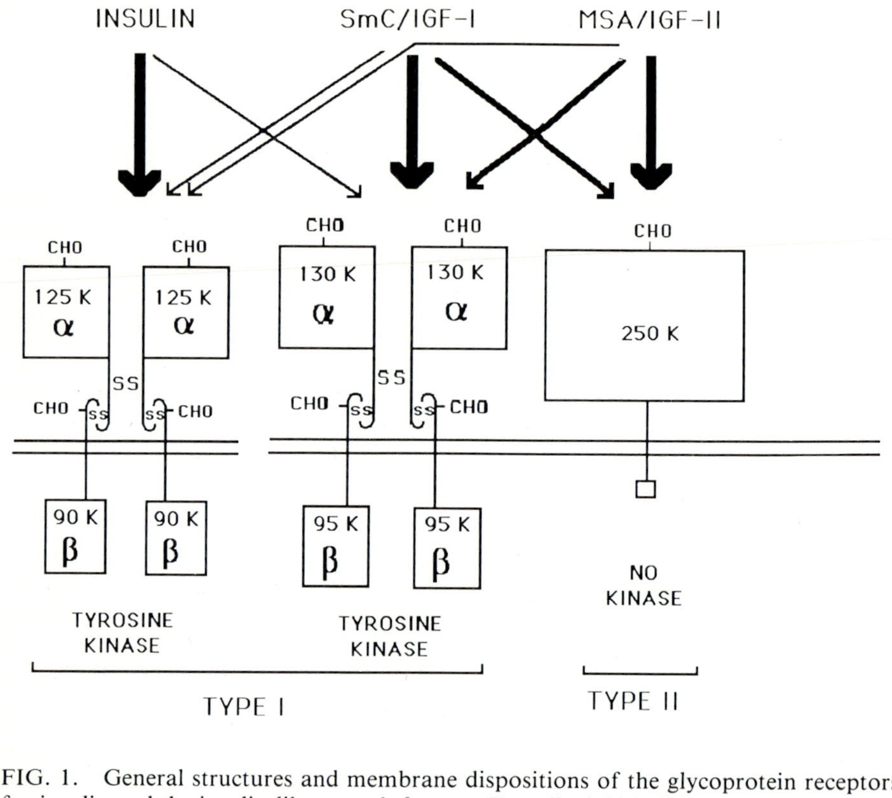

FIG. 1. General structures and membrane dispositions of the glycoprotein receptors for insulin and the insulin-like growth factors. Two disulphide-linked heterotetrameric type I receptors exhibiting highest affinity for either insulin or IGF-I contain intrinsic tyrosine kinase activities within their β-subunit cytoplasmic domains. The type II receptor contains no intrinsic kinase activity and binds IGF-II and IGF-I, but not insulin. This receptor is identical to the Man-6-P receptor. Receptor masses shown are apparent molecular masses calculated from electrophoretic analysis in the presence of SDS.

elicit bio-signals through either of the type I receptors, but do so at concentrations that yield similar receptor occupancies. The type I receptors possess heterotetrameric $\alpha_2\beta_2$ subunit structures (Massague et al 1980, Massague & Czech 1982) and intrinsic ligand-activated tyrosine kinase activities (Kasuga et al 1983) that are essential for their biological functions (Ebina et al 1987). In contrast, the type II IGF receptor preferentially binds IGF-II ($K_d = 1$ nM) with higher affinity than IGF-I, but has no affinity for insulin. This receptor consists of a single polypeptide chain (Czech 1982) lacking detectable kinase activity (Corvera et al 1986), and its physiological role is unclear.

Properties of the type I receptors

The successful isolation of cDNA clones for the human insulin receptor (Ullrich et al 1985, Ebina et al 1985) demonstrates that a single receptor polypeptide serves as the precursor for the mature receptor shown in Fig. 1. The immature proreceptor has a predicted relative molecular mass of 152 000 to 154 000. Biochemical data suggest that the insulin receptor is synthesized as a disulphide-linked polypeptide dimer, which is subsequently glycosylated, proteolytically processed into two subunits, α and β, and capped with terminal sugars before insertion into the plasma membrane (Ronnet et al 1984). Hormone-binding develops in the proreceptor and appears to require glycosylation, as tunicamycin blocks acquisition of binding activity (Ronnet et al 1984). However, once the proreceptor acquires the ability to bind insulin, N-linked oligosaccharide removal with endoglycosidase H does not diminish the proreceptor's ability to bind insulin (Olson & Lane 1987). Thus, glycosylation may initiate conformational changes in the receptor that are ultimately responsible for generating binding activity.

The α-subunit of the insulin receptor contains 37 predicted cysteine residues, 26 of which reside within a stretch of 160 amino acids. Similar cysteine-rich domains have been observed in the low density lipoprotein (LDL) and epidermal growth factor (EGF) receptors. Such cysteine-rich regions within the binding domains of transmembrane receptors may be associated with processes other than ligand binding. These may include intermolecular cross-linking between proteins, potentially resulting in receptor patching and capping. The high density of cysteines within the predicted α-subunit sequence probably provides the sulphydryl groups that form the class I and class II disulphides linking α and β subunits into the receptor heterotetrameric structure shown in Fig. 1. Within the β-subunit, a deduced stretch of 23 amino acids encodes the only region sufficiently hydrophobic to form a transmembrane domain (Fig. 1). Three basic amino acids immediately on the C-terminal side of this membrane-spanning sequence are characteristic of a 'stop-transfer signal' used to anchor membrane proteins.

The insulin receptor β-subunit cytoplasmic domain has significant homology with the family of proteins known to encode tyrosine kinases. The similarity includes the consensus sequence Gly-Xaa-Gly-Xaa-Xaa-Gly and a lysine situated 22 amino acids toward the C-terminus from this sequence, which is crucial in tyrosine and serine/threonine kinases for ATP-binding (Ebina et al 1987). A prominent characteristic of tyrosine kinases is their ability to autophosphorylate. The insulin receptor tyrosine at position 960 is surrounded by a sequence similar to that around the major phosphorylated tyrosine (number 1173) in the EGF receptor (Ullrich et al 1984). The sequence that includes the major phosphorylation site in v-src (tyrosine 1150 in the insulin receptor) is also highly conserved within the insulin receptor. Tyrosine sites in this and the C-terminal region of the insulin receptor (tyrosines 1316 and 1322) appear to be major autophosphorylation sites (Tornqvist et al 1987).

Isolation of a cDNA for the human placental IGF-I receptor (Ullrich et al 1986) revealed major structural similarities with the insulin receptor. The deduced primary amino acid structure, hydropathy, and structural domain distribution are highly conserved between these two receptors. The highest level of sequence identity between the receptor proteins resides within the tyrosine kinase domain (84%). Regions of IGF-I receptor α-subunit flanking cysteine-rich domains show 64–67% homology with the insulin receptor. The cysteine-rich domain itself shows less sequence identity (48%), despite conservation of 24 out of 26 cysteine residues between the two receptors. Fifteen of 26 potential asparagine-linked glycosylation sites in the IGF-I receptor are similarly positioned within the extracellular domain of the insulin receptor. The C-terminal tails within the cytoplasmic domains of the receptors are the most divergent.

Interestingly, in the case of both the insulin and IGF-I receptors, tyrosine phosphorylation leads to marked stimulation of their cytoplasmic tyrosine kinase activities. Tyrosine phosphorylation of these receptors, and the consequent activation of their tyrosine kinase activities, may be mediated by an intermolecular reaction or by intramolecular phosphorylation. An example of the former is the phosphorylation and activation of the insulin receptor by the src kinase (Yu et al 1985). The importance of tyrosine autophosphorylation in the regulation of receptor kinase activity and function is demonstrated by experiments in which mutagenesis of insulin receptor tyrosines 1150 and 1151 to phenylalanines resulted in inhibition of both tyrosine kinase activity and insulin-stimulated glucose uptake (Ellis et al 1986). An important aspect of insulin and IGF-I receptor activation through tyrosine autophosphorylation is that kinase activation is retained even after dissociation of ligand from the receptor. Thus, a critical step in the deactivation of insulin and IGF signals may be the dephosphorylation of the cytoplasmic domains of these receptors.

Properties of the type II IGF receptor

Based on the common ability to bind IGFs, the IGF-II receptor might be expected to be structurally similar to the insulin and IGF-I receptors. However, in contrast to the heterotetrameric structures of the insulin and IGF-I receptors, the type II IGF receptor appears to exist as a single, large glycoprotein with almost all of its mass in the extracellular domain (Fig. 1). The rat (MacDonald et al 1988) and human (Morgan et al 1987) IGF-II receptors were shown to have high sequence similarity to the bovine Man-6-P receptor (Lobel et al 1988). That the same receptor protein binds both ligands was demonstrated in our laboratory by affinity purification on pentamannosyl-6-phosphate (PMP)-Sepharose, followed by affinity labelling with ^{125}I-IGF-II (MacDonald et al 1988). As illustrated in Fig. 2, purified Man-6-P receptor is readily affinity labelled under the experimental conditions described. Consistent with the conclusion that both ligands bind the same receptor glycoprotein, we found that Man-6-P (1–5 mM) increased the affinity of the receptor for ^{125}I-IGF-II. Thus the binding sites for Man-6-P and IGF-II on this receptor exhibit positive cooperativity (MacDonald et al 1988).

The Man-6-P/IGF-II receptor may differ from the other growth factor receptors illustrated in Fig. 1 with respect to signalling function. We found that IGF-II elicits a biological response in H-35 cells even after blockade of its receptor with anti-receptor immunoglobulin (Mottola & Czech 1984). Although some reports have suggested selective signalling by this receptor in certain cell types (Hari et al 1987), most of the published data seem to indicate that metabolic and growth effects of IGF-II are mediated through IGF-I or insulin receptors, not IGF-II/Man-6-P receptors.

We have investigated the presence of Man-6-P receptors in chicken and frogs and compared their ligand-binding properties to the IGF-II/Man-6-P receptor found in rats. Receptors were affinity purified on PMP-Sepharose by incubating detergent extracts from rat placenta, chicken liver and frog liver overnight at 4 °C with PMP-Sepharose. Fractions were then eluted from the resin with buffer alone, buffer with 5 mM Man-1-P, buffer with 5 mM glucose 6-phosphate, and finally buffer with 5 mM Man-6-P. One portion of each fraction was electrophoresed on SDS-polyacrylamide gel electrophoresis (PAGE), then silver-stained. A second fraction was incubated with 2 nM ^{125}I-IGF-II, cross-linked with disuccinimidyl suberate, and run on SDS-PAGE. The silver-stained $M_r \approx 250\,000$ band was present in the Man-6-P-eluted fraction from each species, but only the receptor purified from rat could be affinity labelled by ^{125}I-IGF-II (Clairmont & Czech 1989). These data imply further that in chicken, a species known to possess IGF-II, this peptide mediates biological responses through another receptor, presumably the type I IGF receptor structure. Experiments designed to switch extracellular IGF-II/Man-6-P receptor

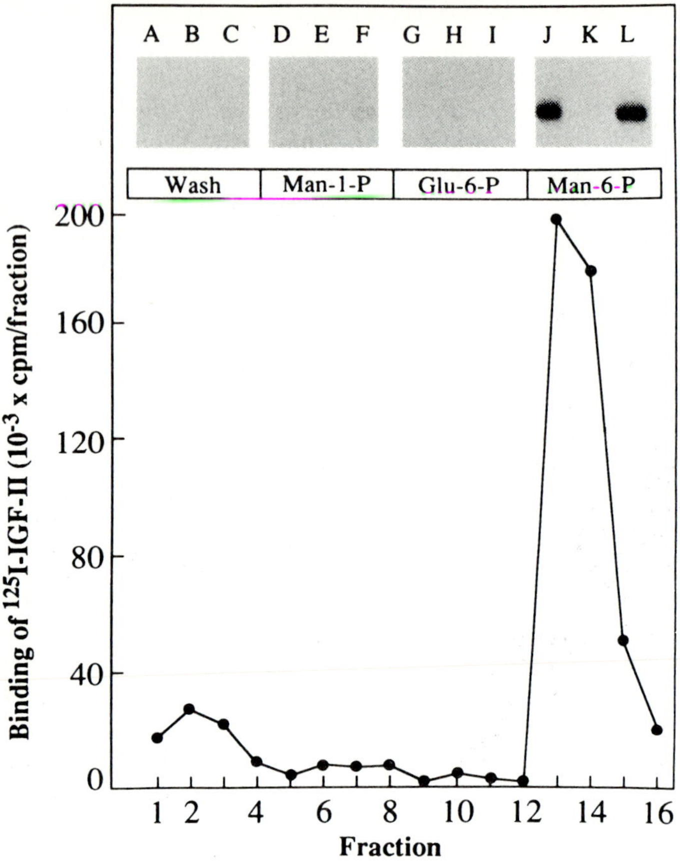

FIG. 2. Chromatography of the type II IGF receptor on immobilized pentamannosyl-6-phosphate (PMP). Rat placental plasma membranes were solubilized in 50 mM Hepes, pH 7.4, 0.15 M NaCl, 5 mM β-glycerophosphate, 0.05% Triton X-100 (1.5 mg of protein in 3 ml). This extract was incubated for 1 hour at 3 °C with 1 ml of PMP-Sepharose. After collecting the flow-through material, the column was washed with 1 ml of buffer, and then eluted sequentially with four 1 ml portions each of buffer containing 5 mM mannose 1-phosphate (Man-1-P), glucose 6-phosphate (Glu-6-P), or Man-6-P (see bar). Fractions (1 ml) were collected, and duplicate 0.1 ml portions were assayed for ^{125}I-IGF-II binding activity. Top. Affinity cross-linking of ^{125}I-IGF-II to material eluted from the column. Portions (50 μl) of pooled fractions 1 to 4 (wash lanes A to C), fractions 5 to 8 (Man-1-P, lanes D to F), fractions 9 to 12 (Glu-6-P, lanes G to I), or fractions 13 to 16 (Man-6-P, lanes J to L) were incubated for 2 hours at 3 °C with 3 nM ^{125}I-IGF-II alone (lanes C, F, I and L). The samples were treated with 0.25 mM disuccinimidyl suberate for 30 minutes at 3 °C and then subjected to electrophoresis on SDS-PAGE. An autoradiogram of the dried gel shows the M_r 250 000 band specifically eluted by Man-6-P. Labelling of this band is displaced by excess IGF-II but not insulin, consistent with the properties of the high affinity type II IGF receptor (reproduced from MacDonald et al 1988).

domain segments between species should be useful in determining the IGF-II-binding domain of the mammalian IGF-II/Man-6-P receptor.

It has been found that soluble forms of the type I IGF receptor are present in the circulation (Kiess et al 1987). Studies in our laboratory were conducted to test the hypothesis that the soluble IGF-II/Man-6-P receptor exists as the extracellular domain of truncated cell membrane receptors. Antipeptide antibodies were prepared in rabbits against deduced peptide sequences in the extracellular and cytoplasmic domains of the rat IGF-II/Man-6-P receptor. Serum IGF-II/Man-6-P receptors retained the ability to be recognized by antibodies to the extracellular domain peptide, but were not recognized by antibody to the cytoplasmic domain peptide. These data are consistent with the idea that this serum receptor is identical to the extracellular portion of the surface receptor, but devoid of the cytoplasmic domain. It seems likely that the serum IGF-II/Man-6-P receptor also lacks the transmembrane segment, but this is not proven. The results suggest that the serum IGF-II/Man-6-P receptor is a cleavage product of a cell membrane receptor precursor, but altered mRNA splicing cannot be eliminated as a possible mechanism for the production of truncated receptor.

Work in our laboratory (Oka et al 1984, Oka & Czech 1986) and others (Wardzala et al 1984) has revealed a marked increase in the number of plasma membrane IGF-II/Man-6-P receptors at the expense of receptors in intracellular pools in response to insulin action. This action of insulin to cause redistribution of IGF-II/Man-6-P receptors to the cell surface membrane reflects a general paradigm in cell biology. Experiments in our laboratory have shown that at least three other receptor proteins undergo similar membrane redistribution reactions in response to insulin. These include the transferrin receptor, the LDL receptor, and the α-2 macroglobulin receptor. In each case, an increased number of these receptor proteins can be directly monitored on the surface of intact target cells upon addition of the hormone. Furthermore, other growth factors that activate specific receptor kinases also appear to modulate the membrane distribution of these proteins. For example, we found that addition of PDGF (platelet-derived growth factor), EGF, IGF-I or insulin to A431 cells in culture caused a rapid expression of transferrin receptors on the cell surface, as monitored by specific anti-receptor antibody (Davis & Czech 1986). All the membrane receptors modulated by insulin in this manner are known to recycle constitutively between the plasma membrane and endosomal membranes. Thus it would appear that insulin and other growth factors modulate one or more steps in the recycling process. Importantly, we have also been able to demonstrate that membrane redistribution of IGF-II/Man-6-P and transferrin receptors is associated with increased uptake of their respective ligands. Thus the receptors for IGF-II and transferrin appear to be regulated by insulin and the other growth factors for the purpose of enhancing cellular uptake of these ligands.

It has been hypothesized that the phosphorylation of the IGF-II/Man-6-P receptor is involved in regulating the movement and subcellular distribution of this molecule. This hypothesis is based on the finding that addition of insulin to adipocytes or H-35 hepatoma cells, which causes a rapid redistribution of receptors among intracellular and plasma membranes, simultaneously decreases the estimated specific activity of [^{32}P]phosphate in plasma membrane IGF-II/Man-6-P receptors. Importantly, IGF-II/Man-6-P receptors isolated from insulin-treated H-35 cells or adipocytes were selectively phosphorylated *in vitro* by a specific serine/threonine kinase, casein kinase II, to a greater extent than receptors isolated from control cells. Thus, the insulin-regulated phosphorylation sites on the IGF-II receptor appear to serve as substrates *in vivo* for casein kinase II or an enzyme with similar substrate specificity (Corvera et al 1988a). A relationship between IGF-II/Man-6-P receptor phosphorylation and its movement among cellular membranes is also consistent with the finding that the highly phosphorylated form of the receptor co-purifies with the heavy chain of clathrin, suggesting that receptor phosphorylation may be involved in the concentration of receptors in clathrin-coated structures (Corvera et al 1988b).

A model which exemplifies a current hypothesis on the role of phosphorylation in IGF-II/Man-6-P receptor transit and on the effects of insulin on the phosphorylation and distribution of the receptor is shown in Fig. 3. According to this model, the distribution of IGF-II/Man-6-P receptors between endosomal and plasma membranes is modulated by an increased rate of exocytosis and a decreased rate of endocytosis in response to insulin. The latter effect is hypothesized to be related to the insulin-mediated dephosphorylation of plasma membrane IGF-II/Man-6-P receptors. The identification of the precise sites that are phosphorylated in this receptor *in vivo*, and their modification by site-directed mutagenesis will be required to directly test this hypothesis.

Insulin receptor signalling mechanisms

A major action of insulin is the modulation of metabolic pathways leading to increased synthesis of protein, glycogen and fat. The signalling system or systems which mediate these anabolic responses result in the regulation of key controlling enzymes in their respective metabolic pathways. In those cases where detailed molecular information is available, these insulin-regulated enzymes appear to undergo dephosphorylation on serine/threonine residues in response to the hormone. Thus, dephosphorylation of glycogen synthase increases the activity of the enzyme leading to increased glycogen deposition (Roach et al 1977), whereas dephosphorylation of hormone-sensitive lipase inhibits its activity and the hydrolysis of triglyceride stores (Stralfors et al 1984). Similarly, pyruvate dehydrogenase is dephosphorylated and activated in response to insulin (Denton et al 1975), leading to increased synthesis of acetyl CoA and fatty acids. Taken

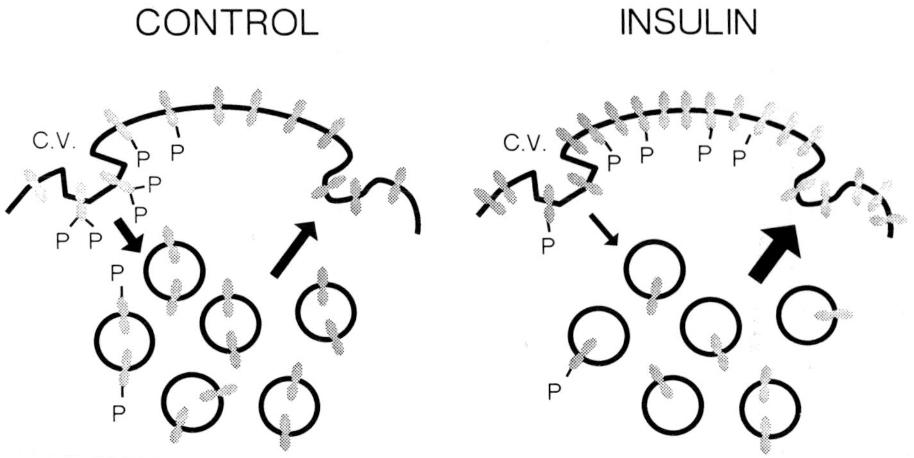

FIG. 3. Hypothetical relationship between insulin-stimulated cell surface IGF-II/Man-6-P receptor number and receptor phosphorylation state. In control adipocytes, most IGF-II/Man-6-P receptors (90% or more) reside in intracellular membrane fractions, whereas addition of insulin causes a marked redistribution of receptors to the plasma membrane. Associated with this up-regulation of IGF-II/Man-6-P receptors, the phosphorylation state of a subpopulation of these receptors in the plasma membrane fraction decreases substantially. This subpopulation of receptors appears to be associated with the clathrin-enriched, detergent-insoluble aspect of isolated plasma membranes. A direct causal relationship between the regulation of IGF-II/Man-6-P receptor phosphorylation and its membrane dynamics has not yet been demonstrated.

together, these and other examples suggest that a central paradigm of insulin action is the dephosphorylation of important regulatory enzymes in intermediary metabolism.

Insulin action also leads to rapid stimulation of protein phosphorylation on serine/threonine residues in target cells within minutes of exposure to the hormone. This stimulation of protein phosphorylation by insulin in ^{32}P-labelled cells is quantitatively more extensive than the enzyme dephosphorylations described above, and can be easily visualized by electrophoretic analysis of whole cell extracts. Several major proteins are targets of this effect, including the S6 ribosomal protein, ATP citrate lyase, acetyl CoA carboxylase, and several unidentified species. The insulin receptor is also serine-phosphorylated in response to insulin-binding in intact cells. Although increased serine/threonine phosphorylation of cellular proteins is a major effect of insulin, in no instance has the insulin-stimulated phosphorylation been yet shown to actually modulate the phosphorylated protein's activity. It is possible that these phosphorylations regulate aspects of protein function, for example, turnover or cellular compartmentalization, that are not yet understood. Alternatively, they may be physiologically silent, reflecting insulin-mediated changes in kinase or phosphatase activities that regulate less abundant, unidentified proteins.

The relatively large magnitude of the effect of insulin to enhance protein phosphorylation suggests its potential importance in the mechanism of insulin action. Results from our laboratory and others provide compelling evidence that multiple serine/threonine kinases are indeed rapidly activated by insulin. The stimulated kinase activities are stable to cell lysis under carefully controlled conditions which usually include the presence of phosphatase inhibitors. Furthermore, the magnitude of the activation of these kinases due to insulin action is often observed to increase as the kinase is partially purified. The insulin-activated serine/threonine kinases reported in the literature are listed with their potential substrates in Fig. 4.

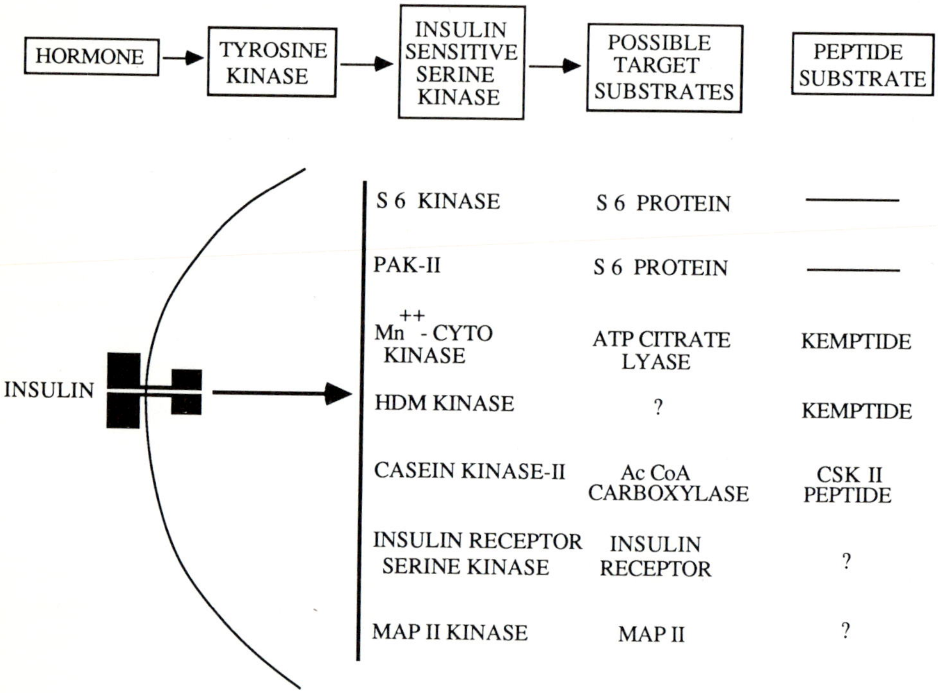

FIG. 4. Insulin receptor signalling involves activation of many serine/threonine kinases. It is now established that insulin enhances the activity of at least seven serine/threonine kinases in target cells. The possible physiological substrates listed are identified cellular proteins that exhibit elevated phosphorylation states in response to insulin in intact cells. They are listed adjacent to the kinase activity most likely to catalyse their insulin-stimulated phosphorylation *in vivo*, based on *in vitro* substrate specificity determinations or identification of phosphorylation site sequences. Peptide substrates for the kinases, where known, are listed to the right.

Two cell-free preparations have been described which contain insulin-sensitive serine/threonine kinase activities associated with a membrane fraction. The first relates to the effect of insulin on intact cells that leads to rapid serine phosphorylation of the insulin receptor itself (Smith et al 1988). Under certain conditions in cell-free systems increased receptor phosphorylation on serine residues has been observed. Moreover, Smith et al (1988) have demonstrated that a serine kinase activity is able to associate with and phosphorylate the insulin receptor during purification from human placental membranes. More recently, we have demonstrated that purification of the human placental insulin receptor on an insulin-agarose affinity column yields a highly purified insulin receptor preparation which retains insulin-sensitive serine kinase activity (Lewis et al 1989). This insulin-sensitive kinase activity phosphorylates the insulin receptor on one threonine and several serines, including serine 1293/1294, *in vitro*. Serine 1293/1294 was also shown to be the major site phosphorylated on the insulin receptor *in vivo* in insulin-treated cells. These data suggest that a serine kinase present in the receptor preparation represents a physiologically relevant enzyme involved in regulating the receptor *in vivo*. A second membrane-bound insulin-sensitive serine kinase is segregated from plasma membranes upon sucrose density gradient centrifugation and appears to be associated with high density microsomes (HDM) (Yu et al 1987a). This kinase activity is characterized by its preferential phosphorylation of histone V-S on serine and, to a lesser extent, threonine. In contrast to the insulin receptor, the HDM kinase activity does not adsorb to wheat germ agglutinin-agarose, indicating that it may not be substantially glycosylated (Yu et al 1987a).

Insulin and other growth factors markedly increase the phosphorylation of ribosomal protein S6 in ^{32}P-labelled cells. The activities of at least two kinases, protease-activated kinase II (Traugh & Pendergast 1986) and S6 kinase (Tabarini et al 1987, Erikson & Maller 1986, Ballou et al 1988), are stimulated in extracts of insulin-treated cells. S6 kinase is cyclic nucleotide-, phospholipid- and Ca^{2+}-independent. The enzyme catalyses the incorporation of up to 5 moles of phosphate per mole of S6 and is highly specific for the S6 protein. Relative molecular masses of 67 000 and 70 000 for purified S6 kinases have been reported from bovine liver (Tabarini et al 1987) and Swiss 3T3 cells (Ballou et al 1988). In addition, a higher M_r of 92 000 was obtained for the kinase purified from *Xenopus* oocytes (Erikson & Maller 1986). The activation of S6 kinase or kinases occurs within minutes at nanomolar concentrations of insulin, paralleling the time course and dose response for insulin-stimulated S6 phosphorylation *in vivo*.

Another novel insulin-stimulated Mn^{2+}-dependent serine kinase of estimated M_r 50 000–60 000 from rat adipocytes has been characterized (Yu et al 1987b). This cytosolic kinase activity was stimulated two-fold in insulin-treated cells using Kemptide (Leu-Arg-Arg-Ala-Ser-Leu-Gly) as substrate. A 4-fold stimulation of adipocyte Mn^{2+}-dependent cytosolic kinase by insulin was observed after

DEAE-Sephacel and molecular sieve chromatography. The activity of this insulin-stimulated kinase is higher in the presence of Mn^{2+} than with Mg^{2+}, is cyclic nucleotide-, phospholipid-, and Ca^{2+}-independent, and appears to be unrelated to insulin-stimulated S6 and protease-activated kinases. Adipocyte Mn^{2+}-dependent cytosolic kinase catalyses the phosphorylation of histone, Kemptide and ATP citrate lyase, but not ribosomal protein S6 and casein. Peptide mapping and radiosequence analysis of ATP citrate lyase phosphorylated *in vitro* in the presence of this kinase activity showed the phosphorylated site to be identical to that observed in intact cells in response to insulin (Yu et al 1989).

A unique cytosolic kinase implicated in insulin action is casein kinase II, an enzyme distributed widely among eukaryotic organisms. The activity of casein kinase II, as evidenced by phosphorylation of its specific substrate peptides, has been found to be enhanced by insulin in differentiated 3T3-L1 adipocytes (Sommercorn et al 1987), and by insulin and insulin-like growth factor in BALB/c-3T3 cells (Klarlund & Czech 1988). The measurements were made in cell extracts and in partially purified preparations of the kinase. The magnitude of the activation is a modest 30–160%. The activation of casein kinase II by insulin is in accord with some studies of potential cellular targets of the enzyme. Acetyl CoA carboxylase from insulin-stimulated ^{32}P-labelled adipocytes exhibits a 31% increased phosphorylation on a unique tryptic peptide in the 240 kDa subunit that is also phosphorylated upon incubation of the purified enzyme with casein kinase II *in vitro* (G. Hardy, personal communication).

Interesting results have been reported on yet another cytosolic kinase activity. Treatment of serum-starved 3T3-L1 adipocytes with insulin results in a 50–200% stimulation of a soluble kinase activity catalysing the specific *in vitro* phosphorylation of microtubule-associated protein 2 (MAP-2) (Sturgill & Ray 1986, Ray & Sturgill 1987). Incorporation of phosphate into MAP-2 occurs principally on serine and to a lesser extent on threonine residues. MAP-2 and S6 kinase activities can be resolved by phosphocellulose chromatography. Most important, S6 kinase activity appears to be enhanced upon its phosphorylation by the MAP-2 kinase (Sturgill et al 1988).

The mechanism or mechanisms whereby activation of membrane-bound or cytosolic serine/threonine kinases by insulin occurs is not known. Although a number of tyrosine-phosphorylated proteins have been identified in insulin-treated cells, none has yet been shown to play a clear role in cellular signalling. It is possible that the key substrate for the insulin receptor tyrosine kinase is the receptor β-subunit itself. Thus, receptor phosphorylation may be a requisite step for the allosteric interaction of the receptor with an effector system which mediates signalling pathways. Another hypothesis suggests that a phosphorylation cascade initiated by the tyrosine phosphorylation of a serine/threonine kinase or kinase regulator links the activated insulin receptor tyrosine kinase to this

Multiple subfamilies of carbohydrate recognition domains in animal lectins

Kurt Drickamer

Department of Biochemistry and Molecular Biophysics, Columbia University, 630 West 168th Street, New York, NY 10032, USA

Abstract. Calcium ion-dependent carbohydrate recognition domains (CRDs) are found in a range of proteins including receptors for serum glycoproteins and proteoglycans of the extracellular matrix. These C-type CRDs have homologous amino acid sequences characterized by the presence of certain invariant residues. Analysis of the genes for five of the proteins reveals that in each case the CRD-coding sequence is separated from the rest of the gene by an intron. The genes fall into two groups: those in which the coding sequence for the CRD is interrupted by two introns, and those in which the coding sequence is contained in a single exon. The sequences of domains in each category are consistent with the suggestion that the different gene structures reflect early evolutionary divergence of two subfamilies of C-type CRDs in animal lectins. However, carbohydrate-binding specificity does not directly parallel the evolutionary categorization. Comparison of the primary structures of CRDs in each subfamily which have related binding specificities may help to identify residues involved in ligating carbohydrates. This type of analysis is being extended by the use of bacterial expression systems to investigate in greater detail the binding characteristics of the CRDs.

1989 Carbohydrate recognition in cellular function. Wiley, Chichester (Ciba Foundation Symposium 145) p 45–61

The wide distribution of complex carbohydrates in animal tissues, particularly in strategic places such as at the surface of cells, in the extracellular matrix, and in the serum, suggests that oligosaccharides may be involved in communication between cells and their environment. The immense number of structures that can be made from a relatively small number of saccharide units and the multiplicity and specificity of the biosynthetic enzymes which assemble these structures are also consistent with the suggestion that information is encoded in oligosaccharides. These considerations suggest, in turn, that there must be biological molecules, presumably proteins, that can decode the information found in complex carbohydrate structures. Animal lectins have the potential to play this role; they are non-enzymic, non-immune proteins which selectively bind to specific carbohydrate structures.

An increasing number of animal lectins are being recognized and characterized at the structural level, and they can be classified according to their amino acid sequence similarities and other properties, as summarized in Table 1. In a number of cases, the ability to bind carbohydrates can be assigned to a portion of the lectin, designated a carbohydrate recognition domain (CRD). As discussed later, C-type (calcium ion-dependent) CRDs are found in proteins with diverse primary structures and biological functions. Similarly, S-type (thiol-dependent) lectins with a variety of molecular masses appear to have carbohydrate recognition domains that share certain critical residues. The CRDs found in these two types of lectins have been reviewed in detail (Drickamer 1988). The ligand-binding portions of the two types of mannose 6-phosphate receptor, although closely related to each other, do not appear to be similar to either the C- or S-type lectins (Dahms et al 1987). Similarly, serum amyloid protein is a member of the structurally unrelated pentraxin family of proteins (Pepys & Baltz 1983). The C-type animal lectins are the focus of this paper.

TABLE 1 Classification of animal lectins

Family name	*Examples*	*Ca^{2+} required?*	*Form of cysteines*	*Location*
C-type	Receptors which mediate glycoprotein endocytosis	Yes	Disulphides	Plasma membrane
	Mannose-binding proteins			Serum and liver
	Invertebrate lectins			Haemolymph
	Proteoglycan core proteins			Extracellular matrix
	Lymphocyte IgE receptor			Plasma membrane
	Pulmonary surfactant apoprotein			Lung fluid
S-type	14 kDa soluble β-galactoside lectins	No	Sulphydryls	Cytoplasm, cell surface
	35 kDa ribonuclear protein (CBP-35)			Nucleus
Man-6-P	250 kDa mannose 6-phosphate receptor	No	Disulphides	Plasma and intracellular membranes
	46 kDa mannose 6-phosphate receptor	(Yes)		
Pentraxins	Serum amyloid protein	Yes	Disulphides	Serum

Characteristics of C-type animal lectins

There are two prototypes for the C-type animal lectins. The first is a set of membrane receptors which mediate endocytosis of serum glycoproteins. The hepatic asialoglycoprotein receptor was the first of these to be described, following the demonstration that glycoproteins are cleared from mammalian circulation upon removal of their terminal sialic acid residues. This receptor has a selective, high affinity for carbohydrate structures with terminal galactose residues, and is often referred to as the rat hepatic lectin. The endocytosis of receptor-bound asialoglycoproteins via coated pits and vesicles, and the routing of these ligands to lysosomes (where they are degraded) have been the subjects of intensive investigations (Ashwell & Harford 1982). A structurally related receptor in chicken liver, the chicken hepatic lectin, mediates transfer of serum glycoproteins from the blood to the liver which is triggered after removal of galactose from typical N-linked oligosaccharides to expose *N*-acetylglucosamine. More recently, a fucose lectin from Kupffer cells has been described (Hoyle & Hill 1988). All these proteins share the general structure outlined in Fig. 1, in which cytoplasmic tails, membrane anchors and neck regions make up the N-terminal portion of each molecule, while the carbohydrate recognition domains are located at the C-termini. These CRDs share considerable structural similarity despite the fact that the proteins have distinct saccharide-binding specificities.

The second major group of C-type lectins are soluble mannose-binding proteins which have been isolated from mammalian liver and serum (Taylor & Summerfield 1986). Structural analysis of two homologous forms of the protein synthesized by rat hepatocytes, and a single form found in human liver reveals that the C-terminal portion of each constitutes a CRD that is homologous with the CRDs of the membrane lectins (Drickamer et al 1986). Much of the remainder of each mannose-binding protein consists of collagen-like sequences. It has been shown that at least one form of rat mannose-binding protein can fix complement, suggesting that this may form part of a 'preimmune' protection pathway by which pathogens that display mannose on their surfaces are recognized (Ikeda et al 1987). Comparison of all the CRDs from the membrane and soluble Ca^{2+}-dependent animal lectins reveals the presence of 18 invariant residues found in a conserved pattern within the 130-amino acids domains (see Fig. 1).

Several soluble lectins from invertebrates have also been found to contain homologous CRDs. Lectins from sea urchins and barnacles consist of CRDs only, with no accessory domains. The CRD in the lectin isolated from the haemolymph of *Sarcophaga peregrina* seems to be located at the N-terminal end of the molecule; the function of the 90 amino acids which follow on the C-terminal side is not known (Muramoto & Kamiya 1986, see Drickamer 1988 for further discussion).

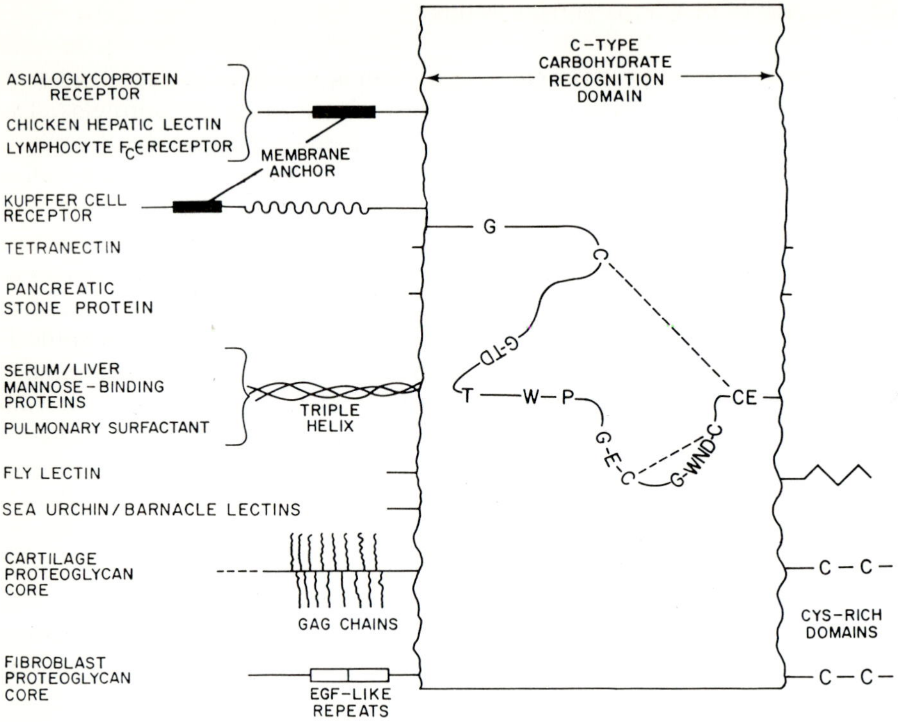

FIG. 1. Summary of structural features of C-type animal lectins. Invariant residues found in the common, Ca^{2+}-dependent carbohydrate recognition domains of a number of animal lectins are shown, flanked by schematic diagrams of the special effector domains (if any) found in individual members of the family. GAG, glycosaminoglycan; EGF, epidermal growth factor. A more detailed discussion may be found in Drickamer (1988, 1989).

Using the conserved residues as a guide, it has been possible to identify portions of other proteins which might be CRDs similar to those found in the known C-type animal lectins. Four examples have come to light so far; in two cases the prediction of carbohydrate recognition activity has been borne out experimentally. The first example is the PS-A apoprotein of pulmonary surfactant (Haagsman et al 1987). This protein is very similar in primary structure to the mammalian soluble mannose-binding proteins. Affinity chromatography has been used to demonstrate Ca^{2+}-dependent binding to a number of monosaccharides but not to amino sugars.

In contrast, the overall organization of the core protein of cartilage proteoglycan is considerably different from that of any of the C-type animal lectins so far discussed. This proteoglycan is a large, multidomain molecule that contains regions specialized for covalent attachment of glycosaminoglycans and for binding to hyaluronic acid, amongst other functions (Doege et al 1987). A possible CRD located near the C-terminus of the molecule was expressed *in vitro* and shown to interact with immobilized carbohydrate (Halberg et al 1988). Competitive studies with free monosaccharides demonstrate a binding preference for fucose and galactose, although some interaction with glucose and mannose was also detected. Fibroblast proteoglycan core protein has also been found to contain a C-type CRD (Krusius et al 1987).

Several groups have shown that the IgE receptor from lymphocytes is organized much like the membrane lectins discussed above, and that the C-terminal domain is homologous with the Ca^{2+}-dependent CRDs of these molecules (Ikuta et al 1987, Luden et al 1987). It has also been noted that the pancreatic stone (thread) protein has the sequence features characteristic of C-type CRDs (de Caro et al 1987). However, carbohydrate-binding activity has not yet been reported in either of these cases.

Subclassification of C-type CRDs based on gene structures

The CRDs in the proteins discussed in the preceding sections are found in association with a range of other effector domains. To determine if the association of a common recognition domain with a range of other functional protein segments has come about as a result of exons shuffling during evolution, we analysed the structures of the genes encoding two proteins with C-type CRDs: the major form of the rat liver asialoglycoprotein receptor and one form of the rat serum/liver mannose-binding protein (Leung et al 1985, Drickamer & McCreary 1987). In both genes, the exons which encode the CRDs of these proteins are separated from the rest of the gene by large introns. However, in the former case the CRD-coding sequences are interrupted by two additional introns, whereas in the latter case the entire CRD is transcribed and translated from a single exon (see Fig. 2). This suggests that it may be possible to classify C-type CRDs into subfamilies according to the presence or absence of introns within the gene segments coding for these domains.

It is now possible to compare the gene structures of several other C-type CRDs with those of these two examples. The intron patterns for the human lymphocyte IgE F_c receptor (Suter et al 1987) and core protein of chicken cartilage proteoglycan (Tanaka et al 1988) have been established. The organization of these genes closely resembles the pattern seen for the asialoglycoprotein receptor: in each case, two interrupting exons are present in almost precisely the same positions as in the asialoglycoprotein receptor gene (Fig. 2). On the other hand, the gene for the human pulmonary surfactant apoprotein PS-A contains an uninterrupted exon for the CRD, and thus resembles the rat mannose-binding protein gene (White et al 1985).

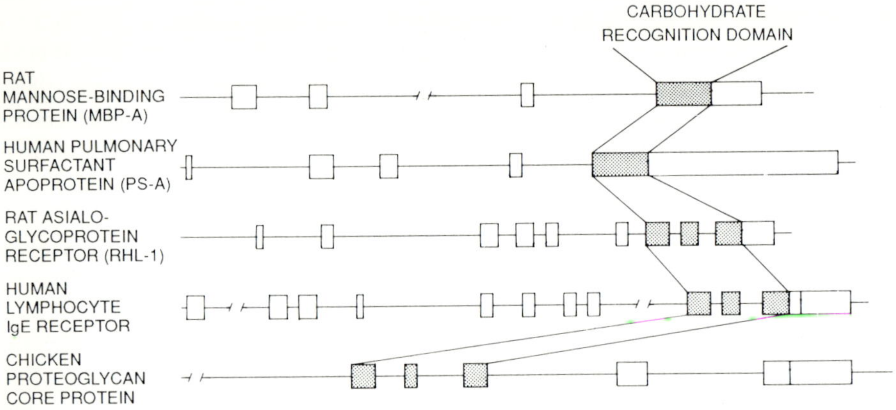

FIG. 2. Comparison of gene structures of C-type CRDs. Two introns that interrupt the coding sequences for the CRDs of the asialoglycoprotein receptor, the lymphocyte IgE receptor and the proteoglycan core protein are not found in the genes for the mannose-binding protein and pulmonary surfactant apoprotein PS-A.

From an evolutionary point of view, these results are consistent with the proposal that two subfamilies of C-type CRDs diverged early in evolution. Arguments presented by Blake (1985) suggest that the ancestral gene is likely to have contained the two introns, and that these have been lost in the case of the subfamily related to the mannose-binding protein. It is convenient to denote the two subfamilies as C(+) and C(−) CRDs to signify the presence or absence of introns in their genes.

Subclassification of C-type CRDs based on sequence comparisons

In addition to definition of highly conserved residues, comparison of the galactose-binding CRDs in the various known mammalian asialoglycoprotein receptors with the CRDs of the mannose-binding proteins allows identification of residues that are uniquely associated with each of these groups. A convenient way of finding such residues is to search for positions at which a particular amino acid is always found in the galactose-binding domains, but never in mannose-binding domains. Such residues, and those defined in the converse way, are shown in Fig. 3. Like the highly conserved residues found in all the C-type CRDs, the mannose- and galactose-characteristic residues are distributed throughout the CRDs, although they are somewhat more concentrated in the N-terminal portions of the domains.

Examination of other CRDs for the presence of these characteristic residues reveals that some of the domains are more closely related to the galactose-binding domains, while others are more like the mannose-binding domains. The results

```
Ala-Glu-Ile-Asn-Thr-Leu-Lys-Ser-Lys-Leu-Glu-Leu-Thr-Asn-Lys-Leu-His-Ala-Phe-Ser-Met-Gly-Lys-Lys-Ser-Gly-Lys-Lys-Phe-Phe-   MBP-A
                                                        Lys                 Ser                             Lys     Phe    Man
                                                                                                    Gly                    INVARIANT
    Cys Gln                         Asn Gly                 Cys Cys Pro     Asn Trp Val                 Ser Cys     Trp    Gal
Ser-Cys-Gln-Met-Ala-Ala-Leu-Arg-Gly-Asn-Gly-Ser-Glu-Arg-Ile-Cys-Cys-Pro-Ile-Asn-Trp-Val-Glu-Tyr-Glu-Gly-Ser-Cys-Tyr-Trp-   RHL-1

Val-Thr-Asn-His-Glu-Arg-Met-Pro-Phe-Ser-Lys-Val-Lys-Ala-Leu-Cys-Ser-Glu-Leu-Arg-Gly-Thr-Val-Ala-Ile-Pro-Arg-Asn-Ala-Glu-   MBP-A
                        Met                     Lys Ala Leu                                 Ala     Pro Arg Asn Ala        Man
                                                            Cys                                                            INVARIANT
    Phe                         Trp                     Tyr                     Ala His Leu     Val         Ser     Glu    Gal
    Phe-Ser-Ser-Ser-Val-Lys-Pro-Trp-Thr-Glu-Ala-Asp-Lys-Tyr-Cys-Gln-Leu-Glu-Asn-Ala-His-Leu-Val-Val-Val-Thr-Ser-Trp-Glu-   RHL-1

    Glu-Asn-Lys-Ala-Ile-Gln-Glu-Val-Ala-Lys-Thr-Ser-Ala-Phe-Leu-Gly-Ile-Thr-Asp-Glu-Val-Thr-Glu-Gly-Gln-Phe-Met-Tyr-Val-   MBP-A
    Glu Asn     Ala Ile                 Lys         Ala Phe Leu     Ile                     Thr Glu         Phe            Man
                                                                Gly     Thr Asp                                            INVARIANT
    Gln     Phe             His                         Trp         Leu                     Gly     Trp Lys Trp         Gly    Gal
Glu-Gln-Arg-Phe-Val-Gln-Gln-His-Met-Gly-Pro-Leu-Asn-Thr-Trp-Ile-Gly-Leu-Thr-Asp-Gln-Asn-Gly-Pro-Trp-Lys-Trp-Val-Asp-Gly-   RHL-1

Thr-Gly-Gly-Arg-Leu-Thr-Tyr-Ser-Asn-Trp-Lys-Lys-Asp-Glu-Pro-Asn-Asp-                        His-Gly-Ser-Gly-Glu-Asp-Cys-Val-   MBP-A
    Gly                                         Glu     Asn                                         Ser                    Man
Thr                                 Trp             Pro                                     Gly         Glu     Cys        INVARIANT
        Tyr                 Lys                     Gln     Asp     Trp     Gly His             Gly                        Gal
Thr-Asp-Tyr-Glu-Thr-Gly-Phe-Lys-Asn-Trp-Arg-Pro-Gly-Gln-Pro-Asp-Asp-Trp-Tyr-Gly-His-Gly-Leu-Gly-Gly-Gly-Glu-Asp-Cys-Ala-   RHL-1

Thr-Ile-Val-Asp-Asn-Gly-Leu-Trp-Asn-Asp-Ile-Ser-Cys-Gln-Ala-Ser-His-Thr-Ala-Val-Cys-Glu-Phe-Pro-Ala                        MBP-A
                Asn                                         Ser                     Phe                                    Man
                    Gly     Trp Asn Asp         Cys                             Cys Glu                                    INVARIANT
                Asp                                                 Arg Trp                                                Gal
His-Phe-Thr-Thr-Asp-Gly-His-Trp-Asn-Asp-Asp-Val-Cys-Arg-Arg-Pro-Tyr-Arg-Trp-Val-Cys-Glu-Thr-Glu-Leu-Gly-Lys-Ala-Asn        RHL-1
```

FIG. 3. Residues characteristic of mannose and galactose CRDs. The sequences of the CRD of one form of rat mannose-binding protein (MBP-A) and one form of the rat asialoglycoprotein receptor (RHL-1) are shown with the invariant residues found in all the C-type CRDs. Also shown are residues present in all the known galactose-binding domains (two rat and two human asialoglycoprotein receptors) but in none of the mannose-binding domains (two rat and one human mannose-binding proteins), and residues found only in mannose- but not galactose-binding domains.

of these comparisons are summarized in Table 2. Classification of C-type CRDs based on this criterion of sequence similarity is closely congruent with the classification based on intron structure (where this information is available). For example, the presence of 14 of 43 galactose-specific residues and none of the mannose-specific residues in the CRD from the proteoglycan core protein and 17 of the 43 galactose-specific residues and only one mannose-specific residue in the lymphocyte IgE receptor is consistent with classification, based on gene organization, of these proteins as C(+)-type CRDs. In other words, the protein sequences of the C-type CRDs reflect the same evolutionary divergence observed in the gene structures. Although the gene structure of the Kupffer cell lectin has not yet been established, analysis of subfamily-characteristic residues strongly suggests that the Kupffer cell lectin belongs to the C(+) subfamily: it contains 18 of the galactose-specific residues found in the protypical C(+) CRDs and only one of the mannose-specific residues found in C(−) CRDs (see Table 2).

TABLE 2 Classification of C-type CRDs into subgroups

Protein in which CRD is found	*Introns present*	*Characteristic residues*		*Monosaccharide binding specificity*				
		Man	*Gal*	*Gal*	*Fuc*	*Man*	*GlcNAc*	*GalNAc*
Asialoglycoprotein receptors:								
RHL-1	+	0	43	+	−		−	+
RHL-2/3		0	43					
Proteoglycan core protein	+	0	14	+	+	±	−	−
Lymphocyte IgE receptor	+	1	17					
Kupffer cell lectin		1	18	+	+	+	−	+
Chicken hepatic lectin		8	14	−	±	±	+	−
Mannose-binding proteins:								
MBP-A	−	31	0	−	+	+	+	−
MBP-C		31	0					
Pulmonary surfactant apoprotein A	−	9	3	+	+	+	−	−

The functional significance of the protein sequence similarities seen by this type of comparison is not as readily apparent. While some of the conservation of residues seen within each category may reflect an absolute requirement for these amino acids in forming particular binding sites, some may simply indicate that the time for divergence within each evolutionary subfamily has not been long enough for complete randomization. Further analysis of these factors requires consideration of the diverse binding specificities of the various C-type CRDs.

Binding specificity of subfamilies of C-type CRDs

A crude summary of carbohydrate-binding specificity, defined at the level of monosaccharides, is also presented in Table 2. It should be emphasized that these preferences have been defined in different ways for each of the various CRDs, so the comparison is only an approximate one. Nevertheless, some important patterns can be discerned. For example, there is some resemblance in binding characteristics within each of the evolutionary subfamilies: all the C(+) CRDs bind galactose and none binds *N*-acetylglucosamine, whereas all the C(−) domains bind mannose and fucose. Equally important, however, is the fact that some specificities can be shared by members of the two branches of the CRD family. For example, fucose and mannose are ligands for the proteoglycan core protein and Kupffer cell lectin CRDs as well as for the C(−) CRDs, and pulmonary surfactant apoprotein shares galactose-binding activity with the C(+) CRDs. This is shown schematically in Fig. 4.

By combining the ligand-binding information with the sequence comparisons discussed earlier, it is possible to gain a clearer picture of the role of some of the conserved residues within each evolutionary subgroup. For example, residues found in all the C(+) CRDs known to bind galactose (rather than those found in just the asialoglycoprotein receptors) are the ones most likely to be involved in ligating this sugar.

A particularly informative set of comparisons can be made with the chicken hepatic lectin. This protein has a function very similar to the mammalian asialoglycoprotein receptor; it is also a membrane receptor involved in endocytosis. However, it is most closely related to the mannose-binding proteins in ligand specificity—it binds *N*-acetylglucosamine, mannose, and fucose—and lies somewhere in the middle between the two C-type CRD subfamilies in terms of characteristic residues (see Table 2). This pattern can be explained by postulating that the CRD has arisen (by divergent evolution) from the C(+) subfamily, but has gained the ability to bind ligands such as *N*-acetylglucosamine by becoming (through convergent evolution) similar to members of the C(−) subfamily. The viability of this hypothesis can eventually be tested by analysis of the exon structure of the gene for this protein when it is isolated. If it holds

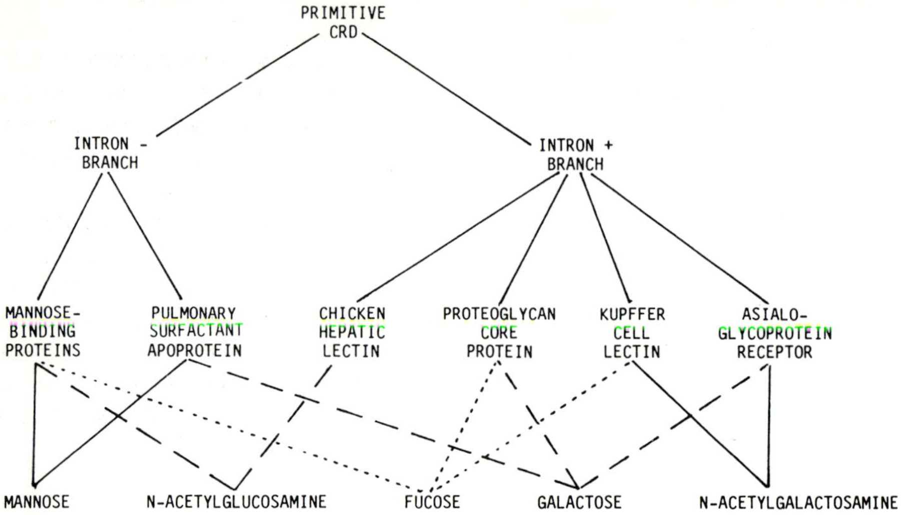

FIG. 4. Partial evolutionary scheme for C-type CRDs. The first and second levels of divergence within the C-type CRD family are shown in the upper portion of the figure. At the bottom, binding specificities of the CRDs, defined at the level of monosaccharides, are summarized.

up to this test, the theory predicts that residues found both in this C(+) CRD and in the C(−) domains are highly likely to be responsible for conferring the ability to bind ligands such as *N*-acetylglucosamine. This prediction will be testable by appropriate site-specific mutagenesis experiments.

In order to explain more precisely the conserved or convergent presence of certain residues in individual subgroups of binding domains, it will be particularly important to take into account binding specificities of the CRDs at higher resolution than the monosaccharides discussed up to this point. A possibility which has not yet been extensively considered is that there may be two distinct saccharide-binding sites in a single C-type CRD. That this might be the case has been indicated by results of the analysis of the interaction between the isolated CRD of the chicken hepatic lectin and unbranched and polyantennary ligands (Loeb & Drickamer 1988). The very much higher affinity of this domain for oligosaccharides with multiple terminal *N*-acetylglucosamine residues is most readily explained by the presence of multiple binding sites within the monomeric domain. Some techniques that will make possible more extensive analysis of the interactions between the C-type CRDs and oligosaccharide ligands are described in the next section.

Tools for future comparisons of CRDs

In order to extend our understanding of ligand-binding to the C-type CRDs, methods have been developed for expression of the domains in bacteria. One of the soluble mannose-binding proteins has been selected for the preliminary studies. It is possible to express this binding domain in the absence of the collagen-like portion of the molecule using a vector which directs the secretion of the protein into the periplasmic space. The protein produced in this manner is not correctly folded, but can be renatured following denaturation in guanidine hydrochloride in the presence of reducing agent. The renatured material is readily purified by affinity chromatography on mannose-Sepharose (Drickamer 1989).

Mannose-binding protein CRD prepared in this way has been used to probe for high affinity ligands. Nitrocellulose blots of glycoproteins separated by SDS-polyacrylamide gel electrophoresis can be stained with iodinated CRD (Fig. 5). The results confirm the specificity of binding demonstrated by more conventional methods. The best ligand for the domain is the oligomannosidic structure of yeast invertase. The CRD is also shown to bind neoglycoproteins which contain terminal mannose or *N*-acetylglycosamine, whereas it does not bind to a galactose neoglycoprotein. The CRD also shows specificity for terminal *N*-acetylglycosamine by binding to degalactosylated α_1-acid glycoprotein, but not to material containing terminal galactose or sialic acid.

In future, it will be possible to extend this approach to investigate the ligand specificities of CRDs using neoglycolipids. Oligosaccharides obtained by enzymic or chemical release from glycoproteins (N- and O-linked) are conjugated to phosphatidylethanolamine to form neoglycolipids. These neoglycolipids can be resolved by established high-performance thin-layer chromatographic methods, and the chromatograms can be probed with radioactive binding protein (Tang & Feizi 1987, see Feizi, this volume). The advantage of this method is that it allows identification of natural oligosaccharide ligands which are available in only very limited quantities. Furthermore, these do not have to be initially pure, as the high resolution of the chromatographic technique allows separation of even closely related structural isomers. Although blotting of intact glycoproteins can provide useful information about the specificity of certain binding domains, this approach suffers from the fact that the oligosaccharides attached to most glycoproteins are heterogeneous. Analysis of individual species of oligosaccharides in the form of neoglycolipids is clearly preferable because binding preferences for each form of oligosaccharide present in the mixture can be ascertained.

As expression systems for other CRDs are developed, these methods will allow us to determine which have shared binding specificities. This approach has the advantage of allowing comparisons to be made under identical conditions using the same ligands. In combination with the type of sequence comparisons discussed in this paper these results will provide important insights into how a wide range of distinct carbohydrate-binding capabilities have evolved from one primordial CRD.

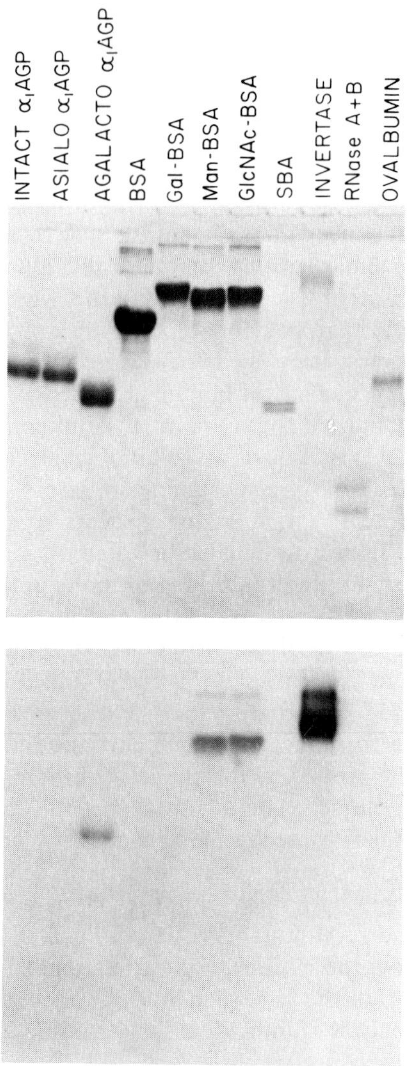

FIG. 5. Blotting of glycoproteins with expressed CRD. CRD from the mannose-binding protein was expressed in *Escherichia coli*, purified and radioiodinated. Glycoproteins were separated on SDS-polyacrylamide gels and either detected with Coomassie blue (upper panel) or transferred to nitrocellulose (lower panel) and blotted with the radiolabelled CRD. α_1AGP, α_1-acid glycoprotein; BSA, bovine serum albumin; SBA, soyabean agglutinin. Reprinted with permission from *Transactions of the Biochemical Society*.

Acknowledgements

This work was supported by Grant GM37903 from the National Institutes of Health, and a Career Scientist Award from the Irma T. Hirschl Foundation.

References

Ashwell G, Harford J 1982 Carbohydrate-specific receptors of the liver. Annu Rev Biochem 51:531–554

Blake CCF 1985 Exons and the evolution of proteins. Int Rev Cytol 93:149–185

deCaro AM, Bonicel JJ, Rouimi P, deCaro JD, Sarles H, Rovery M 1987 Complete amino acid sequence of an immunoreactive form of human pancreatic stone protein isolated from pancreatic juice. Eur J Biochem 168:201–207

Dahms NM, Lobel P, Breitmeyer J, Chirgwin JM, Kornfeld S 1987 46 kD Mannose 6-phosphate receptor—cloning, expression, and homology to the 215 kD mannose 6-phosphate receptor. Cell 50:181–192

Doege K, Sasaki M, Horigan E, Hassell JR, Yamada Y 1987 Complete primary structure of the rat cartilage proteoglycan core protein deduced from cDNA clones. J Biol Chem 262:17757–17767

Drickamer K 1988 Two distinct classes of carbohydrate-recognition domains in animal lectins. J Biol Chem 263:9557–9560

Drickamer K 1989 Demonstration of carbohydrate-recognition activity in diverse proteins which share a common primary structure motif. Biochem Soc Trans 17:13–15

Drickamer K, McCreary V 1987 Exon structure of a mannose-binding protein gene reflects its evolutionary relationship to the asialoglycoprotein receptor and nonfibrillar collagens. J Biol Chem 262:2582–2589

Drickamer K, Dordal MS, Reynolds L 1986 Mannose-binding proteins isolated from rat liver contain carbohydrate-recognition domains linked to collagenous tails. J Biol Chem 261:6878–6886

Feizi T 1989 Glycoprotein oligosaccharides as recognition structures. In: Carbohydrate recognition in cellular function. Wiley, Chichester (Ciba Found Symp 145) p 62–79

Haagsman HP, Haegood S, Sargeant T et al 1987 The major lung surfactant protein, SP-28-36, is a calcium dependent, carbohydrate-binding protein. J Biol Chem 262:13877–13880

Halberg DF, Proulx G, Doege K, Yamada Y, Drickamer K 1988 A segment of the cartilage proteoglycan core protein has lectin-like activity. J Biol Chem 263:9486–9490

Hoyle GW, Hill RL 1988 Molecular cloning and sequencing of a cDNA for a carbohydrate binding receptor unique to rat Kupffer cells. J Biol Chem 263:7487–7492

Ikeda K, Sannoh T, Kawasaki N, Kawasaki T, Yamashina I 1987 Serum lectin with known structure activates complement through the classical pathway. J Biol Chem 262:7451–7454

Ikuta K, Takami M, Kim CW et al 1987 Human lymphocyte Fc receptor for IgE: sequence homology of its cloned cDNA with animal lectins. Proc Natl Acad Sci USA 84:819–823

Krusius T, Gehlsen KR, Rouslahti E 1987 A fibroblast chondroitin sulfate proteoglycan core protein contains lectin-like and growth factor-like sequences. J Biol Chem 262:13120–13125

Leung JO, Holland EC, Drickamer K 1985 Characterization of the gene encoding the major rat liver asialoglycoprotein receptor. J Biol Chem 260:12523–12527

Ludin C, Hofstetter H, Sarfati M et al 1987 Cloning and expression of the cDNA coding for a human lymphocyte IgE receptor. EMBO (Eur Mol Biol Organ) J 6:109–114

Loeb JA, Drickamer K 1988 Conformational changes in the chicken receptor for endocytosis of glycoproteins. J Biol Chem 263:9752–9760
Muramoto K, Kamiya H 1986 The amino acid sequence of a lectin of the acorn barnacle *Megabalanus rosa*. Biochim Biophys Acta 874:285–295
Pepys MB, Baltz ML 1983 Acute phase proteins with special reference to C-reactive protein and related proteins (pentraxins) and serum amyloid A protein. Adv Immunol 34:141–211
Suter U, Bastos R, Hofstetter H 1987 Molecular structure of the gene and the 5′-flanking region of the human lymphocyte immunoglobulin-E receptor. Nucleic Acids Res 15:7295–7308
Tanaka T, Har-El R, Tanzer ML 1988 Partial structure of the gene for chicken cartilage proteoglycan core protein. J Biol Chem 263:15831–15835
Tang PW, Feizi T 1987 Neoglycolipid micro-immunoassays applied to the oligosaccharides of human milk galactosyltransferase detect blood group related antigens on both O-linked and N-linked chains. Carbohydr Res 161:133–143
Taylor ME, Summerfield JA 1986 Mammalian mannose-binding proteins. Clin Sci (Lond) 70:539–546
White RT, Damm D, Miller J et al 1985 Isolation and characterization of the human pulmonary surfactant apoprotein gene. Nature (Lond) 317:361–363

DISCUSSION

Ruoslahti: When you express the carbohydrate recognition domain (CRD) in a prokaryotic system, does it fold correctly? Do the disulphide bonds form?

Drickamer: The protein does not fold correctly in the prokaryotic system; it precipitates. But by every physical criterion we can apply, renaturing in guanidine does produce the normal conformation.

Ruoslahti: Has the normal conformation been documented by activity?

Drickamer: It has been determined by activity and by crude physical measurements, such as circular dichroism.

Barondes: Does the fusion protein also have carbohydrate-binding activity, or do you have to chop it out after you make it?

Drickamer: It is a fusion protein only to the extent that it has six extra amino acids at the N-terminus. We hoped that by expressing the binding domain with a bacterial signal sequence we would see both secretion into the periplasm and correct folding. But it didn't fold correctly. However, being directed to the periplasm by the sequence signal does seem to make the protein more stable; when we express it in the cytoplasm it tends to be degraded.

Rademacher: A glycoprotein may be assumed to have an interactive relationship with its own attached carbohydrate (i.e. proteins may bind their own carbohydrates extremely tightly). How then does one differentiate between such a carbohydrate-binding domain and one which functions as a lectin, binding carbohydrates on another protein. In other words, if you applied your approaches to the IgG molecule, which has a carbohydrate-binding domain that binds its own sugars tightly, would it be defined as a lectin? Are you not simply mapping the hydrophobic surface of proteins to which carbohydrates tend to bind? Extending your techniques to all glycoproteins, would we always find a carbohydrate-binding patch which the carbohydrate covers?

Drickamer: I don't know how to exclude that. But the simple mannose-binding proteins don't have any carbohydrates, so they are not designed to bind their own carbohydrate.

Carver: Have you searched data banks for sites on known glycoproteins that are homologous with the sugar-binding sites on your lectins?

Drickamer: Yes, but algorithms which search for distributed homologies are much improved now. Using our crude method, we think we have found most of the proteins which contain all the critical residues. A very small subset of those residues might be present in other proteins.

Lee: Many experiments have been done on Gal/GalNAc-binding proteins. Those proteins have N-linked polysaccharides which usually contain sialic acids. If sialic acid is removed, they bind to each other and cease to function normally.

Feizi: Dr Rademacher, is the binding of immunoglobulin to its own sugar chain really tight? Can you demonstrate that by methods such as those Dr Drickamer uses?

Rademacher: This high affinity binding is suggested by experiments which show that the oligosaccharides on the IgG molecule are not recognized by lectins and various exoglycosidases. If you don't denature the IgG molecule, it doesn't bind to lectin affinity columns.

Feizi: Could it be that the oligosaccharide is merely buried within the folded protein?

Rademacher: Even if the sugar was only exposed 1% of the time the equilibrium would still be shifted and binding would occur. This is an important point; the IgG sugars are never exposed. So the evidence, at least in terms of its ability to interact with other molecules, is that they are fixed and never exposed to the surface.

Feizi: The homology of tetranectin with carbohydrate-binding proteins is intriguing. Is tetranectin a carbohydrate-binding protein?

Drickamer: There is no evidence for lectin activity yet. We are simply predicting from its sequence that tetranectin will bind sugars. It binds to plasminogen, but I don't know whether its function is modulatory.

Stanley: To what extent do you think this is the tip of the iceberg? For example, would you expect to discover many more mannose-binding lectins if you made affinity columns using Man_5, Man_6, Man_7 etc?

Drickamer: If you plot how many proteins have been discovered every year that appear to have a C-type CRD, we are not saturated. By that type of argument, there must be a lot of them out there.

Stanley: Do you have evidence suggesting that you are missing the ones that have a greater specificity for naturally occurring carbohydrates?

Drickamer: We don't have direct evidence, other than the fact that there are proteins which were not known to be lectins that now turn out to be. The proteoglycan core protein is a subtle example because it does not exhibit tight binding to the 'wrong' monosaccharide ligands which we use on affinity columns. However, because we had reason to suspect that the core protein of proteoglycan has carbohydrate-binding activity, we could look for it. There are probably many other proteins which bind carbohydrates but don't stick tightly to affinity columns of the type we are using.

Jessell: Does the primary structure tell you anything about the calcium dependence? Do you see classical calcium-binding domains?

Drickamer: We don't expect to see EF hands, because the average binding constant for the two apparent sites is minimal and EF hands have micromolar affinities. Amongst the 18 invariant residues, eight or nine are potential calcium ligands—glutamic and aspartic acids, glutamines and asparagines—but they are not in a typical pattern.

Fukuda: A number of amino acids could be involved in carbohydrate binding.

Drickamer: That's why I want somebody to solve the crystal structure!

Fukuda: Do you want to answer this by crystallography rather than by mutagenesis?

Drickamer: We are doing some mutagenesis work, but we have to be careful: the domain is too big for us to start randomly changing amino acids.

Carver: You mentioned a multivalent ligand that showed a high (nanomolar) affinity for the chicken hepatic lectin CRD.

Drickamer: That was asialo-agalacto-orosomucoid, which is multivalent in two respects—it has polyantennary oligosaccharides and multiple oligosaccharides.

Carver: The affinity could be a reflection of the structure rather than of an additional interaction per monomeric domain.

Drickamer: I am over-simplifying the results as an introduction; the binding-site almost certainly binds more than a single sugar residue.

Carver: The same kind of sequence comparison for plant lectins gives the reverse story. The structurally important things are conserved, such as a common fold. Even for the so-called mannose-binding lectins there's quite a variation in the binding sites, because the interaction is with the backbone. The only side chain that's involved is one that is conserved for calcium binding. I wonder how many of your homologous residues are conserving a scaffolding rather than a common sugar-binding site.

Drickamer: It's not the reverse of what we are seeing because we don't know what our domains are. The primary interpretation I would like to place on the domains is essentially an evolutionary one: the fact of their conservation in clearly defined subfamilies. There is a secondary hypothesis that some of those domains form binding sites with related sugar selectivity.

Carver: In plant lectins a different monosaccharide specificity is evolved on an identical scaffold. So differentiating Man and Gal specificities the way you have and suggesting that those are evolving independently is contrary to what we find in plant lectins.

Hart: Have you seen any sequence homologies with glycosidases or with the two glycosyltransferases that have been sequenced?

Drickamer: My conclusion is that Nature has evolved carbohydrate recognition functions independently many times. We cannot see any relationship between our C-type domains and plant lectins, bacterial binding proteins, glycosidases or tranferases. Our search suggests that these various groups are not divergent.

Ruoslahti: Is there any reason to believe that these domains might be binding something other than carbohydrates? Nature, having invented a good binding domain, can use it in different ways.

Drickamer: There is no evidence for that.

Kobata: The model for your chicken hepatic lectin looks like C1q, a subunit of the first complement, and you said that it fixes complement. Does it contain a collagen-like domain?

Drickamer: It is the rat mannose-binding protein that fixes complement. The collagen-like domain is in every respect collagenous; it has hydroxyprolines, glucosyl and galactosyl hydroxylysines, and is sensitive to collagenase digestion. So it is almost certainly triple helical. The opsonic function is probably dependent upon that collagen-like domain.

Kobata: I am asking because of the multiple valency of the lectin. The collagen-like arm may provide flexibility in the arrangement of the binding sites. If so, the sugar structures required for the epitope may not be as strictly fixed as they are for the galactose-binding protein, where Dr Lee showed that 2,4,2-triantennary sugar chains are important (Townsend et al 1986).

Reference

Townsend RR, Hardy MR, Wang TC, Lee YC 1986 Binding of N-linked bovine fetuin glycopeptides to isolated rabbit hepatocytes—Gal/GalNAc hepatic lectin discrimination between Galβ(1,4)GlcNAc and Galβ(1,3)GlcNAc in a triantennary structure. Biochemistry 25:5716–5725

Glycoprotein oligosaccharides as recognition structures

Ten Feizi

MRC Clinical Research Centre, Watford Road, Harrow, Middlesex HA1 3UJ, UK

Abstract. A series of observations—the pronounced changes in the expression and distribution of oligosaccharide antigens during embryonic development, cell differentiation and oncogenesis, the prominence of these changing structures (oncodevelopmental antigens) on the receptor for epidermal growth factor, and the stimulation of receptor autophosphorylation following their perturbation with antibodies—has suggested that the oligosaccharides of growth factor receptors and complementary lectins may be intimately involved in molecular recognition events in growth and differentiation processes. For elucidating oligosaccharide recognition by diverse cellular and secreted proteins and microbial adhesins, a new technique has been developed which involves the overlay of immobilized oligosaccharide probes (neoglycolipids) derived from glycoproteins and other sources. New insights have been gained into carbohydrate recognition by several mammalian lectins, and a novel receptor system has been discovered in *Escherichia coli* isolated from patients with urinary tract infections. This new technique seems ideal for elucidating oligosaccharide recognition in diverse biological settings, and for 'quality control' of the sugar chains of recombinant glycoproteins engineered for the purpose of administration to man.

1989 Carbohydrate recognition in cellular function. Wiley, Chichester (Ciba Foundation Symposium 145) p 62–79

With the use of monoclonal antibodies (Feizi 1985, Hakomori 1985, Thorpe et al 1988) and the considerable advances in the techniques of carbohydrate structural analysis (Hounsell 1987, Kobata 1988), we have become increasingly aware of the pronounced changes in the display and positional patterning of various oligosaccharides during embryogenesis and morphogenesis, the continuum of changes in their expression during cell differentiation, and the regular and predictable changes during oncogenesis. Many of the structural changes, particularly those in the backbone and peripheral domains of the oligosaccharides of glycoproteins and glycolipids, are manifest as antigenic changes. Other more recently characterized structures in the backbone and peripheral domains of O-linked oligosaccharides of fetal glycoproteins are potential antigens (Hounsell et al 1988). Collectively such transiently expressed antigens are referred to as oncodevelopmental antigens (Feizi 1985), many of which are related to the major blood group antigens, A, B, H and Lewis (Watkins 1980, Kabat 1982). All this indicates that there is a remarkable programming of glycosyltransferase activities during the development and differentiation processes. The challenge we now face is to understand the roles of this diversity of structure and the relevance of the changes in the growth and differentiation processes.

Oligosaccharides: ligands in growth regulating networks?

Might oligosaccharides be the missing links in the molecular recognition events in cell growth and differentiation? Two observations have suggested to us that they may be. The first is that one of the major carriers of blood group-related oncodevelopmental antigens is the epidermal growth factor (EGF) receptor (reviewed by Feizi & Childs 1987a). The second is that perturbation of the blood group A-active oligosaccharide chains of the solubilized EGF receptor from A431 cells stimulates receptor autophosphorylation. This conclusion was reached after autophosphorylation experiments by Yarden & Schlessinger (1985) using the solubilized EGF receptor and two monoclonal anti-receptor antibodies which subsequently were shown (Gooi et al 1985) to be directed at the blood group A-related oligosaccharide structures on the receptor glycoprotein. Doubtless, the A431 cell line is derived from a blood group A individual, and alternative oligosaccharide structures would be predicted to occur on EGF receptors from cells lacking blood group A. We believe (Feizi & Childs 1987a) that the importance of these observations is that the reaction of sugar chains of the receptor with complementary antibodies elicited an effect that is associated with receptor activation. Whether this effect is due to receptor aggregation or to allosteric effects on the protein moiety is not yet known. In either case, we interpret the results as being an important clue to the role of oligosaccharides (including the blood group structures in cells that express them normally) in growth regulation. By analogy with the growth-modulating effects of exogenously added plant lectins on animal cells, we proposed that endogenous lectins may react with the sugar chains of growth factor receptors and tune their response to growth factors (Feizi & Childs 1987a). Our thesis is that oligosaccharides of cell surface glycoproteins and glycolipids, and complementary lectins, may constitute networks of functionally coupled receptor systems that transmit signals within and across cell membranes (Fig. 1).

A report by Defize et al (1988) is in strong support of this idea. They have identified and separated two cell populations in A431 cells: the first is blood group A-positive and the second blood group A-negative. These two glycosylation variants differ both in the number of high affinity receptors for

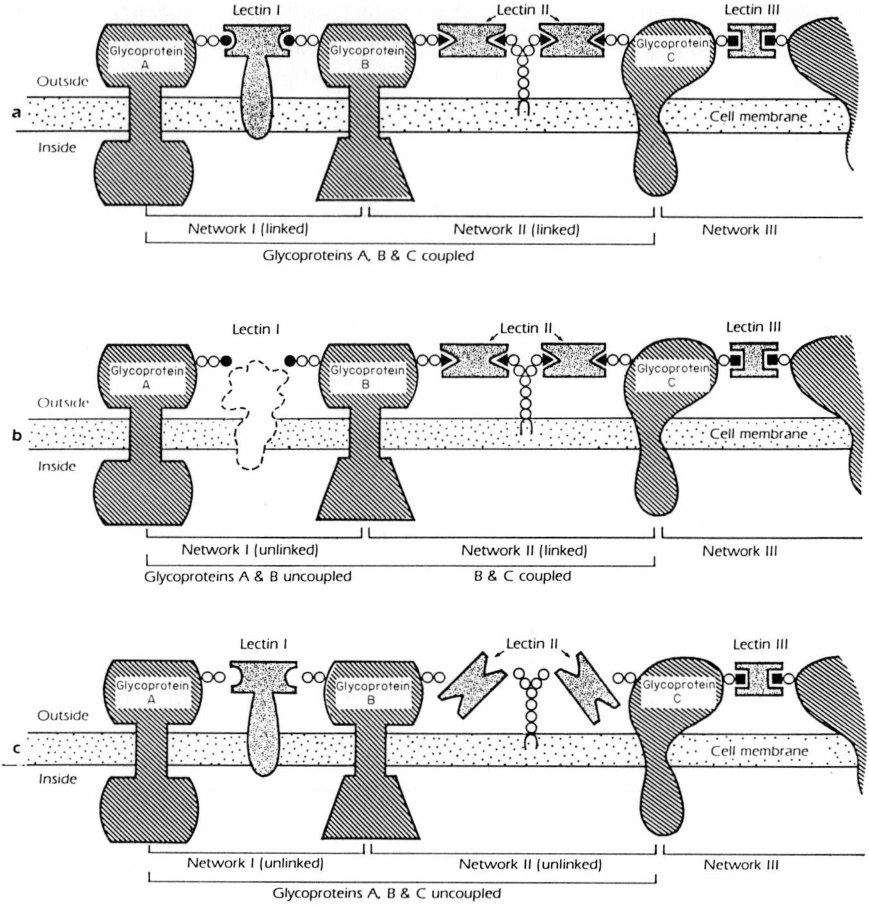

FIG. 1. Diagram of a cell membrane depicting a hypothetical network based on specific oligosaccharide–lectin interactions as a means of coupling different glycoprotein receptors. (a) Glycoproteins and glycolipids with like carbohydrate structures are linked by appropriate lectins. Linkage of glycoconjugates with different carbohydrate structures is achieved through the mediation of lectins of different binding specificities, e.g. glycoprotein A with glycoprotein C via networks I and II. (b) Network I is disrupted by the absence or inactivity of lectin I. (c) Networks I and II are both unlinked because the glycoconjugates are inappropriately glycosylated. Lectin I is represented as a membrane-associated protein, and lectins II and III are shown as soluble proteins; the three lectins have different oligosaccharide specificities. The lectins in this scheme might include microbial carbohydrate-binding proteins. Lectin–sugar networks need not be confined to plasma membranes but might operate in intracellular membranes and even in the cytosol. ○, ▲, ■, ●, monosaccharides linked in various ways; ∩ , ceramide moiety of a glycolipid. (From Feizi & Childs 1987a).

EGF and in their biological responses to added EGF. These observations suggest that a blood group A recognition system may be operating in this cell line. This requires further investigation.

A second series of relevant observations are those concerning the glycoprotein receptor for the insulin-like growth factor II (IGF-II). The first of these is the discovery from the molecular biology studies of Morgan et al (1987) that the IGF-II receptor and the cation-independent receptor for oligosaccharides bearing 6-phosphorylated mannose residues (Lobel et al 1987) are one and the same molecule. This topic is further discussed by Dr M. Czech (this volume). To accommodate these developments, an updated scheme for our proposed receptor 'cross-talk' mechanism is shown in Fig. 2 with a growth factor receptor (glycoprotein B) depicted as a multifunctional glycoprotein with a combining site for the growth factor and another for the sugar chains of those glycoproteins (e.g. glycoprotein A) with which it needs to communicate. We have proposed (Feizi & Childs 1987b) that such a mechanism, dependent on correct glycosylation, would be a versatile and adjustable means of communication between growth factor receptors and associated proteins in the same cells, and could explain many of the cooperative effects that have been observed among growth factor receptors.

These ideas should be testable in the near future, but first we need to have a sensitive means of identifying proteins that recognize the sugar chains of glycoproteins. Such a method has now been introduced and has given new information in several oligosaccharide recognition systems.

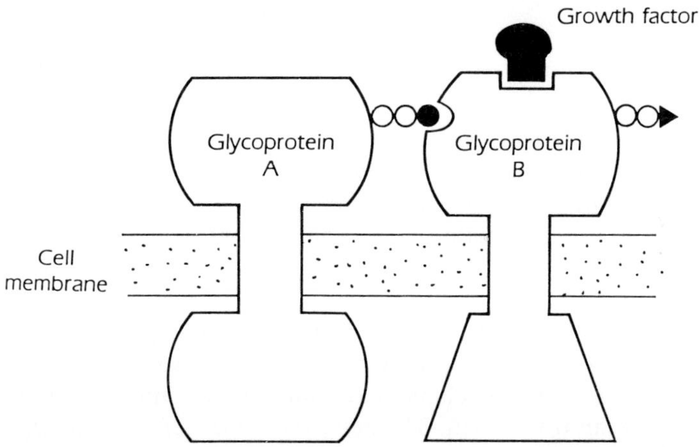

FIG. 2. Diagram depicting part of a hypothetical growth-regulating network involving oligosaccharide recognition. The growth factor receptor is shown here as a bifunctional glycoprotein (B) with a combining site for the growth factor and a 'lectin-like' site for an oligosaccharide structure that occurs on a second growth-regulatory glycoprotein (A). (From Feizi 1988).

Neoglycolipids as oligosaccharide probes for ligand-binding assays

The principle is to release oligosaccharides from glycoproteins, to conjugate these to a lipid and to use the resulting neoglycolipids as probes of oligosaccharide recognition after immobilization on insoluble matrices, e.g. silica gel or plastic plates, or after incorporation into liposomes (Tang et al 1985). Any oligosaccharide structure might be used in this system, including those synthesized chemically. Reduced oligosaccharides (such as those released from mucin-type glycoproteins by alkaline borohydride degradation) which are unreactive in conjugation reactions, can be converted into reactive oligosaccharide aldehydes by mild periodate oxidation (Tang et al 1985). For the idea and guidance in the oxidation experiments, I am indebted to Dr Y. C. Lee.

The lipid of choice was phosphatidylethanolamine dipalmitoyl (PPEADP, 1), because preliminary experiments have shown superior reactivities of conjugates using this lipid compared with those using several other readily available lipids. Conjugation is achieved by reductive amination, and the feasibility of this approach has been shown by the potent antigenicities of O-linked

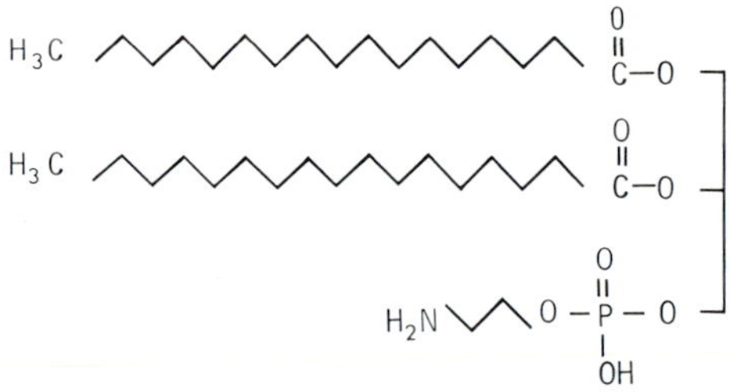

(1)

oligosaccharides released from mucin-type glycoproteins (Tang et al 1985) and from galactosyltransferase (Tang & Feizi 1987), and of sulphated oligosaccharides released from keratan sulphate by using endo-β-galactosidase (Tang et al 1986).

The conditions for conjugation of reducing oligosaccharides to PPEADP have been improved (Stoll et al 1988) and a series of neoglycolipids have been prepared from N-linked chains derived from structurally characterized glycoproteins (Mizuochi et al 1989), such as high mannose-type (oligomannosidic) oligosaccharides with five to eight mannose residues from ribonuclease B, and a spectrum of complex-type chains from human and mouse immunoglobulin G and from human transferrin. The neoglycolipids were chromatographed on silica gel plates. In collaboration with Dr A. M. Lawson, we have identified (Mizuochi et al 1989) individual neoglycolipid bands by liquid secondary ion mass spectrometry directly from the chromatograms using the procedure of Kushi & Handa (1985). The reactivities of the neoglycolipids with various carbohydrate-binding proteins are being investigated by overlay assays as used with natural glycosphingolipids (after Magnani et al 1980), as described below.

Ribonuclease B

```
                 ┌
                 │ Man(α1-6)
                 │          \
                 │            Man(α1-6)
                 │          /           \
± Man(α1-2)n     │ Man(α1-3)              Man(β1-4)GlcNAc (β1-4)GlcNAc
                 │                      /
                 │            Man(α1-3)
                 └
```

n = 0, 1, 2 or 3

Human IgG, mouse IgG and transferrin

```
                                         ± GlcNAc(β1-4)
[ ± Gal(β1-4) ] GlcNAc(β1-2)Man(α1-6)          |         ± Fuc(α1-6)
                                     \         |                |
                                       Man(β1-4)GlcNAc(β1-4)GlcNAc
                                     /
[ ± Gal(β1-4) ] GlcNAc(β1-2)Man(α1-3)
```

Neoglycolipid overlays with the plant lectin concanavalin A and animal lectins

Growth modulation in animal cells by exogenously added plant lectins was mentioned earlier. As a first step in the study of lectins with growth-modulating properties, we assessed the reactivity of concanavalin A (Con A) with neoglycolipids derived from N-linked chains of glycoproteins (Stoll et al 1988). The specificity and sensitivity of this approach is illustrated for neoglycolipids from ribonuclease B (Fig. 3, panel I). The neoglycolipid mixture (2.5 μg total carbohydrate per lane) was chromatographed on silica gel plates and overlaid with ^{125}I-labelled Con A. Binding was detected by autoradiography (lane CA). The same lane was later stained by orcinol reagent to reveal the oligosaccharide bands (lane O). In accordance with previous knowledge, Con A bound to all four high mannose-type probes. This binding was specifically inhibited by α-methylmannoside (lane CA′). Also in accordance with previous knowledge, binding to biantennary oligosaccharide probes was weaker than to the high mannose-type probes. These results are described in detail in Childs et al (1989).

In collaboration with Dr K. Drickamer, we (Childs et al 1989) have investigated oligosaccharide recognition by the recombinant carbohydrate recognition domain of rat mannose-binding protein, rMBP-A (see Drickamer, this volume). This protein has been found to react preferentially with neoglycolipids from N-linked chains of the complex-type that have two non-reducing terminal *N*-acetylglucosamine residues. It reacts only weakly in the presence of a single non-reducing terminal galactose residue and with the high mannose-type oligosaccharides, and not at all with analogues that have two

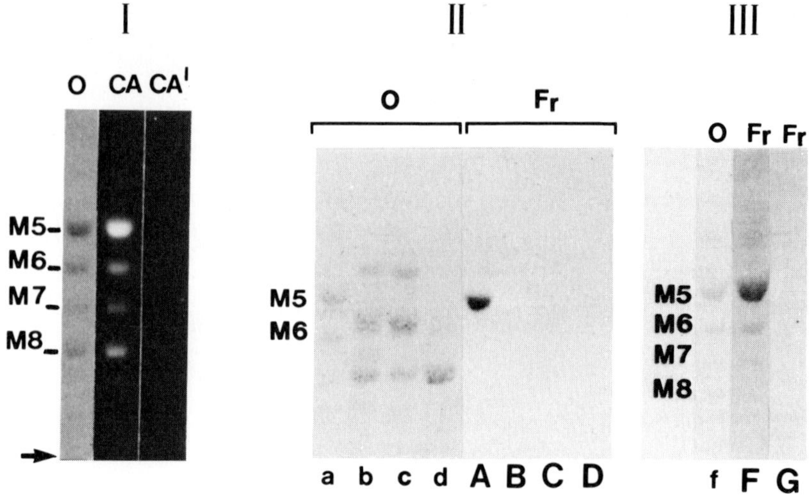

FIG. 3. Reactivities of the plant lectin Con A and of *E. coli* with neoglycolipids derived from glycoprotein oligosaccharides. Panel I, a mixture of neoglycolipids (2.5 μg carbohydrate per lane) derived from the oligosaccharides of ribonuclease B were chromatographed on silica gel plates using chloroform/methanol/water 105:100:28 (by volume) and overlaid with ^{125}I-labelled Con A. Lane CA shows binding of Con A detected by autoradiography and lane O shows the same lane after chemical staining with orcinol to reveal the neoglycolipid bands. Lane CA′ shows lack of Con A binding in the presence of 200 mM α-methylmannoside. Panel II, neoglycolipids (5 μg carbohydrate per lane) derived from the oligosaccharides of ribonuclease B (lanes A, a) and from sialidase-treated oligosaccharides of human and mouse IgG (lanes B, b and C, c respectively) and transferrin (lanes D, d) were chromatographed as above and overlaid with ^{14}C-labelled *E. coli* strain C600 isolated from a patient with urinary tract infection. Binding was detected by fluorography (Fr) and the same plate was stained with orcinol (O) to reveal the neoglycolipid bands. Panel III, ^{14}C-labelled *E. coli* were overlaid onto a chromatogram containing ribonuclease B oligosaccharides (10 μg carbohydrate per lane), in the absence (lane F) and in the presence (lane G) of 300 mM mannose. Lanes F and G show results of fluorography (Fr). Lane f shows the same lane as F stained with orcinol. Arrow indicates position of sample application. M_5 to M_8 indicate positions of the five high mannose-type oligosaccharide probes from ribonuclease B. Panel I from Stoll et al (1988) and panels II and III from Rosenstein et al (1988).

terminal galactose residues. Additional experiments with simple oligosaccharide probes have revealed a reactivity with non-reducing terminal fucose residues, as in the sequence Fuc(α1–4)[Gal(β1–3)]GlcNAc. Similar studies have been performed with natural mannose-binding proteins of rat and human sera, and with bovine serum conglutinin (collaborative studies with Dr T. Kawasaki, Professor P. J. Lachmann and Dr S. Thiel). The results (Childs et al 1989, Loveless et al 1989, Mizuochi et al 1989) indicate that the binding specificities of the four mammalian proteins are related, but not identical. Another common feature is that the recognition units for these proteins are small, possibly not much longer than monosaccharides.

Neoglycolipid overlays with *Escherichia coli*

The new technique using neoglycolipids lends itself well to studies of oligosaccharide recognition by microbial adhesins, particularly to identification of individual oligosaccharide structures among the diverse chains on glycoproteins. In studies with type 1 fimbriated *E. coli* from patients with urinary tract infections, specific adhesion has been visualized to selected neoglycolipids derived from glycoprotein oligosaccharides (Rosenstein et al 1988). Thus, when a series of oligosaccharide probes from ribonuclease B, human and mouse immunoglobulin G and from transferrin (5 μg carbohydrate per lane) were evaluated for reactivity with ^{14}C-labelled *E. coli* (Fig. 3, panel II), binding was detected to only the high mannose-type $Man_5GlcNAc_2$ (Man_5) probe (panel Fr, lane A). This is in accordance with previous studies (Firon et al 1982, Neeser et al 1986) which showed that cell attachment by type 1 fimbriated *E. coli* can be inhibited by free mannose and by certain oligosaccharides and glycopeptides having the non-reducing terminal sequence Man(α1–3)Man(β1–4)GlcNAc. This sequence is present on the Man_5

TABLE 1 Designations and structures of oligosaccharides, and the binding reactions of their neoglycolipids with *E. coli* strain C600 isolated from a patient with urinary tract infection

Designation[a]	*Oligosaccharide*	*Reactivity of neoglycolipid*
Lactose	Gal(β1–4)Glc	+ + + +
LNT	Gal(β1–3)GlcNAc(β1–3)Gal(β1–4)Glc	+ + + +
LNFP-I (H)	Gal(β1–3)GlcNAc(β1–3)Gal(β1–4)Glc / Fuc(α1–2)	+ +
LNDFH-I (Le^b)	Gal(β1–3)GlcNAc(β1–3)Gal(β1–4)Glc / | Fuc(α1–2) Fuc(α1–4)	+
(A) tetra	GalNAc(α1–3)Gal(β1–4)Glc | Fuc(α1–2)	+
(A) hexa	GalNAc(α1–3)Gal(β1–3)GlcNAc(β1–3)Gal(β1–4)Glc / Fuc(α1–2)	±

[a]Blood group A, H and Le^b activities, where present, are shown in parentheses. Taken from Rosenstein et al (1988).

structure and on some of the higher oligomannosyl structures (Man_6–Man_8) from ribonuclease B. The predicted faint reactivities with Man_6–Man_8 structures cannot be discerned in Fig. 3, panel II. However, when 10 μg carbohydrate material was applied per lane, as in Fig. 3, panel III, some binding to these higher structures was detected, and binding to all four bands was specifically inhibited in the presence of 300 mM mannose.

When a series of neoglycolipids derived from human milk oligosaccharides were used as probes in overlay experiments with several *E. coli* strains (Rosenstein et al 1988), a new adhesive specificity was revealed (Table 1). This adhesion is unrelated to the presence of type 1 fimbriae and appears to involve the lactose [Gal(β1–4)Glc]-related sequence in association with lipid, because it is not inhibited by free oligosaccharides. A related adhesive specificity has been observed among gonococci by Stromberg et al (1988) using a series of natural,

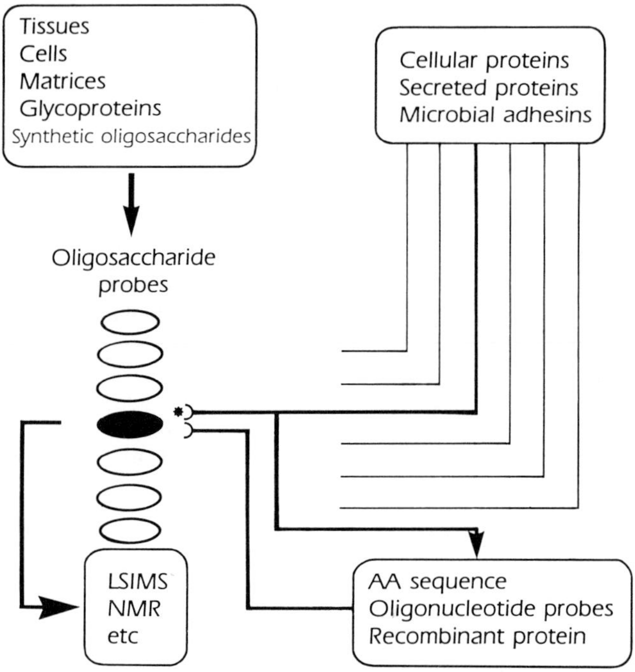

FIG. 4. A general strategy for elucidating oligosaccharide recognition systems. Whole tissues, isolated cells, extracellular matrices or individual glycoproteins, in addition to desired synthetic oligosaccharides, are used to generate neoglycolipids which are used in overlay experiments with radiolabelled proteins. Individual cellular or secreted proteins or microbial adhesins (asterisked) with carbohydrate recognition properties are singled out and characterized at the level of protein and gene structure. The oligosaccharides recognized are characterized by state-of-the-art methods of structural analysis (LSIMS, liquid secondary ion mass spectrometry; NMR, nuclear magnetic resonance spectroscopy).

epithelium-derived glycolipids. Since the lactose sequence is contained in the lipid-linked, membrane-associated domains of the majority of host cell glycolipids, we propose that this type of adhesive specificity may play an important role in the invasion of damaged epithelial cells, where the saccharide–lipid junction may be exposed.

When the lactose-containing sequence is modified by additional monosaccharides (including the blood group monosaccharides) binding of the *E. coli* is markedly impaired (Table 1). Thus it is predicted that the secretor gene and the genes coding for the blood group enzymes, or other glycosyltransferases whose levels change in epithelial cells during differentiation, proliferation and maturation, markedly influence binding and hence susceptibility to invasion.

There are many potential applications of neoglycolipids in the field of infectious diseases. We envisage banks of diverse oligosaccharide probes derived from natural sources and from synthetic oligosaccharides. These may be used to identify and map the oligosaccharide receptors for pathogens and their individual adhesins. Such studies could form the basis of novel drug designs targeted at the prevention of the initial events of infection.

Future perspectives

I wish to highlight the wide biomedical potential of this new method. Figure 4 shows a general strategy for elucidating novel oligosaccharide recognition systems in diverse settings. It is envisaged that, as appropriate, whole tissues, isolated cells, extracellular matrices as well as individual glycoproteins will be used as sources of oligosaccharide probes, and that these will be supplemented by probes derived from synthetic oligosaccharides. By the neoglycolipid overlay approach, it should be possible to identify specific carbohydrate-binding proteins from among cellular and secreted proteins and microbial proteins. The proteins that bind to particular neoglycolipid bands would be eluted, sequenced and eventually produced by protein molecular engineering. In parallel, the oligosaccharide recognition units would be identified by state-of-the-art analyses and new profiling methods under development by Drs E. F. Hounsell and M. S. Stoll and associates in our laboratory.

There is a great potential for neoglycolipids in the 'quality control' and selection of recombinant glycoproteins produced by molecular engineering for administration to man. It is well appreciated that particular sugar chains of glycoproteins are important determinants of the compatibility, the clearance rates and *in vivo* bioactivities of some glycoproteins. Strategies are envisaged, as illustrated in Fig. 5, whereby oligosaccharide probes from recombinant glycoproteins generated from alternative cellular sources are subjected to the

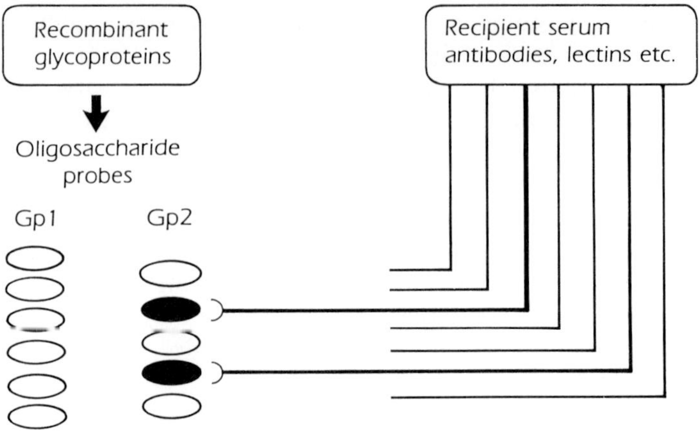

FIG. 5. A strategy for 'quality control' of the oligosaccharides of recombinant glycoproteins. Alternative glycoproteins, designated as Gp1 and Gp2, are assessed with respect to the compatibility of their oligosaccharides with antibodies in recipient sera and their reactivities with sequence-specific monoclonal antibodies and lectins, by neoglycolipid overlays. Thus incompatible oligosaccharides reactive with antibodies in recipient sera, or, conversely, oligosaccharides with desired structures, are specifically and conveniently identified.

new profiling methods which would be extended to include probing with recipient sera and with sequence-specific monoclonal antibodies and lectins. In this way, glycoproteins decorated with undesirable oligosaccharides could be readily identified and discontinued, and those with the desired structures accepted for large-scale production.

Acknowledgements

The author acknowledges the work and discussions with colleagues supported by the MRC, the Cancer Research Campaign, the Leukaemia Research Fund and the Arthritis and Rheumatism Council which have contributed to many of the concepts in this article.

References

Childs RA, Drickamer K, Kawasaki T, Thiel S, Mizuochi T, Feizi T 1989 Neoglycolipids as probes of oligosaccharide recognition by recombinant and natural mannose-binding proteins of the rat and man. Biochem J, in press

Czech MP, Lewis RE, Corvera S, 1989 Multifunctional roles of the receptors for insulin and the insulin-like growth factors. In: Carbohydrate recognition in cellular function. Wiley, Chichester (Ciba Found Symp 145) p 27–44

Defize LHK, Arndt-Jovin DJ, Jovin TM et al 1988 A431 cell variants lacking the blood Group A antigen display increased high affinity epidermal growth factor-receptor number, protein-tyrosine kinase activity, and receptor turnover. J Cell Biol 107:939–949

Drickamer K 1989 Multiple subfamilies of carbohydrate recognition domains in animal lectins. In: Carbohydrate recognition in cellular function. Wiley, Chichester (Ciba Found Symp 145) p 45–61

Feizi T 1985 Demonstration by monoclonal antibodies that carbohydrate structures of glycoproteins and glycolipids are onco-developmental antigens. Nature (Lond) 314: 53–57

Feizi T 1988 Oligosaccharides in molecular recognition. Biochem Soc Trans 16:930–934

Feizi T, Childs RA 1987a Carbohydrates as antigenic determinants of glycoproteins. Biochem J 245:1–11

Feizi T, Childs RA 1987b Growth regulating network? Nature (Lond) 329:678

Firon N, Ofek I, Sharon N 1982 Interaction of mannose-containing oligosaccharides with the fimbrial lectin of *Escherichia coli*. Biochem Biophys Res Commun 105:1426–1432

Gooi HC, Hounsell EF, Picard JK et al 1985 Differing reactions of monoclonal anti-A antibodies with oligosaccharides related to blood group A. J. Biol Chem 260:13218–13224

Hakomori S 1985 Aberrant glycosylation in cancer cell membranes as focused on glycolipids: overview and perspectives. Cancer Res 45:2405–2414

Hounsell EF 1987 Tate and Lyle Lecture: Structural and conformational characterization of carbohydrate differentiation antigens. Chem Soc Rev 16:161–185

Hounsell EF, Lawson AM, Feeney J et al 1988 Identification of a novel oligosaccharide backbone structure with a galactose residue monosubstituted at C-6 in human foetal gastrointestinal mucins. Biochem J 256:397–401

Kabat EA 1982 Philip Levine Award Lecture: Contributions of quantitative immunochemistry to knowledge of blood group A, B, H, Le, I and i antigens. Am J Clin Pathol 78:281–292

Kobata A 1988 Structures, function and transformational changes of the sugar chains of glycohormones. J Cell Biochem 37:79–90

Kushi Y, Handa S 1985 Direct analysis of lipids on thin layer plates by matrix-assisted secondary ion mass spectrometry. J Biochem 98:265–268

Lobel P, Dahms NM, Breitmeyer J, Chirgwin JM, Kornfeld S 1987 Cloning of the bovine 215-kDa cation-independent mannose 6-phosphate receptor. Proc Natl Acad Sci USA 84:2233–2237

Loveless RW, Feizi T, Childs RA et al 1989 Bovine serum conglutinin is a lectin which binds non-reducing terminal *N*-acetylglucosamine, mannose and fucose residues. Biochem J 258:109–113

Magnani JL, Smith DF, Ginsburg V 1980 Detection of gangliosides that bind cholera toxin: direct binding of ^{125}I-labeled toxin to thin-layer chromatograms. Anal Biochem 109:399–402

Mizuochi T, Loveless RW, Lawson AM et al 1989 A library of oligosaccharide probes (neoglycolipids) from *N*-glycosylated proteins reveals that conglutinin binds to certain complex type as well as high-mannose type oligosaccharide chains. J Biol Chem, in press

Morgan DO, Edman JC, Standring DN et al 1987 Insulin-like growth factor II receptor as a multifunctional binding protein. Nature (Lond) 329:301–307

Neeser J-R, Koellreutter B, Wuersch P 1986 Oligomannoside-type glycopeptides inhibiting adhesion of *Escherichia coli* strains mediated by type 1 pili: preparation of potent inhibitors from plant glycoproteins. Infect Immun 52:428–436

Rosenstein IJ, Stoll MS, Mizuochi T, Childs RA, Hounsell EF, Feizi T 1988 Oligosaccharide probes reveal a new type of adhesive specificity in *Escherichia coli* from patients with urinary tract infection. Lancet II: 1327–1330

Stoll MS, Mizuochi T, Childs RA, Feizi T 1988 Improved procedure for the construction of neoglycolipids having antigenic and lectin-binding activities from reducing oligosaccharides. Biochem J 256:661–664

Stromberg N, Deal C, Nyberg G, Normark S, So M, Karlsson K-A 1988 Identification of carbohydrate structures that are possible receptors for *Neisseria gonorrhoeae*. Proc Natl Acad Sci USA 85:4902–4906

Tang PW, Feizi T 1987 Neoglycolipid micro-immunoassays applied to the oligosaccharides of human milk galactosyltransferase detect blood group related antigens on both O- and N-linked chains. Carbohydr Res 161:133–143

Tang PW, Gooi HC, Hardy M, Lee YC, Feizi T 1985 Novel approach to the study of the antigenicities and receptor functions of carbohydrate chains of glycoproteins. Biochem Biophys Res Commun 132:474–480

Tang PW, Scudder P, Mehmet H, Hounsell EF, Feizi T 1986 Sulphate groups are involved in the antigenicity of keratan sulphate and mask i antigen expression on their poly-*N*-acetyllactosamine backbones. Eur J Biochem 160:537–545

Thorpe SJ, Bellairs R, Feizi T 1988 Developmental patterning of carbohydrate antigens during early embryogenesis of the chick: expression of antigens of the poly-*N*-acetyllactosamine series. Development 102:193–210

Watkins WM 1980 Biochemistry and genetics of the ABO, Lewis and P blood group systems. In: Harris H, Hirschhorn K (eds) Advances in human genetics. Plenum Publishing Co, New York, vol 10:1–136, 379–385

Yarden Y, Schlessinger J 1985 The EGF receptor kinase: evidence for allosteric activation and intramolecular self-phosphorylation. In: Growth factors in biology and medicine. (Ciba Foundation Symp 116) Pitman, London, p 23–45

DISCUSSION

Ruoslahti: Have you demonstrated that you can take a mixture of bacterial extract, do the overlay assay, and have it work specifically as illustrated in Fig. 4?

Feizi: So far we have worked with whole bacteria that have been metabolically labelled. If one were to make an extract, one could hope to identify the adhesive protein itself.

Hughes: The thin-layer chromatogram that had been overlaid with bacteria (Fig. 3, panel III) showed remarkable specificity for $Man_5GlcNAc_2$ (Man_5): binding to the $Man_6GlcNAc_2$ species was practically non-existent.

Feizi: This is according to prediction, because the non-substituted Man(α1–3)Man(β1–4)GlcNAc sequence would be present on only a subpopulation of the Man_6 structures (Liang et al 1980).

Hughes: Do you always find the same binding specificity in different strains of *E. coli*, or is there a range of bacterial strains or substrains with different $Man_nGlcNAc_2$ oligosaccharide-binding specificities?

Feizi: We have not studied that yet.

Sharon: I was delighted to see that you confirmed our findings and those of Neeser et al (1986) that Man_5 is a good inhibitor of cell attachment by type 1 fimbriated *E. coli*. In hapten inhibition experiments, Man_5 is about 20–30 times better than methyl α-mannoside (Firon et al 1983).

We studied several strains of *E. coli*, mainly with hydrophobic glycosides which are strong inhibitors. We also looked at some strains with branched oligosaccharides, synthetic methyl glycosides prepared by T. Ogawa (Saitama, Japan); they did not have the $GlcNAc_2Asn$ structure. Neeser et al (1986) used different strains of *E. coli* and a different assay system. Although they found essentially the same results with Man_5, there were some differences with the Man_6 oligosaccharide. We also found that aromatic α-mannosides are powerful inhibitors of the mannose-specific *E. coli* lectin (Firon et al 1987). Our data seem to indicate that all *E. coli* that are 'mannose-specific' have the same specificity for different oligomannosides.

In most bacteria, the lectin or lectin-like proteins are on the outer surface, usually in the form of fimbriae. The fimbriae can be easily isolated by shaking the bacteria followed by differential centrifugation. The isolated fimbriae show the same pattern of carbohydrate specificity as the fimbriated bacteria. Interestingly, examination of the binding of purified fimbriae with different specificities, isolated from infectious strains of *E. coli*, to sections of human kidney (Korhonen et al 1986) and of rat brain (Parkkinen et al 1988) gave patterns supporting the widely held belief that binding of the parent bacteria to complementary cell surface sugars plays an important role during bacterial invasion.

Ruoslahti: Dr Feizi, have you tried to get mammalian cells to bind to, say, fibronectin, as we showed several years ago?

Feizi: We are preparing to do binding studies with whole cells.

Sharon: Dr Ruoslahti, you tried some years ago to overlay with mammalian cells proteins separated by polyacrylamide gel electrophoresis.

Ruoslahti: Yes, but that was with adhesion proteins, such as fibronectin and vitronectin. That works well. Thus, in principle one can do this with mammalian cells.

Sharon: One can do it with bacteria on polyacrylamide gels.

Fukuda: Dr Feizi, can you separate your neoglycolipid by HPLC as well?

Feizi: We are investigating that.

Fukuda: When you conjugate with reduced oligosaccharides you need to use periodate, which probably destroys the sialic acid.

Feizi: This is something we have been discussing with Dr Lee. With reduced oligosaccharides we need to refine the oxidation procedure to protect the sialic acid. I don't know whether we shall completely overcome this. We may have to be resigned to losing a proportion of the sialic acid residues.

Hart: Do your coupling procedures destroy O-acetylated sialic acid? At what pH is your reductive deamination done?

Feizi: It is virtually neutral. If the O-acetylated sialic acid were on a reducing oligosaccharide it would not be destroyed during conjugation. If it were on an O-linked chain, it would have been destroyed already in the course of releasing the chain from protein by the alkaline borohydride degradation procedure.

Watkins: Or released by hydrazinolysis, because this is done under alkaline conditions and leads to loss of both N- and O-acetyl groups.

Feizi: Yes; this also applies to N-linked chains released by hydrazinolysis. Such oligosaccharides should ideally be released using endoglycosidases.

Finne: Your result with the mannose-binding protein was perhaps unexpected. Is it possible that sometimes when you conjugate the oligosaccharides to the carrier lipid you change the conformation of the oligosaccharides in such a way that you get the 'wrong' result?

Feizi: That is something we have to consider. But so far we have not observed any spurious reactions using L-α-phosphatidylethanolamine dipalmitoyl (PPEADP).

Rademacher: The implication is that the mannose-binding protein is binding *N*-acetylglucosamine. Could you comment on the purification of those sugars? Human IgG contains a substantial amount of oligomannose which could co-run.

Feizi: It is most unlikely that the reactions we have observed with IgG oligosaccharides represent lectin binding to trace amounts of oligomannosidic (high mannose) oligosaccharides, because the oligomannosidic oligosaccharides from ribonuclease B reacted extremely weakly.

Rademacher: Was that binding confirmed by treatment with jack bean β-hexosaminidase?

Feizi: We have observed that while the lipid conjugates of oligosaccharides with terminal galactose residues derived from transferrin do not bind to the mannose-binding protein, the conjugates of the β-galactosidase-treated oligosaccharides bind strongly (Childs et al 1989).

Rademacher: Have you blotted a preparation of group A streptococci with the mannose-binding protein?

Feizi: No.

Rademacher: Have you blotted neoglycolipids using whole serum?

Feizi: Yes; for example, whole sera containing monoclonal antibodies have been used successfully (Tang et al 1985, 1986).

Rademacher: The assumption is that the monoclonal antibody is purely to the sugar and there is no peptide contibution to the binding. What is the minimum binding affinity detected by your method?

Feizi: We need to do proper measurements. At this stage, we can say that it is analogous to the minimum binding affinity required for immunofluorescence or Western blotting. The overall affinity is high because the ligands are presented in a multivalent state. With monovalent oligosaccharide ligands the affinity may be too low for binding in this type of assay. The virtue of this method is that the chromatography gives rise to clustering of the oligosaccharides. Reactivities with antibodies and lectins are strongly influenced by cooperative effects of multivalency.

Barondes: Can you use glycopeptides or are they too messy?

Feizi: We have not yet made neoglycolipids using glycopeptides.

Barondes: What is the largest oligosaccharide you can resolve?

Feizi: Up to a decasaccharide, so far.

Hughes: You have used the carbohydrate-binding domains of the lectin, as isolated by Kurt Drickamer. Might the specificity of the whole protein be different?

Feizi: I didn't go into this, but we have also worked with the whole mannose-binding protein from rat serum, and there is a subtle difference in reaction pattern (Childs et al 1989). It is not yet clear whether the collagen domain of the mature protein somehow influences the reactivity of the carbohydrate recognition domain, or whether there is more than one type of carbohydrate recognition domain in the natural protein; i.e. it is a heteropolymer, not a homopolymer.

Hughes: Kurt would have picked up the latter case in his sequence search.

Drickamer: It is known that there are two different genes that encode highly homologous mannose-binding proteins. It is not clear whether the natural protein is a mixture of homopolymers of the two different gene products, or a heteropolymer and, if it is the latter, what the proportions are in the circulation. Until one clarifies these issues, as Ten says, one can't distinguish any effect of the additional structure of the intact protein, not present in the recombinant binding domain, from the effect of the presence of several non-identical proteins.

Watkins: How do you think the endogenous lectins tune the response of the epidermal growth factor (EGF) receptor through an interaction with A determinants? Are you suggesting that there are endogenous *N*-acetylgalactosamine-binding lectins in group A individuals? Would that not be rather dangerous?

Feizi: I am suggesting that endogenous lectins (possibly those with blood group A specificities in A431 cells) might, for example, influence the extent of aggregation of the EGF receptor on the cell membrane and thus tune the level of responsiveness of the receptor glycoprotein to EGF. There is already a good example of an *N*-acetylgalactosamine-specific lectin—one of the so-called soluble β-galactoside-binding proteins in rat lung (Leffler & Barondes 1986). It reacts more strongly with blood group A structures than with the unsubstituted β-galactosyl backbone. Whether this lectin normally associates with the blood group A structures in intact membranes must be studied. Lectins of this type have all the hallmarks of cytoplasmic proteins; they lack the signal sequence found on secreted and membrane-associated proteins, although some people claim to have also detected them on the cell surface. Cells do break down and these are very stable proteins. It is therefore conceivable that the released proteins find their way to the cell surface. There may, of course, be other families of membrane-associated lectins yet to be discovered.

Watkins: If it were to have any physiological effect, the lectin would have to be fairly widely distributed.

Feizi: Soluble lectins of this family occur in most cell types.

Fukuda: We usually look at binding in solution. Your assay involves an interaction between solid-phase carbohydrate and the protein in solution. That could account for differences in the specificity of the interaction.

Feizi: The same reservation applies to overlay assays with natural glycolipids. It is true that we don't know how these structures are orientated on the silica gel plate.

Fukuda: It's nice to complement such experiments with studies using liposomes, which may be similar to the membrane.

Feizi: Binding assays using liposomes are rather messy. We are interested in incorporating the neoglycolipids into cells. That might be easier to work with.

Muramatsu: Can your reagents adhere to plastic disks? If so, how many experiments could you perform with 1 mg of oligosaccharide?

Feizi: Thousands of microtitre wells can be coated with neoglycolipids derived from that amount.

Kobata: I have a comment on the mannose-binding lectins of bacteria. We have analysed several glycoproteins purified from the epithelial cells of large intestine and found that none of them contains oligomannosidic oligosaccharides. The sugars are mostly highly branched, with elongated outer chains. So I wonder what that bacterial lectin is doing in the intestine, because it has no ligand on the epithelium.

Feizi: We studied *E. coli* from patients with severe urinary tract infections. These bacteria adhere to epithelial cells of the urinary tract.

Sharon: Dr Kobata, are there no hybrid-type oligosaccharides either?

Kobata: We could not find them.

Fukuda: We have some preliminary data on colonic cell glycopeptides. Poly-*N*-acetyllactosamine is the major type of oligosaccharide. There are no oligomannosidic oligosaccharides, but many O-linked oligosaccharides are present.

Hughes: Perhaps the 'mannose-binding' protein binds to poly-*N*-acetyllactosamine structures.

Sharon: Or perhaps the mannose-binding protein is really an *N*-acetylglucosamine-binding protein.

Feizi: The presence of non-reducing terminal galactose residues on complex-type oligosaccharides inhibits the binding of the rat lectin. Terminal *N*-acetylglucosamine seems to be the ligand of choice.

Hughes: It could be a poly-*N*-acetyllactosamine chain with *N*-acetylglucosamine at the non-reducing terminus.

Feizi: Branched poly-*N*-acetyllactosamine structures with terminal *N*-acetylglucosamine residues are certainly candidate ligands. We find that neoglycolipids derived from chitin oligosaccharides, such as chitobiose [GlcNAc(β1-4)GlcNAc], react well.

Finne: We observe a similar absence of oligomannosidic glycans in small intestinal epithelial cells (Finne et al 1989). It appears that we do not want to have *E. coli* that bind mannose in our intestinal tract.

Sharon: You really don't need much. The Tamm-Horsfall glycoprotein has only one oligomannose chain (Dall'Olio et al 1988) and yet it binds very well to mannose-specific *E. coli*. It may have a role in infection, for example as a vehicle for clearing bacteria from the urine.

References

Childs RA, Drickamer K, Kawasaki T, Thiel S, Mizuochi T, Feizi T 1989 Neoglycolipids as probes of oligosaccharide recognition by recombinant and natural mannose-binding proteins of the rat and man. Biochem J, in press

Dall'Olio F, de Kanter FJJ, van den Eijnden DH, Serafini-Cessi F 1988 Structural analysis of the preponderant high-mannose oligosaccharide of human Tamm-Horsfall glycoprotein. Carbohydr Res 178:327–332

Finne J, Breimer ME, Hansson GC, Karlsson K-A, Leffler H, Vliegenthart JFG, van Halbeek H 1989 Novel polyfucosylated N-linked glycopeptides with blood group A,H,X and Y determinants from human small-intestinal epithelial cells. J Biol Chem, in press

Firon N, Ofek I, Sharon N 1983 Carbohydrate specificity of the surface lectins of *Escherichia coli*, *Klebsiella pneumoniae* and *Salmonella typhimurium*. Carbohydr Res 120:235–249

Firon N, Ashkenazi S, Mirelman D, Ofek I, Sharon N 1987 Aromatic alpha-glycosides of mannose are powerful inhibitors of the adherence of type 1 fimbriated *Escherichia coli* to yeast and intestinal epithelial cells. Infect Immun 55:472–476

Korhonen TK, Virkola R, Holthofer H 1986 Localization of binding sites for purified *Escherichia coli* P fimbriae in the human kidney. Infect Immun 54:328–332

Leffler H, Barondes SH 1986 Specificity of binding of three soluble rat lung lectins to substituted and unsubstituted mammalian β-galactosides. J Biol Chem 261:10119–10126

Liang C-J, Yamashita K, Kobata A 1980 Structural study of the carbohydrate moiety of bovine pancreatic ribonuclease B. J Biochem 88:51–58

Neeser N-R, Koellreutter B, Wuersch P 1986 Oligomannoside-type glycopeptides inhibiting adhesion of *Escherichia coli* strains mediated by type 1 pili: preparation of potent inhibitors from plant glycoproteins. Infect Immun 52:428–436

Parkkinen J, Korhonen TK, Pere A, Hacker J, Soinila S 1988 Binding sites in the rat brain for *Escherichia coli* S fimbriae associated with neonatal meningitis. J Clin Invest 81:860–865

Tang PW, Gooi HC, Hardy M, Lee YC, Feizi T 1985 Novel approach to the study of the antigenicities and receptor functions of carbohydrate chains of glycoproteins. Biochem Biophys Res Commun 132:474–480

Tang PW, Scudder P, Mehmet H, Hounsell EF, Feizi T 1986 Sulphate groups are involved in the antigenicity of keratan sulphate and mask i antigen expression on their poly-N-acetyllactosamine backbones. Eur J Biochem 160:537–545

Binding modes of mammalian hepatic Gal/GalNAc receptors

Y. C. Lee

Department of Biology, The Johns Hopkins University, Baltimore, MD 21218, USA

Abstract. Mammalian Gal/GalNAc receptors show dramatic preference for three-branched oligosaccharide structures over two- or one-branched counterparts. The spatial arrangement of the Gal residues is extremely important for optimal binding. The three terminal Gal residues in the best triantennary ligand are about 15–30 Å apart, and therefore the sugar-combining sites on the receptor may also have the same geometry. The results obtained with synthetic ligands containing Gal or GalNAc are in agreement with this concept. Photoaffinity labelling and GalNAc-lactoperoxidase catalysed iodolabelling of the receptors revealed that the minor subunits (52 and 60 kDa) were more readily labelled than the major subunit (43 kDa). The stoichiometry of binding was determined with synthetic ligands containing GalNAc. The results indicated that each subunit may have two sugar-combining sites. A model for subunit assembly is proposed on the basis of these results and the finding that coexpression of all subunits is necessary for the binding and processing of asialo-orosomucoid in transfected cells, whereas Gal-polylysine can be bound and processed by the cells expressing only the major subunit.

1989 Carbohydrate recognition in cellular function. Wiley, Chichester (Ciba Foundation Symposium 145) p 80–95

Since their discovery (see Ashwell & Morrell 1974), mammalian hepatic Gal/GalNAc receptors have been studied intensively by many investigators (for reviews, see Ashwell & Harford 1982, Breitfeld et al 1985). Although the receptor protein has been solubilized and purified, its sugar-binding specificity has been studied extensively (Lee & Lee 1982), and its molecular biology is reasonably well understood (Drickamer et al 1984, Halberg et al 1987), there are still many unresolved questions about these receptors. They have several interesting characteristics, some of which will be discussed in this paper.

Binding enhancement by clustering of glycosides

In our initial studies of binding of glycoconjugates by the mammalian hepatic carbohydrate-binding receptors, we used ‘neoglycoproteins’—proteins modified with structurally well-defined carbohydrate derivatives (Lee et al 1976). These derivatives were originally designed to provide a series of glycoconjugates, each of which contained a single, definitive carbohydrate structure. Using neoglycoproteins containing only monosaccharide units, we established that high binding affinity for mammalian hepatic carbohydrate receptors can be obtained with proteins modified only with Gal or GalNAc; complex oligosaccharide structures were unnecessary.

The neoglycoproteins can be prepared with different sugar densities, i.e. varying mole ratios of sugars to protein. We found that the logarithm of the binding affinity of the neoglycoproteins by the solubilized rabbit liver Gal/GalNAc receptor increased nearly linearly with a linear increase in the sugar density up to about 30 moles Gal/mole protein, beyond which the affinity reached a plateau (Fig. 1A). Subsequently a similar effect was observed with isolated rabbit hepatocytes: the binding affinity increased logarithmically up to about 20 moles Gal/mole protein before a plateau was reached (Fig. 1B). Thus, we observed that an increase in the sugar density on a protein causes more affinity increase than can be expected on a statistical basis (Connolly 1982). This idea was tested with a series of simple model compounds (1–3) based on tris(hydroxymethyl)aminomethane, 'Tris' (Lee 1978). Studies of the binding of

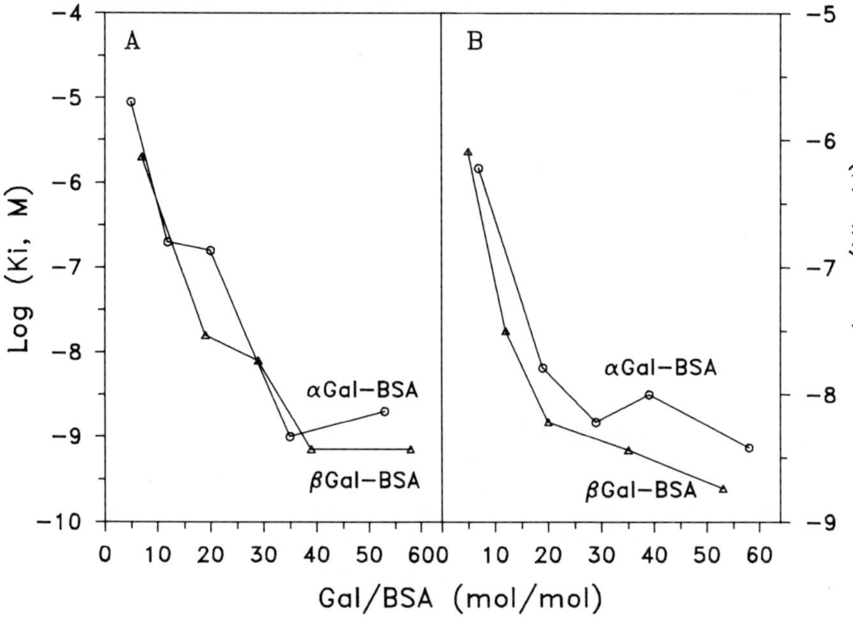

FIG. 1. Inhibition of ^{125}I-asialo-orosomucoid binding to solubilized rabbit Gal/GalNAc receptor (A) and intact rabbit hepatocytes (B) by Gal-BSA of different sugar densities. The inhibitory effect was expressed as K_i (M).

these simple synthetic compounds by the receptor on the isolated hepatocytes indicated that 'clustering' of glycosides does enhance the binding affinity more than can be accounted for by simple increases in statistical chances (Connolly et al 1982). Interestingly, this phenomenon becomes much less prominent when a detergent-solubilized receptor is used with these synthetic glycosides.

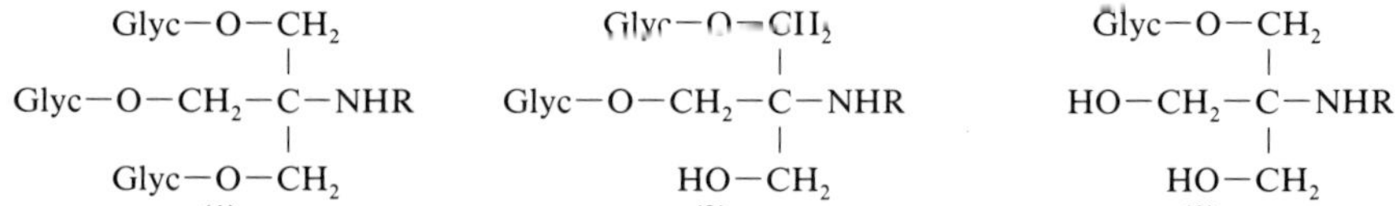

(1) (2) (3)

Glyc = glycosyl residue,

R = e.g. $-\overset{O}{\overset{\|}{C}}(CH_2)_6NH_2$

Encouraged by these results, and by a fortunate collaboration with Dr J. Lönngren's group in Sweden, we studied binding of a series of synthetic oligosaccharides which are close analogues of natural N-glycoside chains, varying in complexity up to an undecasaccharide of tetraantennary (four-branched)

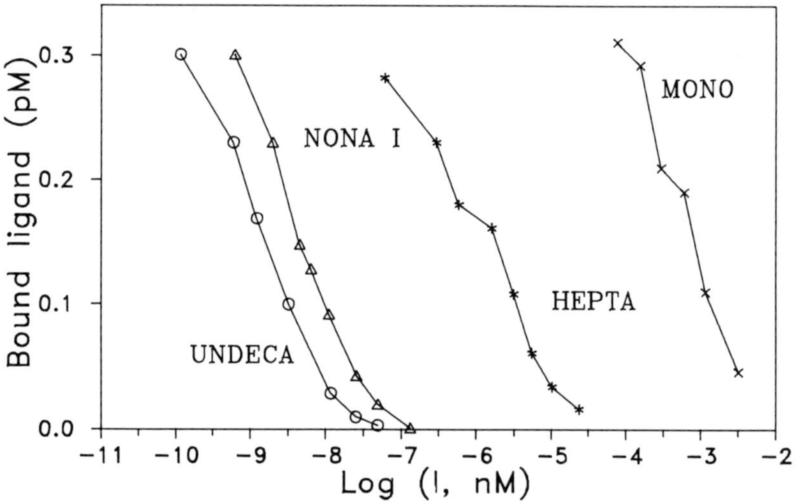

FIG. 2. Inhibition of binding of ^{125}I-asialo-orosomucoid to rabbit hepatocytes by mono- (MONO), bi- (HEPTA) and tri- (NONA I) antennary oligosaccharides (Lee et al 1983).

structure. To our astonishment, increases in the number of branches of these oligosaccharides from one to two to three (Fig. 2) resulted in increases of binding affinities of nearly a thousandfold, and a millionfold, respectively (Lee et al 1983, 1984). This is perhaps one of the most striking 'cooperative effects' found in biology. The change from a triantennary structure to a tetraantennary structure did not result in any dramatic increase in binding affinity.

More detailed studies revealed that the number of branches does not alone determine the binding affinity of these oligosaccharides (see Table 1). Considerable differences in affinities were seen among the biantennary as well as triantennary oligosaccharides. Through our collaboration with Dr Klaus Bock, who did conformational analyses on some of these oligosaccharides, it was revealed that the binding affinities of these compounds are closely related to their conformation (Lee et al 1984). For example, in the most strongly bound triantennary structure, NONA I, the three inter-Gal distances are about 15Å,

TABLE 1 Some synthetic analogues of natural oligosaccharides

Structure		K_d
Gal(β1–4)GlcNAc(β1–2)Man	MONO	283 μM
Gal(β1–4)GlcNAc(β1–2)Man(α1–6) \ Man / Gal(β1–4)GlcNAc(β1–2)Man(α1–3)	HEPTA	13.2 μM
Gal(β1–4)GlcNAc(β1–2)Man(α1–6) \ Man / Gal(β1–4)GlcNAc(β1–2)Man(α1–3) / Gal(β1–4)GlcNAc(β1–4)	NONA I	1.85 nM
Gal(β1–4)GlcNAc(β1–6) \ Gal(β1–4)GlcNAc(β1–2)Man(α1–6) \ Man / Gal(β1–4)GlcNAc(β1–2)Man(α1–3)	NONA II	81.3 nM
Gal(β1–4)GlcNAc(β1–6) \ Gal(β1–4)GlcNAc(β1–2)Man(α1–6) \ Man / Gal(β1–4)GlcNAc(β1–2)Man(α1–3) / Gal(β1–4)GlcNAc(β1–4)	UNDECA	1.3 nM

22Å, and 25Å in the minimal energy conformation. The affinity of this oligosaccharide is only slightly less than that of the tetraantennary structure (UNDECA), and is nearly as strong as that of asialo-orosomucoid (ASOR), one of the most tightly bound naturally-derived ligands for this receptor. From these findings, it is reasonable to propose that the functional receptor unit contains three Gal/GalNAc-combining sites (Lee et al 1983, 1984), which will be referred to as the 'golden triangle'.

The stereochemical requirement for optimal binding appears to be quite stringent. For example, an analogue of NONA I in which the third Gal(β1–4)GlcNAc unit is linked to the Man(α1–6) branch rather than to the Man(α1–3) (NONA II in Table 1) showed a much lower binding affinity. The inter-Gal distances in this compound are somewhat larger than those described above.

An even more subtle structural alteration can bring about a large change in binding affinity. A new triantennary N-glycoside has been found in bovine fetuin in which one of the three terminal Gals is linked (β1–3) rather than (β1–4) to GlcNAc (Table 2). Conformational analysis revealed that in this variant oligosaccharide one of the Gal residues is a few ångstroms away from its position in the classic structure (K. Bock, personal communication; N. Rao, unpublished results), as shown in Fig. 3. The new oligosaccharide, however, binds to the rat hepatocytes more than hundredfold less tightly than the classic triantennary oligosaccharide of NONA I-type structure (Townsend et al 1986).

TABLE 2 Two types of triantennary glycoproteins in fetuin

Gal(β1–4)GlcNAc(β1–2)Man(α1–6) \\ Man(β1–4)-R / Gal(β1–4)GlcNAc(β1–2)Man(α1–3) / Gal(β1–4)GlcNAc(β1–4)	Gal(β1–4)-TRI
Gal(β1–4)GlcNAc(β1–2)Man(α1–6) \\ Man(β1–4)-R / Gal(β1–4)GlcNAc(β1–2)Man(α1–3) / Gal(β1–3)GlcNAc(β1–4)	Gal(β1–3)-TRI

R = GlcNAc(β1–4)GlcNAcβ-Asn

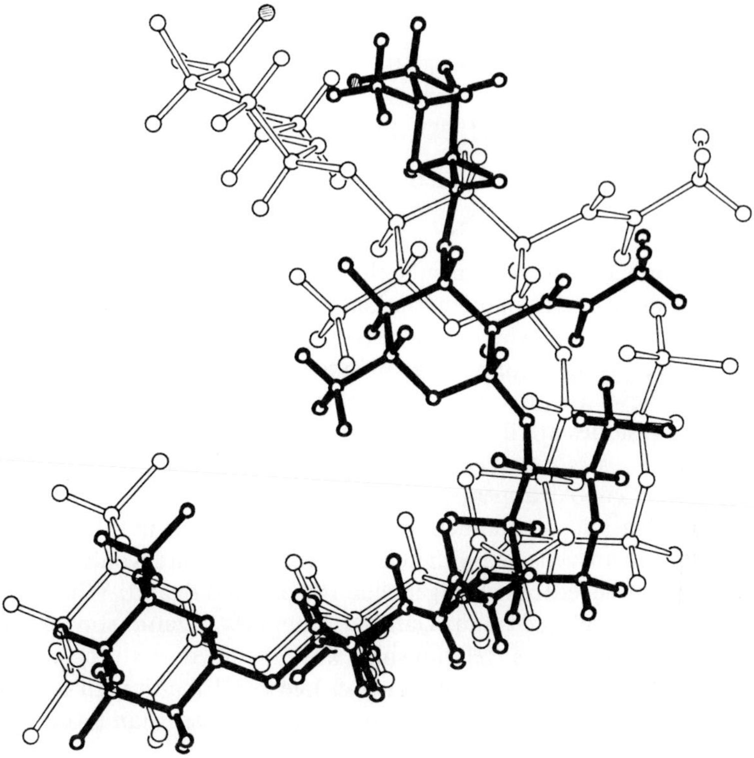

FIG. 3. The minimal energy conformation of the two oligosaccharides shown in Table 2, Gal(β1–4)-TRI (light) and Gal(β1–3)-TRI (dark) (N. Rao, unpublished results).

Synthetic cluster glycosides

Although these oligosaccharides of natural analogues proved to be invaluable, their syntheses are laborious. Because neoglycoproteins bearing only Gal or GalNAc derivatives were efficient ligands, it seemed reasonable to construct synthetic ligands in which Gal or GalNAc residues were arranged to occupy the vertices of the golden triangle without construction of complete branched oligosaccharides. A series of synthetic branched compounds (4) mimicking the branched oligosaccharide structure were constructed using Asp (sometimes extended with Gly) to which 6-aminohexyl glycosides (the only glycosyl residues) were linked through its carboxyl groups (Lee et al 1984, Lee & Lee 1987b).

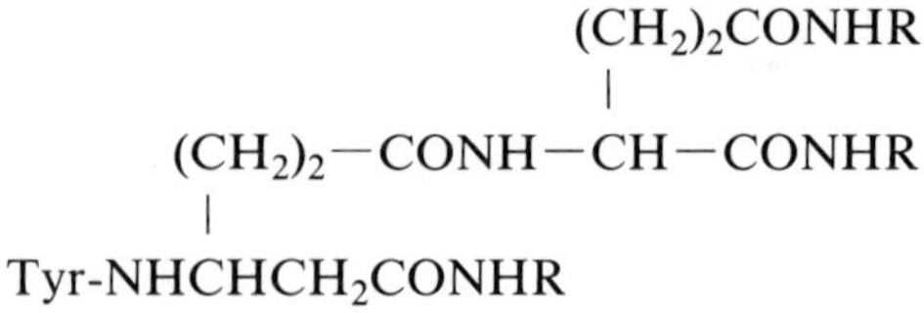

R = $(CH_2)_6$—O—Lac or $(CH_2)_6$—O—GalNAc

(4)

The results were in good agreement with those obtained with oligosaccharides. Synthetic compounds that can orient the Gal residues at the vertices of the golden triangle are good ligands, whereas those that cannot are inferior ligands, using both isolated lectins and isolated hepatocytes (Table 3).

Such a synthetic design (termed 'shorthand synthesis' because only the essential sugar residues are incorporated in the structure) provided interesting opportunities to test the concept of the golden triangle (see Fig. 4). The monosaccharide GalNAc appears to be about tenfold stronger than Gal in its inhibition of ^{125}I-asialo-orosomucoid. If both Gal residues in the synthetic di-branched structure are replaced with GalNAc, the new ligand would be expected to bind about a hundredfold stronger than the original ligand. When all three Gal residues are replaced with GalNAc, the new ligand should bind a thousandfold tighter. This relationship was observed with both the rat hepatocytes and the isolated receptor (Lee & Lee 1987b), although the binding affinities with the isolated protein are always much lower than those with the

TABLE 3 Binding of synthetic GalNAc derivatives[a]

	I_{50} (μM)	
	Soluble receptor	*Hepatocytes*
Monosubstituted ligands		
LacAH[b]	190[c]	40
GalNAcAH	5	4
Disubstituted ligands		
Asp(LacAH)$_2$	7	0.6
Asp(Gly)$_2$(GalNAcAH)$_2$	0.08	0.004
Trisubstituted ligands		
TyrGluβ(LacAH)Glu(LacAH)$_2$	3	0.05
TyrGluβ(GalNAcAH)Glu(GalNAcAH)$_2$	0.004	0.0002[c]

[a]Expressed as 50% inhibition point, using ^{125}I-asialo-orosomucoid as labelled ligand, and receptors and cells derived from rat liver (Lee & Lee 1987b).
[b]AH = 6-aminohexyl.
[c]Determination of I_{50} values of extremely low or high affinities are subject to greater errors than at intermediate affinities.

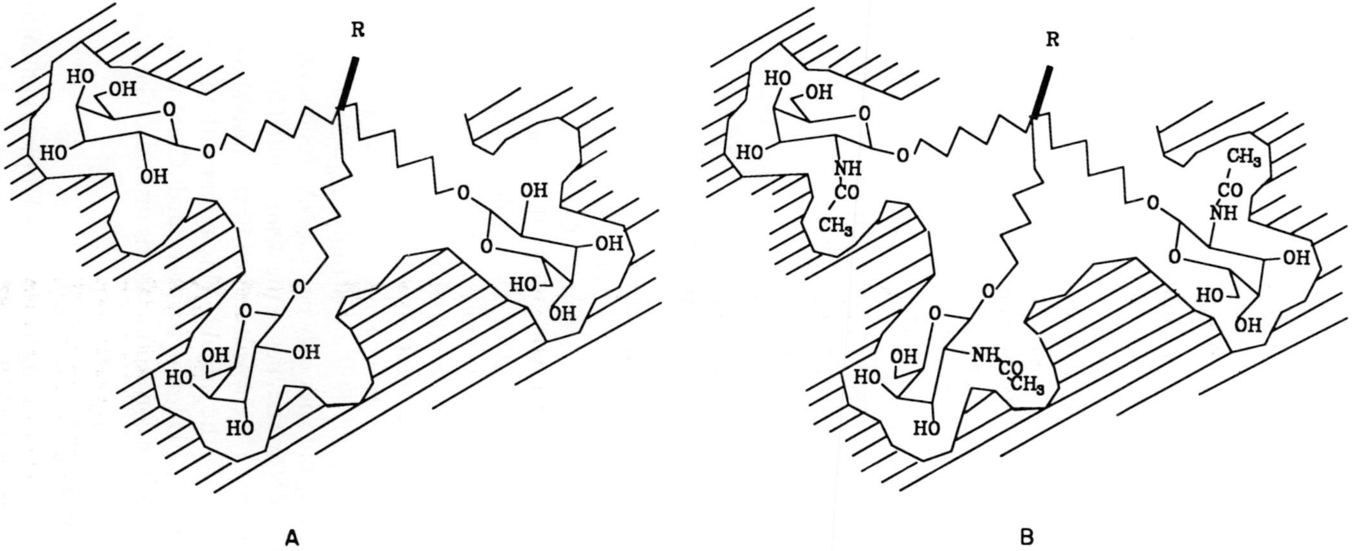

FIG. 4. Models of trivalent binding. A, binding of Gal-ligand; B, binding of GalNAc-ligand. $K_{d,GalNAc}/K_{d,Gal} \simeq 10$; $(K_{d,GalNAc})^2/(K_{d,Gal})^2 \simeq 100$; $(K_{d,GalNAc})^3/(K_{d,Gal})^3 \simeq 1000$.

hepatocytes (Table 3). These results indicate that Gal and GalNAc bind to the same combining site. Furthermore, it is evident that the clustering effect is also manifest in the solubilized system, albeit less dramatically than in the isolated rat liver hepatocytes. That the simple clustered glycosides based on Tris showed very little 'cluster' effect with solubilized rabbit receptor protein (Connolly et al 1982) may be due to their low binding affinities, which tend to cause underestimation of binding under the experimental protocols used.

The existence of a clustering effect also raises the possibility that the golden triangle, or a modified arrangement of it, may exist in the solubilized receptor proteins for these high affinity GalNAc ligands. This is corroborated in the binding stoichiometry, as described below.

Stoichiometry of binding

The synthetic GalNAc-containing ligands described above are simple and definitive in structure and possess very high binding affinities for rat liver hepatocytes. Using these ligands, we were able to determine the stoichiometry of binding to subunits of the receptor (Lee & Lee 1988). We used the gel filtration technique of Hummel & Dreyer (1962). A suitable amount of detergent-solubilized receptor protein is applied to a Sephadex G-50 column equilibrated with a radiolabelled GalNAc ligand. As it passes through the column, the receptor protein binds the ligand in the column so that the protein peak becomes more radioactive than the background count. A corresponding trough will be seen at a later elution position from which the radioactive ligand has been removed. In these studies, a major subunit of the isolated receptor protein, RHL-1 (43 kDa), as well as a mixture of RHL-1 and the minor subunits, RHL-2/3 (52 and 60 kDa), were used. Two moles of ligand containing one GalNAc residue were bound to one mole of the rat receptor protein subunit, indicating that there are two GalNAc-combining sites per subunit. When a ligand with two GalNAc residues was tested, one mole of the ligand was bound per mole of the subunit. Although this ligand can theoretically be bound by one or both GalNAc residues, the binding affinity ($K_d = 5 \times 10^{-7}$ M) was much higher than that of the ligand with only one GalNAc residue ($K_d = 5 \times 10^{-5}$ M), suggesting that both GalNAc residues in the disubstituted ligand are bound by the subunit. The binding of the trisubstituted ligand also showed that two or three GalNAc are bound by the subunit. This agrees with the observation that the solubilized receptors bind the trisubstituted GalNAc ligand about a thousandfold tighter than the corresponding trisubstituted Gal ligand (Table 3).

Results of affinity labelling

Another approach for investigating receptor subunit organization is to use affinity labelling. We have devised a method to convert the 6-OH group of Gal into 6-NH_2 by oxidation with galactose oxidase followed by reductive amination with benzylamine. On removal of the benzyl group, the amino group is exposed and available for modification (Lee & Lee 1986). We modified a mixture of desialylated triantennary glycopeptides from bovine fetuin by this method, and attached a 4-azido-2-nitro-benzoyl group to the newly formed 6-NH_2 group. The reagent thus prepared was used for photoaffinity labelling of rat hepatocytes (Lee & Lee 1987a).

The cells were allowed to bind the reagent in the cold, and the excess reagent was washed away. Ultraviolet illumination of the cell suspension revealed the labelling of the Gal/GalNAc receptors. Surprisingly the minor subunits (52 and 60 kDa) were much more efficiently labelled than the major subunit (43 kDa) (Table 4). This could suggest that the surface and interior receptors consist of different subunits; the preferential labelling of the minor subunits might be due to their preponderance on the plasma membrane. The following experiments showed that this was not the case. Intracellular receptors can be accessed by macromolecules after treating the cells with 0.05% digitonin (Weigel & Oka 1983). The digitonin-treated hepatocytes were labelled by the same reagent under similar conditions and the results were the same as the labelling of the surface receptors: minor subunits were preferentially labelled over the major subunit.

In another approach, lactoperoxidase (LPO) was modified with GalNAc. The idea behind this was that GalNAc-LPO bound to the receptor would generate an active iodination agent nearby and selective labelling of the surface receptors might be achieved. The labelling with GalNAc-LPO also showed that the minor subunits were labelled most efficiently. With purely chemical labelling, using

TABLE 4 Labelling of Gal/GalNAc receptor subunit[a]

	^{125}I-Distribution among subunits (%)[b]		
Labelling methods	*43 kDa*	*52 kDa*	*60 kDa*
Rat hepatocytes			
Photolabelling (− digitonin)	22	51	27
Photolabelling (+ digitonin)	16	60	24
GalNAc-LPO[c]	6	53	41
Chloramine-T	63	27	10
Solubilized lectin			
GalNAc-LPO	63	26	11
Coomassie (PAGE)	75	16	9

[a]Lee & Lee (1987a)
[b]Determined after polyacrylamide gel electrophoresis (PAGE).
[c]Lactoperoxidase modified with GalNAc derivatives.

chloramine-T, all the subunits were labelled. All subunits were also uniformly labelled when the receptor protein was prepared in Triton X-100 and labelled with the photoaffinity reagent or GalNAc-LPO, indicating that there are no inherent differences in reactivity among the subunits.

The importance of the minor subunits

These results are extremely interesting in view of the results obtained with the cells transfected with cDNAs of RHL-1 and RHL-2/3. Co-transfection of hepatoma tissue culture (HTC) cells lacking endogenous Gal/GalNAc receptors with both RHL-1 and RHL-2/3 was necessary to bind asialo-orosomucoid (McPhaul & Berg 1986). Further studies revealed that those transfected HTC cells capable of binding and processing asialo-orosomucoid expressed both RHL-1 and RHL-2/3. The cells expressing only RHL-1 were not capable of binding and processing asialo-orosomucoid (Braiterman et al 1989). Interestingly, the cells expressing only RHL-1 can bind Gal-polylysine (5), albeit less tightly ($K_d \approx 10^{-9}$ M) than rat hepatocytes ($K_d \approx 10^{-10}$ M), and are capable of processing Gal-polylysine. Thus, although RHL-1 can be biosynthesized and transported in the absence of RHL-2/3, it cannot bind ASOR without the coexpression of RHL-2/3.

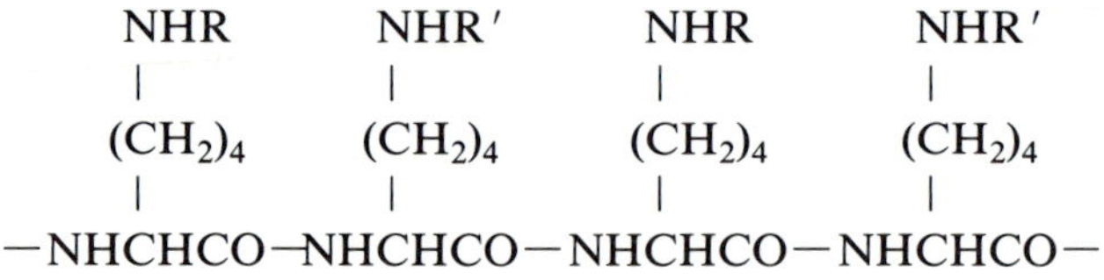

R = CO(CH$_2$)$_5$—S—Gal, R′ = COMe

(5)

N-Acetylated Gal-polylysine

The following hypothesis can be developed from these results. In the absence of RHL-2/3, RHL-1 may aggregate, but only in such a way that the triangles formed from their sugar-combining sites deviate considerably from the golden triangle needed for high affinity binding of ASOR (Fig. 5). However, Gal-polylysine (5) has flexible arms and high sugar density (100 Gal per 250 lysine residues) and may form a different receptor–ligand complex which still possesses a high affinity. The results from the photo- and other affinity labelling showed that the subunits corresponding to RHL-2/3 may have sugar-combining sites of higher binding affinity than RHL-1 or which are more accessible than RHL-1. It may be speculated that when RHL-2/3 subunits are present, a heteropolymer is formed in which two of the sugar-combining sites are from either RHL-2

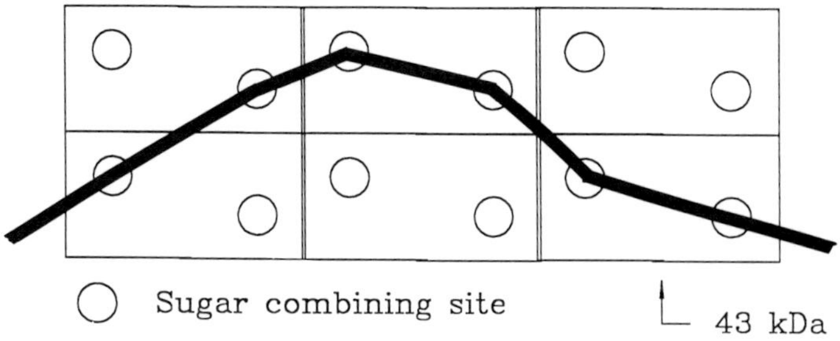

FIG. 5. A homopolymer model (top view) for association of RHL-1 (43 kDa) and binding of Gal-polylysine.

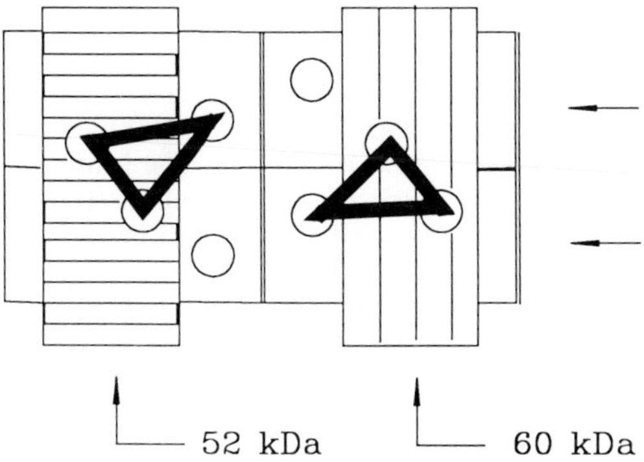

FIG. 6. A heteropolymer model (top view) for association of RHL-1 (43 kDa) and RHL-2/3 (52 and 60 kDa) and binding of triantennary oligosaccharides.

or RHL-3, and the last site comes from RHL-1 (Fig. 6). In this mode, RHL-2/3 serves as cement for RHL-1, enabling formation of the golden triangle with sugar-combining sites of both RHL-2/3 and RHL-1. Preferential labelling of RHL-2/3 can be explained by the selective affinity for the subunits providing two combining sites. If the homopolymer or heteropolymer can coexist in the plasma membrane, the apparently contradictory findings of the cross-linked homopolymer (Halberg et al 1987) and the cross-linked heteropolymer (Bischoff et al 1988) may be reconcilable.

Acknowledgements

The author wishes to thank the following persons for their work, which made this article possible: L. T. Braiterman, S. C. Chance, D. T. Connolly, M. R. Hardy, A. L. Hubbard, K. Kawaguchi, R. T. Lee, P. Lin, W. R. Porter, N. Rao and R. R. Townsend (in alphabetical order). Research support from NIH (DK 09970) and NSF (DCB 8509638) is gratefully acknowledged.

References

Ashwell G, Harford J 1982 Carbohydrate-specific receptors of the liver. Annu Rev Biochem 51:531–554

Ashwell G, Morrell AG 1974 Role of surface carbohydrates in hepatic recognition and transport of circulating glycoproteins. Adv Enzymol 41:99–128

Bischoff J, Libresco S, Shia M, Lodish HF 1988 The H1 and H2 polypeptides associate to form the asialoglycoprotein receptor in human hepatoma cells. J Cell Biol 106:1067–1074

Braiterman LT, Chance SC, Porter WR, Lee YC, Townsend RR, Hubbard AL 1989 The major subunit of the rat asialoglycoprotein receptor can function alone as a receptor. J Biol Chem 264:1682–1688

Breitfeld PP, Simmons CF, Strous GJAM, Genze H, Schwartz AL 1985 Cell biology of the asialoglycoprotein receptor system: a model of receptor-mediated endocytosis. Int Rev Cytol 97:47–94

Connolly DT, Townsend RR, Kawaguchi K, Bell WR, Lee YC 1982 Binding and endocytosis of cluster glycosides by rabbit hepatocytes. Evidence for a short circuit pathway that does not lead to degradation. J Biol Chem 257:939–945

Connolly DT 1982 Binding properties of the hepatic lectin. Studies using the isolated lectin and intact rabbit hepatocytes. PhD Dissertation, The Johns Hopkins University

Drickamer K, Mamon JF, Binns G, Leung JO 1984 Primary structure of the rat liver asialoglycoprotein receptor. Structural evidence for multiple polypeptide species. J Biol Chem 259:770–778

Halberg DF, Wager RE, Farrell DC et al 1987 Major and minor forms of the rat-liver asialoglycoprotein receptor are independent galactose-binding proteins. Primary structure and glycosylation heterogeneity of minor receptor forms. J Biol Chem 262:9828–9838

Hummel JP, Dreyer WJ 1962 Measurement of protein-binding phenomena by gel filtration. Biochim Biophys Acta 63:530–532

Lee RT, Lin P, Lee YC 1984 New synthetic cluster ligands for galactose/N-acetylgalactosamine-specific lectin of mammalian liver. Biochemistry 23:4255–4261

Lee RT, Lee YC 1986 Preparation of a high-affinity photolabeling reagent for the Gal/GalNAc lectin of mammalian liver—demonstration of galactose-combining sites on each subunit of rabbit hepatic lectin. Biochemistry 25:6835–6841

Lee RT, Lee YC 1987a Affinity labeling of the galactose N-acetylgalactosamine-specific receptor of rat hepatocytes—preferential labeling of one of the subunits. Biochemistry 26:6320–6329

Lee RT, Lee YC 1987b Preparation of cluster glycosides of N-acetyl galactosamine that have subnanomolar binding constants towards the mammalian hepatic Gal/GalNAc specific receptor. Glycoconjugate J 4:317–328

Lee RT, Lee YC 1988 Rabbit and rat hepatic lectin have two sugar combining sites per monomeric unit. Biochem Biophys Res Commun 155:1444–1451

Lee YC 1978 Synthesis of some cluster glycosides suitable for attachment to proteins or solid matrices. Carbohydr Res 67:509–514

Lee YC, Lee RT 1982 In: Horowitz MI (ed) The Glycoconjugates Vol IV. Academic Press, New York, p 57–83

Lee YC, Stowell CP, Krantz MJ 1976 2-Imino-2-methoxyethyl 1-thioglycosides—new reagents for attaching sugars to proteins. Biochemistry 15:3956–3963

Lee YC, Townsend RR, Hardy MR et al 1983 Binding of synthetic oligosaccharides to the hepatic Gal/GalNAc lectin. Dependence on fine-structural features. J Biol Chem 258:199–202

Lee YC, Townsend RR, Hardy MR, Lönngren J, Bock K 1984 Binding of synthetic clustered ligands to the Gal/GalNAc lectin on rabbit hepatocytes. In: Lo TP, Kiu TY, Li CH (eds) Biochemical and biophysical studies of proteins and nucleic acids. Elsevier Science Publishers, New York, p 349–360
McPhaul M, Berg P 1986 Formation of functional asialoglycoprotein receptor after transfection with cDNAs encoding the receptor proteins. Proc Natl Acad Sci USA 83:8863–8867
Townsend RR, Hardy MR, Wong TC, Lee YC 1986 Binding of N-linked bovine fetuin glycopeptides to isolated rabbit hepatocytes—Gal/GalNAc hepatic lectin discrimination between Galβ(1,4)GlcNAc and Galβ(1,3)GlcNAc in a triantennary structure. Biochemistry 25:5716–5725
Weigel PH, Oka J 1983 The large intracellular pool of asialoglycoprotein receptors functions during the endocytosis of asialoglycoproteins by isolated rat hepatocytes. J Biol Chem 258:5095–5102

DISCUSSION

Feizi: You propose that there is more than one sugar-binding site on each subunit of the rat Gal/GalNAc receptors. Is there any evidence for this at the protein level? Are there two homologous domains per subunit?

Lee: Even for a single site you cannot expect a combining site to consist of a small local region in the sequence. Each sugar-combining site probably uses amino acids from various places in the sequence.

Drickamer: Within the polypeptide chain we have found no evidence for two homologous domains; there is no obvious gene duplication. They may exist in the three-dimensional structure.

Ruoslahti: Do the two ligand sugars compete with one another? You showed that the binding site is homogeneous. It could still be that one sugar binds to a slightly different area within the binding site. Do the galactose and *N*-acetylgalactosamine bind to exactly the same site?

Lee: That's what we think. Using a wide variety of synthetic compounds, we see that the simple sugars inhibit one another.

Fukuda: Has the galactose-binding protein been cloned? Is the sequence known, and how does it fit with the model?

Drickamer: These are the RHL proteins that I described (Drickamer, this volume). Cloning gives us the primary structure.

Ruoslahti: We need data from crystal structures.

Sharon: Many plant lectins that are specific for Gal/GalNAc have a high affinity for hydrophobic glycosides or *N*-dansylgalactosamine. Do your lectins also bind strongly to these compounds?

Lee: The hepatic lectins, mammalian as well as chicken, do have a little affinity for hydrophobic glycosides, but not as much as concanavalin A.

Sharon: Have you tested sugars with other substituents on the amino group?

Lee: Yes; bulky substituents weaken the binding, but not enough to influence the overall picture.

Hindsgaul: Are the inter-galactose distances measured to O4 of the galactose?

Lee: At that time we used C4 to calculate inter-galactose distances. They are not firm numbers.

Hindsgaul: You draw the binding site as a flat triangle, but if you are binding from the back of the galactose, it must be in a pocket.

Lee: Those experiments were done in 1982. We now realize that one cannot only go by the inter-sugar distances. For the two triantennary oligosaccharide chains from fetuin, not only the distances but also the orientations of the sugar are important. Two pairs of galactoses can have very different orientations but the same inter-C4 distances.

Feizi: Does this rat hepatic lectin also bind glucose?

Lee: Yes, but glucose is only bound if there is an amidine in the aglycone. We analysed the binding requirements by modifying each hydroxyl in turn. For this lectin, the 3-OH groups are essential: one cannot modify the 3-OH groups and retain binding. However, I believe the binding energy to have many components. In this example, if you have glucose instead of galactose, you lose the contribution from the binding of an axial hydroxyl group. But, provided there is an amidine group in the aglycone, binding still occurs. The binding of a glycosyl unit should be regarded as a spectrum of energy contributions, along the same lines as Jeremy Carver suggested for the averaged ensemble of various conformations (Carver et al, this volume).

Sharon: If you have the same aglycone with galactose, do you have much stronger binding?

Lee: Yes. This is more clearly shown with the simple glycosides than with neoglycoproteins (Lee et al 1982).

Nagai: Is the Gal-polylysine highly basic?

Lee: No. We convert the ε-amino groups of polylysine into amides with a galactoside which contains a carboxyl group in its aglycone. Those amino groups not modified with a galactose derivative are protected by N-acetylation. (See the formula on p 90.)

Nagai: The N-acetylation must render the polylysine hydrophobic.

Lee: Yes, but fortunately we have a series of compounds; for example, we have both galactose-and mannose-polylysine. We use mannose-polylysine as a control compound and can therefore account for the non-specific binding.

Ruoslahti: Are there any humans or animals that lack the hepatic lectin, and, if so, what's the phenotype?

Lee: I don't think such a phenotype has been found in any animal. However, an established cell line, HepG2, has an aberration in the number of these receptor proteins and in their distribution.

Ruoslahti: What would one predict the phenotype to be?

Drickamer: I presume it would be lethal. Liver damage from cirrhosis or hepatoma results in a lower clearance activity and accumulation of asialoglycoproteins in the circulation. That implies that there is an active neuraminidase present.

Feizi: If immunoglobulins accumulate, a severe hyperviscosity might be predicted.

Sharon: Although there was a claim in the literature that immunoglobulins are cleared from the circulatory system by the Ashwell–Morell pathway, this is apparently not the case.

Hughes: A common problem with therapeutic engineered glycoproteins is their rapid loss from the circulation after intravenous injection, due perhaps to underglycosylation and exposure of terminal galactose or mannose residues. Could you suggest a simple, inexpensive reagent that might enhance circulation in these systems?

Lee: Galactose-containing neoglycoproteins and other galactose-containing synthetic compounds suppress the clearance. I cannot say much about the cost, nor about any problems with immunogenicity of these compounds.

Hughes: An attractive candidate would be your galactosylated and acetylated poly-L-lysines, which are presumably rather non-immunogenic.

Lee: Yes, that's one of the hopeful leads. They are not immunogenic. We also make galactose-modified poly-D-lysines, which have a prolonged life in isolated hepatocytes. However, we don't know how they behave *in vivo*.

References

Carver JP, Michnick SW, Imberty A, Cumming DA 1989 Oligosaccharide–protein interactions: a three-dimensional view. In: Carbohydrate recognition in cellular function. Wiley, Chichester (Ciba Found Symp 145) p 18–26

Drickamer K 1989 Multiple subfamilies of carbohydrate recognition domains in animal lectins. In: Carbohydrate recognition in cellular function. Wiley, Chichester (Ciba Found Symp 145) p 45–61

Lee RT, Myers RW, Lee YC 1982 Further studies on the binding characteristics of rabbit liver galactose/N-acetylgalactosamine-specific lectin. Biochemistry 21:6292–6298

General discussion I

Two human lysosomal membrane glycoproteins

Fukuda: We have isolated two human lysosomal membrane glycoproteins, h-lamp-1 and h-lamp-2, which have 389 and 380 amino acids respectively. Eighteen (h-lamp-1) and sixteen (h-lamp-2) N-glycan attachment sites can be identified in the sequence derived by cDNA analysis (Carlsson et al 1988, Fukuda et al 1988a). In the middle section of the protein sequence, we found a strong homology with the hinge region of the immunoglobulin α_1 chain. This section is enriched with proline and serine or threonine and we think it is functionally important. Alternatively, if we put the second and the third cysteines together to make a disulphide loop, a different three-dimensional structure is formed which fits with that of the immunoglobulin superfamily; the tyrosine-X-cysteine sequence found in the immunoglobulin superfamily is present in each C-terminal region of the disulphide loops.

These proteins are probably interesting for two reasons: they contain poly-*N*-acetyllactosamine oligosaccharides; and they are targeted to lysosomes. We found that if the human chorionic gonadotropin α chain is secreted, the protein receives typical complex-type N-linked oligosaccharides. When it is fused with the trans-membrane portion of G protein, however, it acquires poly-*N*-acetyllactosamine oligosaccharides. We think this occurs because the gonadotropin α chain is now bound to the transmembrane portion, and the glycosylation sites are close to the membrane protein (Fukuda et al 1988b). h-lamp-1 has fewer N-glycosylation sites close to the transmembrane portion than h-lamp-2. By glycopeptide analysis we found that h-lamp-2 contains more poly-*N*-acetyllactosamine than h-lamp-1. Therefore, our hypothesis is that the distance to the transmembrane portion is critical for the extension enzyme, (β1–3)-*N*-acetylglucosaminyltransferase.

We isolated h-lamp-1 and h-lamp-2 because, among poly-*N*-acetyllactos-amine-containing glycoproteins in total cell lysates, these two lysosomal membrane glycoproteins are the major ones susceptible to endo-β-galactosidase. We used immunoelectron microscopy to confirm that they are lysosomal membrane glycoproteins. Rabbit antibodies specific to the glycoproteins showed convincing staining of the lysosomal membrane, as did the *Datura stramonium* lectin. When the distribution of lectin binding, presumably to poly-*N*-acetyl-lactosamine, and antibody staining were compared, the patterns were found to be almost identical. These results also support the idea that h-lamp-1 and h-lamp-2 are major carriers for poly-*N*-acetyllactosamine oligosaccharides.

The distribution of these lysosomal membrane glycoproteins is very different from that of mannose 6-phosphate receptors or many other components involved in endocytosis. Where fusion takes place between lysosomes and endosomes, there is always a high density of lysosomal membrane glycoproteins. Lysosomal membrane glycoproteins are synthesized in the Golgi and remain until the endosome fuses with prelysosome. Once fusion occurs, most of the glycoproteins go to the lysosome. In contrast, Man-6-P receptor–enzyme complexes tend to dissociate when they meet the acidic pH of the prelysosome. Dissociated receptors then return to the Golgi, which explains the different distributions of mannose 6-phosphate receptors and lysosomal membrane glycoproteins.

What is the signal for targeting of lysosomal membrane glycoproteins? Because most of the molecules are inside the vesicle, it is likely that a short cytoplasmic segment is of functional significance for this molecule to go to the lysosome. We used a transfection vector based on the SV40 promoter. When COS-1 cells are transfected with the vector, they support the proliferation of vector DNA by providing large T antigen. When the control h-lamp-1 cDNA was inserted in the vector, the resultant proteins went to the lysosome, as expected. This was done by Mark Williams. He deleted the cytoplasmic segment by site-directed mutagenesis of serine to a stop codon and then the protein no longer went to the lysosome. Instead it went to the endoplasmic reticulum and Golgi. Thus it appears that the cytoplasmic segment downstream from serine is important in the targeting of lysosomal membrane glycoproteins. The sequence HAGY (His-Ala-Gly-Tyr) is found at the cytoplasmic segment in both h-lamp-1 and h-lamp-2. Only modification of tyrosine prevents the glycoprotein from going to lysosomes. Other mutations do not influence the traffic of this molecule. I hope to make HAGY famous like the RGD sequence!

Ruoslahti: What did you change the tyrosine to?

Fukuda: We changed tyrosine to histidine. Now we are trying to change it to phenylalanine.

Hughes: What is the evidence that this tetrapeptide HAGY is important? Have you made constructs of non-lysosomal membrane proteins with that sequence in the cytoplasmic domain and shown that these proteins transfer into lysosomes?

Fukuda: We are doing that. First we changed a serine into the stop codon, TGA. This mutation eliminates the cytoplasmic segment and the mutant gene product does not go to lysosomes. Then we modified the amino acids one by one, and only the tyrosine had any influence.

Hart: If that's a signal to keep the protein in the lysosome, why is there so much of that protein on the cell surface?

Fukuda: That is an important question. It is not yet clear how much lysosomal membrane protein is on the cell surface in a given cell. In COS cells, which are tumour cells, it is easy to see these glycoproteins on the cell surface. Lysosomal enzymes are secreted more in metastatic tumour cells than in non-metastatic ones. Therefore we are investigating whether there is any difference in distribution between these tumours. Lippincott-Schwartz & Fambrough (1987) studied the distribution of the chicken protein and found 2–3% in plasma membrane and 90% in lysosome. Griffiths et al (1988) failed to show any on the cell surface. I have found it very hard to locate lysosomal membrane glycoproteins on the cell surface of HeLa and other human cells.

Lee: Lippincott-Schwartz & Fambrough's finding is not only important because of the 2% on the surface. The glycoprotein moves from the surface to the lysosome, which shows that there is communication between these areas.

Gahmberg: I presume that the large amounts of N-linked polysaccharides are important for the protection of the inner surface of the lysosomes. Does tunicamycin degrade the protein, and does it then get to the lysosomes?

Fukuda: I haven't done that experiment. Barriocanal et al (1986) have shown that the half-time is decreased when tunicamycin is added to the cells.

Barondes: Are there other major poly-*N*-acetyllactosamine-containing proteins in the cells you studied?

Fukuda: It is hard to say. In granulocytes, which are phagocytic nucleated cells, these proteins are the major ones containing poly-*N*-acetyllactosamine.

Barondes: What about non-phagocytic cells?

Fukuda: They also contain large amounts of lysosomal membrane glycoproteins. These proteins are in all the nucleated cells we have checked, and are the predominant poly-*N*-acetyllactosamine-containing proteins.

Feizi: We have studied permeabilized granulocytes with antibodies that recognize oligosaccharides of poly-*N*-acetyllactosamine type, such as anti-i (Thorpe & Feizi 1984). In support of your results, there is a huge amount of staining, not only at the surface, but also in the lining of intracellular vesicular structures that resemble lysosomes.

How sure are you that the apparent accumulation of your modified protein in the endoplasmic reticulum and the Golgi isn't a compensatory increased production rather than stagnation?

Fukuda: We have a normal control. The vector containing the normal gene appears to produce a large amount of the proteins, and they go to lysosomes.

Feizi: But if your protein is being rapidly degraded, you may have a compensatory increase in production.

Fukuda: I do not think that is the problem. Rather there may be some difficulty in protein folding. That's why it is important to construct a hybrid protein using the cytoplasmic segment and a reporter molecule. If a fused protein does not go to the lysosome after mutation at the tyrosine residue, that is clear evidence for the signal.

Stanley: Is your sequence seen in the other lysosomal membrane proteins that Howe et al (1988) have sequenced?

Fukuda: Yes. The same kinds of molecules from different species are almost completely homologous. Because we isolated two different types of this lysosomal membrane glycoprotein which have less homology, we can identify the potential targeting signal. For example, lamp-1 molecules from chicken and humans have more than 60% homology, but between human lamp-1 and lamp-2 the homology is less than 40%. However, the HAGY sequence is conserved between these two different types of protein.

Ruoslahti: Have you' looked for this sequence in a protein data bank?

Fukuda: Yes, on a small scale. And we found no homologous sequence in other proteins.

Hart: We all like to imagine that these proteins are acting like some kind of stomach lining, keeping the lysosome from digesting the cell, but the lysosome is loaded with glycosidases and I don't understand how having highly glycosylated proteins is going to solve that.

Fukuda: That's why I think the poly-*N*-acetyllactosamine in these proteins is important. It is hard to digest poly-*N*-acetyllactosamine with exoglycosidases, even with a large quantity of β-galactosidase and β-*N*-acetylglucosaminidase. One hypothesis is that if we can modify poly-*N*-acetyllactosamine attachment sites we might see a decrease of the half-life.

Czech: Might this protein either use, or be part of the segregation mechanism that uses, mannose 6-phosphate to segregate lysosomal enzymes? Does this protein contain mannose 6-phosphate or does it bind mannose 6-phosphate-like proteins?

Fukuda: We haven't looked at that. In a preliminary experiment, we detected almost no phosphorylation, so we don't think mannose 6-phosphate is involved. Others have also found that they couldn't phosphorylate lysosomal membrane glycoproteins.

Czech: If it bound mannose 6-phosphate, it might be part of the mechanism of segregation.

Fukuda: It is likely that the signal is not mannose 6-phosphate and there is no sequence homology between the mannose 6-phosphate receptor and our protein. My speculation is that this protein may be involved in the fusion process because it is found in plasma membranes and late endosomes as well, but I don't have strong evidence.

Finne: What is the functional significance of the hinge region-type of sequence?

Fukuda: I am not sure. We are investigating what happens to this lysosomal glycoprotein if we delete one domain. We could also delete the hinge region and see what happens.

Muramatsu: Is the immunoglobulin-like domain more similar to the J domain or C domain?

Fukuda: I don't have such extensive data. There is a domain which has the tyrosine-X-cysteine sequence, and the loop consists of 73 amino acids. I can't find further homology.

Wang: What is the evidence that this is a trans-lysosomal membrane protein and that the putative signal sequence is exposed in the cytoplasm?

Fukuda: I don't think anyone has done the necessary proteolytic digestion or antibody studies. But it has been shown it is a membrane protein; you can extract by Triton, but not with salt or a high pH.

Wang: There is evidence that the binding of brain hexokinase to mitochondria is mediated by a N-terminal hydrophobic sequence that includes a tyrosine residue (Polakis & Wilson 1985). Limited digestion of the bindable form of hexokinase results in the loss of this tyrosine-containing sequence with concomitant conversion to a non-bindable form of the enzyme. I wonder whether you might have a similar situation here; the critical tyrosine actually mediates the binding of your protein to a receptor on the lysosomal membrane.

Fukuda: That's one possibility. That tyrosine is important is corroborated by other studies. Influenza haemagglutinin is usually on the cell surface, but Lazarovits & Roth (1988) made several mutations, including cysteine to tyrosine, and one of the tyrosine mutants brought this protein into the endosome and lysosome.

Hughes: But did that construct contain the full HAGY sequence which you suggest is the dominant signal for transfer into lysosomes?

Fukuda: No. But all the lysosomal membrane glycoproteins in the cell types we have studied contain this sequence. On the other hand, Pohlmann et al (1988) reported a different lysosomal membrane glycoprotein that doesn't have that sequence except tyrosine. So it is likely that we are dealing with several mechanisms which bring this protein into the lysosome, but a major one involves the HAGY sequence.

Hughes: Isn't it possible that the tyrosine residue is just one part of a conformational feature of lysosomal membrane proteins that is required for lysosomal transfer? Other amino acids may be important but not necessarily adjacent to tyrosine in the primary structure: such residues may be hard to identify.

Fukuda: Yes. As I said, we are trying to put the cytoplasmic segment together with a reporter molecule to see if a hybrid protein goes into the lysosome. If this hybrid protein goes to lysosome and if transfer is abolished by modifying tyrosine in the cytoplasmic sequence in the hybrid protein, I think we shall have final proof.

References

Barriocanal JG, Bonifacino JS, Yuan L, Sandoval IV 1986 Biosynthesis, glycosylation, movement through the Golgi system, and transport to lysosomes by an N-linked carbohydrate-independent mechanism of three lysosomal integral membrane proteins. J Biol Chem 261:16755–16763

Carlsson SR, Roth J, Piller F, Fukuda M 1988 Isolation and characterization of human lysosomal membrane glycoproteins, h-lamp-1 and h-lamp-2. J Biol Chem 263:18911–18919

Fukuda M, Viitala J, Matteson J, Carlsson SR 1988a Cloning of cDNAs encoding human lysosomal membrane glycoproteins, h-lamp-1 and h-lamp-2. J Biol Chem 263:18920–18928

Fukuda M, Guan J-L, Rose JK 1988b A membrane-anchored form but not the secretory form of human chorionic gonadotrophin-α chain acquires polylactosaminoglycan. J Biol Chem 263:5314–5318

Griffiths G, Hoflack B, Simons K, Mellman I, Kornfeld S 1988 The mannose 6-phosphate receptor and the biogenesis of lysosomes. Cell 52:329–341

Howe CL, Granger B-L, Hull M, Green SA, Gable CA, Helenius A, Mellman I 1988 Derived protein sequences, oligosaccharides, and membrane insertion of the 120-kDa lysosomal membrane glycoprotein (lgp120): identification of a highly conserved family of lysosomal membrane glycoproteins. Proc Natl Acad Sci USA 85:7577–7581

Lazarovits J, Roth M 1988 A single amino acid change in the cytoplasmic domain allows the influenza virus hemagglutinin to be endocytosed through coated pits. Cell 53:743–752

Lippincott-Schwartz J, Fambrough DM 1987 Cycling of the integral membrane glycoprotein, LEP100, between plasma membrane and lysosomes: kinetic and morphological analysis. Cell 49:669–677

Pohlmann R, Krentler C, Schmidt B et al 1988 Human lysosomal acid phosphatase: cloning, expression and chromosomal assignment. EMBO (Eur Mol Biol Organ) J 7:2343–2350

Polakis PG, Wilson JE 1985 An intact hydrophobic N-terminal sequence is critical for binding of rat brain hexokinase to mitochondria. Arch Biochem Biophys 236:328–337

Thorpe SJ, Feizi T 1984 Species differences in the expression of carbohydrate differentiation antigens on mammalian blood cells revealed by immunofluorescence with monoclonal antibodies. Biosci Rep 4:673–685

Nucleoplasmic and cytoplasmic glycoproteins

Gerald W. Hart, Robert S. Haltiwanger, Gordon D. Holt, and William G. Kelly

Department of Biological Chemistry, The Johns Hopkins University School of Medicine, 725 N. Wolfe Street, Baltimore, Maryland 21205, USA

Abstract. We have described a new form of protein glycosylation in which *N*-acetylglucosamine is glycosidically linked to the hydroxyl of serine or threonine (*O*-GlcNAc). Unlike most other forms of protein glycosylation, *O*-GlcNAc is predominantly localized in the nuclear and cytoplasmic compartments of cells, where it occurs on important nuclear pore glycoproteins, well-characterized cytoskeletal proteins, as well as on many chromatin proteins, including factors that regulate gene transcription. Gas-phase protein sequencing of three *O*-GlcNAc-modified proteins has identified a common structural feature at sites of *O*-GlcNAc addition. An assay for UDP-GlcNAc:polypeptide *O*-GlcNAc transferase has been developed. The enzyme appears to be membrane-associated, its active site is cytoplasmic, and it has an absolute requirement for Mn^{2+}. We are now purifying this glycosyltransferase, characterizing its substrate specificity, and determining the extent of elongation of attached saccharide moieties. The functions of *O*-GlcNAc remain largely unknown, but it may be important in blocking phosphorylation sites, it may be required for the assembly of specific multiprotein complexes, it might serve as a nuclear transport signal, or it may be directly involved in the active transport of macromolecules across nuclear pores.

1989 Carbohydrate recognition in cellular function. Wiley, Chichester (Ciba Foundation Symposium 145) p 102–118

Only recently, with the development of facile affinity purification techniques (Beyer et al 1981), has it become feasible to use purified glycosyltransferases as probes of saccharide structures and functions. Because there is one specific enzyme for each type of glycosidic linkage, these probes provide exquisite specificity analogous to that of the restriction enzymes used to probe nucleic acid structure and function. As probes of cell surfaces, both sugar nucleotide donors and the enzymes themselves are impermeant to the plasma membrane. Most importantly, unlike lectin probes, the use of radiolabelled sugar nucleotides and purified glycosyltransferases allows for product characterization, and provides a quantitative, biological method for specifically altering carbohydrate structures on cell surfaces or on individual isolated receptors.

During studies in which bovine milk galactosyltransferase was used to probe terminal *N*-acetylglucosamine (GlcNAc) residues on the surfaces of living murine lymphocyte subpopulations (Torres & Hart 1984), we unexpectedly found that 90%, 80% and 50% of the terminal GlcNAc residues on B lymphocytes, T lymphocytes and thymocytes, respectively, occurred as monosaccharide residues O-glycosidically attached to integral membrane proteins. On the other hand, latency studies showed that the vast majority of this *O*-GlcNAc was inside these cells, with B lymphocytes containing approximately 150 million residues per cell. Identification of *O*-GlcNAc was confirmed by insensitivity of galactosyltransferase-labelled material to the enzyme glycopeptide N-glycosidase A, which cleaves nearly all N-linked oligosaccharides, and by kinetic analyses of alkali-induced β-elimination of galactosylated *O*-GlcNAc moieties. In addition, the identity of *O*-GlcNAc was verified by high-resolution gel filtration chromatography, followed by high-pressure liquid chromatography on two different types of column, as well as by sensitivity of the galactosyltransferase-labelled *O*-GlcNAc to β-galactosidases, and by metabolic radiolabelling studies, using [^{3}H] glucosamine.

Because most *O*-GlcNAc-bearing proteins appeared to be intracellular, we investigated their distribution in highly purified subcellular fractions from rat liver (Holt & Hart 1986). About this time, it was reported in an abstract that nuclei contained large amounts of terminal GlcNAc residues (see Schindler et al 1987). Surprisingly, as shown in Fig. 1 and Table 1, we found *O*-GlcNAc to be greatly enriched in nuclear envelope, nuclei and cytosolic fractions of rat liver. *O*-GlcNAc was found on unique arrays of proteins in every organelle

TABLE 1 Distribution of O-linked GlcNAc in subcellular organelles in an 'average' liver cell

Organelle	*O-Linked GlcNAc*[a] *(pmol/mg organelle)*	*Total O-Linked GlcNAc*[b] *(pmol/mg liver)*
Plasma membrane	128	2.5
Cytosol[c]	171	51.3
Golgi	203	4.0
Rough microsome	70	8.4
Nuclei	533	79.9

[a]Calculated from the total radioactivity ([^{3}H]galactose) incorporated in each organelle multiplied by the percentage of *O*-GlcNAc present, measured by β-elimination experiments.
[b]Calculated from the amount of O-linked GlcNAc (pmol) for each organelle multiplied by the percentage of the total protein that the organelles represent in whole liver homogenate (Holt & Hart 1986).
[c]Obtained from non-perfused livers. When perfused livers were used nearly all terminal GlcNAc residues were O-linked GlcNAc.

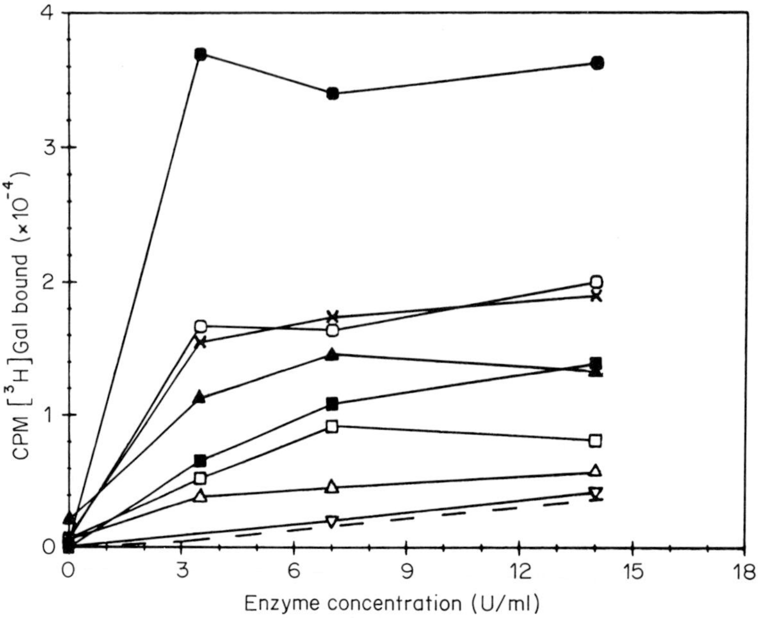

FIG. 1. Galactosylation of terminal GlcNAc residues in rat liver organelles. For each data point, 50 µg of highly purified organelle preparations were solubilized with Nonidet P-40. The samples were incubated with UDP-[^{3}H]galactose and increasing concentrations of galactosyltransferase for 30 min at 37 °C. The products were fractionated on Sephadex G-50 columns (1 × 60 cm) and the quantity of protein-bound [^{3}H]galactose eluting in the column's void volume was determined. ●, nuclear envelope; ○, rough microsome; ×, nuclei; ■, soluble fraction; ▲ Golgi; □, plasma membrane; △, stripped microsome; ▽, mitochondria; —, no membrane. Reproduced from Holt & Hart (1986) with permission.

(Fig. 2), except for mitochondria which contained no detectable terminal GlcNAc moieties. Even though serum contains large amounts of N-linked and 'mucin-type' oligosaccharide chains attached to proteins, *O*-GlcNAc was not found. Based upon the wealth of information concerning the distribution of proteins in organelles of rat liver, it was estimated that in liver cells about 55% and 35% of the total cellular *O*-GlcNAc is localized in nuclei and cytosol, respectively (Table 1), with much smaller amounts present in other organelles (Holt & Hart 1986).

Given the enrichment of *O*-GlcNAc in the nuclear envelope, we decided to investigate this organelle further. Davis & Blobel (1986) described a monoclonal antibody specific to nuclear pores that recognized a protein of 62 kDa (p62). Based upon pulse-chase experiments, earlier studies describing a major wheat germ agglutinin (WGA)-binding protein in nuclear pores (Baglia & Maul 1983),

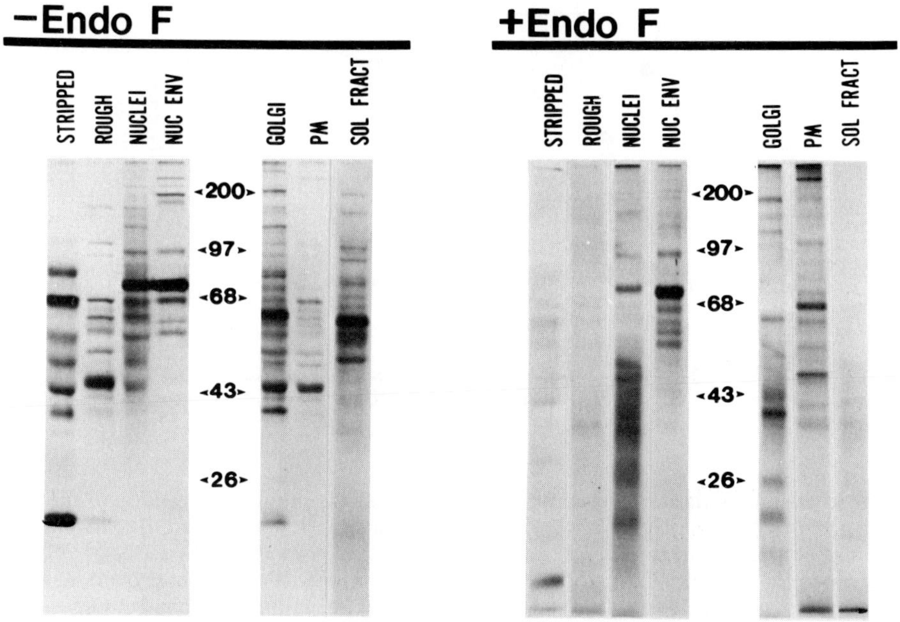

FIG. 2. Galactosyltransferase labels distinct subsets of proteins in each organelle that are insensitive to deglycosylation by glycopeptide N-glycosidase F (Endo F). [^{3}H]Galactosylated organelles were subjected to deglycosylation by Endo F (cleaves N-linked saccharides), or were separated directly by SDS-PAGE (polyacrylamide gel electrophoresis). The treated samples were fractionated on Sephadex G-50 columns, and the Endo F-resistant proteins, eluting in the void volume of the columns, were separated by SDS-PAGE and autoradiography. The migration positions of molecular mass standards ($\times 10^{-3}$) are indicated. Reproduced from Holt & Hart (1986), with permission. Stripped, stripped microsome; rough, rough microsome; nuc env, nuclear envelope; PM, plasma membrane; sol fract, soluble fraction.

the demonstration that p62 bound WGA, and our findings of *O*-GlcNAc in nuclear envelopes, these workers suggested that this nuclear pore protein might be post-translationally modified by *O*-GlcNAc. Concomitantly, Snow et al (1987) also prepared several monoclonal antibodies specific to proteins of the nuclear pore complex. Strikingly, these monoclonal antibodies appeared to recognize the same family of nuclear envelope polypeptides that we found to be modified by *O*-GlcNAc (Holt et al 1987b). Even though these antibodies are monoclonal, they all recognize various subsets of the family of proteins comprising the nuclear pore complex. As shown in Fig. 3, we found that galactosylation of these nuclear pore proteins by UDP-galactose and bovine milk galactosyltransferase abrogated the recognition of these proteins by these antibodies (Holt et al 1987b). In addition, removal of *O*-GlcNAc by hexosaminidase prevented the binding of

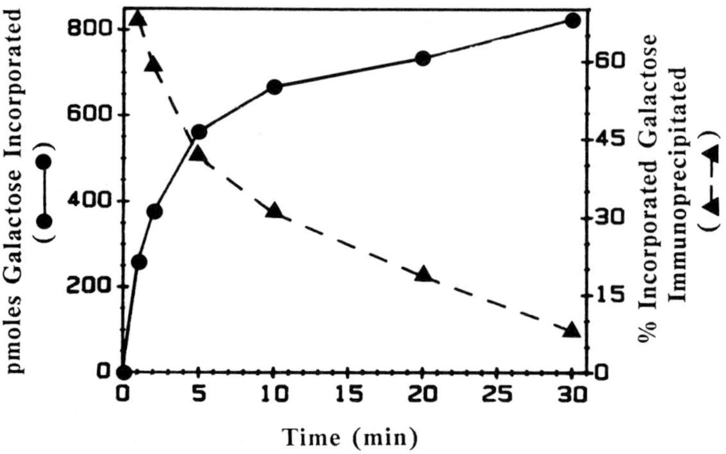

FIG. 3. Galactosylation of nuclear envelopes blocks the immunoadsorption of nuclear pore proteins by monoclonal antibodies. Nuclear envelopes were subjected to increasing extents of galactosylation as described in Fig. 1, and then combined with antibody (RL-2)-Sepharose for immunoadsorption. The antibody supernatants and pelleted material were separated from unincorporated radioactivity by fractionation on Sephadex G-50. The percentage of incorporated galactose immunoadsorbed (--- ▲ ---) is expressed as the amount of [^{3}H]galactose-labelled protein immunoadsorbed by RL-2-Sepharose divided by the total macromolecular radioactivity incorporated into protein at a given time point. The incorporation curve (—●—) is expressed as the pmol [^{3}H]galactose incorporated per 1.5×10^8 nuclear equivalents. Reproduced from Holt et al (1987b) with permission.

TABLE 2 Evidence that *O*-GlcNAc is a major epitope of nuclear pore proteins

*Monoclonal antibodies bind a family of proteins
*Galactosylation abrogates antibody binding
*Hexosaminidase treatment abrogates antibody binding
*Antibodies *do not* bind limit pronase glycopeptides

all these antibodies to nuclear pore glycoproteins to varying extents. Table 2 summarizes the evidence which suggests that *O*-GlcNAc is a major epitope recognized by these antibodies. Although binding requires *O*-GlcNAc, the fact that these antibodies did not bind to limit pronase glycopeptides containing these moieties, and that binding was not inhibited by high concentrations of free GlcNAc or GlcNAc-neoglycoproteins, suggests that the epitope recognized by these antibodies is created by a combination of both peptide structure and *O*-GlcNAc residues.

Further studies also demonstrated that the nuclear pore glycoproteins were multiply glycosylated with as many as ten to twelve *O*-GlcNAc moieties per protein (Table 3). Amino acid analysis of nuclear pore proteins that had been

TABLE 3 Determination of the number of sites of O-linked GlcNAc addition on nuclear pore glycoproteins

Molecular mass of protein (kDa)	*Galactose/protein (mol ± sd)*	*No. of determinations*
180	9.8 ± 0.6	2
100	2.0 ± 1.2	3
63	10.0 ± 1.8	3

subjected to alkali-induced β-elimination in the presence of sodium borotritide, directly demonstrated that the majority of the *O*-GlcNAc was attached to serine residues and a small amount to threonine residues. Immunogold electron microscopy, using one of the *O*-GlcNAc-recognizing nuclear pore-specific monoclonal antibodies (Snow et al 1987), confirmed the biochemical evidence that *O*-GlcNAc is cytoplasmically and nucleoplasmically oriented (Holt & Hart 1986, Holt et al 1987a). Latency and WGA-binding studies by Hanover and colleagues (1987) also demonstrated that *O*-GlcNAc is localized at the cytoplasmic and nucleoplasmic faces of nuclear pores. Further support for the high immunogenicity of these saccharides was obtained by independently isolating nuclear envelope-specific monoclonal antibodies and demonstrating that their epitopes largely consisted of *O*-GlcNAc (Park et al 1987). Several laboratories have directly implicated *O*-GlcNAc at the faces of nuclear pores in nucleocytoplasmic transport of macromolecules (Baglia & Maul 1983, Finlay et al 1987, Yoneda et al 1987, Featherstone et al 1988, Dabauvalle et al 1988). Interestingly, the binding of WGA or antibodies to *O*-GlcNAc appears to prevent transport by a mechanism other than steric blockage. These agents also do not inhibit the binding of nuclear proteins to the pore complex, but rather *O*-GlcNAc appears to be involved in the active translocation process (Newmeyer & Forbes 1988).

Several intrinsic membrane proteins of both rough and smooth endoplasmic reticulum have also been shown to contain cytosolic-oriented *O*-GlcNAc residues (Abeijon & Hirschberg 1988). In addition, a major basic phosphoprotein localized in the tegument (between the capsid and the viral envelope) of human cytomegalovirus has been shown to be modified by *O*-GlcNAc (Benko et al 1988). Such a localization is unprecedented for a glycosylated viral protein. *O*-GlcNAc is also a major form of protein glycosylation in the pathogenic parasite *Schistosoma mansoni* (Nyame et al 1987).

In order to more easily identify a cytosolic protein containing *O*-GlcNAc, and to obtain a readily available source of potential substrates for purification of the polypeptide *O*-GlcNAc glycosyltransferase, we investigated the presence of *O*-GlcNAc in human erythrocytes (Holt et al 1987b). As shown in Fig. 4a, nearly all the terminal GlcNAc in these cells is intracellular, most of it occurring as *O*-GlcNAc. Subcellular fractionation of erythrocytes (Fig. 4b) demonstrated that the major *O*-GlcNAc protein in these cells is an unidentified cytosolic soluble protein of 65 kDa. In addition, several proteins apparently associated with the innerside of the plasma membrane are also modified by *O*-GlcNAc. One of these was purified to homogeneity and shown to be Band 4.1, a well-characterized cytoskeletal protein. Band 4.1 is important in maintaining erythrocyte shape, because it anchors the underlying cytoskeleton to the plasma membrane by interacting with the cytoplasmic tail of glycophorin and the spectrin/actin skeleton (Bennett 1985).

We have used UDP-[^{3}H]galactose and galactosyltransferase to label *O*-GlcNAc moieties on Band 4.1, the 65 kDa erythrocyte protein, and p62 of nuclear pores, and have purified these proteins to homogeneity. Radiolabelled glycopeptides were obtained by proteolysis and/or by cyanogen bromide

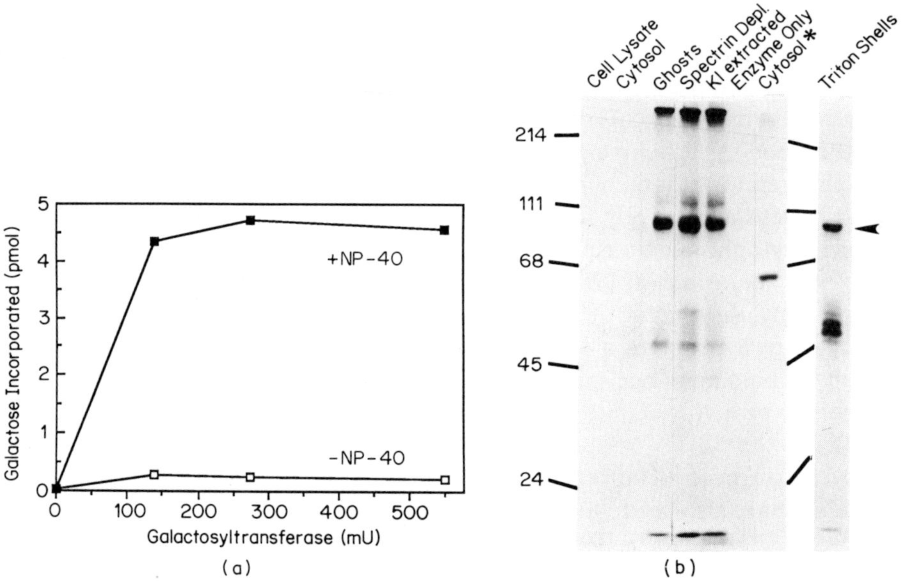

FIG. 4. Human erythrocytes contain cytosolic proteins bearing terminal GlcNAc. (a) Erythrocytes were lysed by the addition of Nonidet P-40 (+NP-40), or were left intact for labelling by galactosyltransferase (−NP-40). (b) Autoradiogram of erythrocyte subcellular fractions galactosylated with UDP-[^{3}H]galactose and bovine milk galactosyltransferase. Approximately 10 μg of cell fraction proteins were labelled and loaded on each lane, except for the Cytosol* fraction in which 500 μg of protein was used. Arrow indicates the migration position of Band 4.1. The migration positions of molecular mass standards ($\times 10^{-3}$) are indicated. Reproduced from Holt et al (1987a) with permission.

treatment, and purified by reverse-phase high-performance liquid chromatography. Glycopeptides purified by two different solvent systems were then subjected to gas-phase protein sequencing to identify the nature of the glycosylation sites. Table 4 shows sequences of glycopeptides containing sites of *O*-GlcNAc attachment for Band 4.1, the 65 kDa protein and the p62 nuclear pore protcin, and tentatively assigns the sites of attachment of the saccharide. The common structural features include the presence of an acidic amino acid adjacent to the glycosylation site and a proline within a distance of one to three amino acids. In addition, all these proteins appear to have long stretches of serine and/or threonine repeats. Based upon this sequence of the p62 nuclear pore protein, a probe was synthesized that was subsequently used to isolate and sequence a cDNA encoding this protein (D'Onofrio et al 1988).

Pulse-chase studies on the biosynthesis of the p62 nuclear pore protein suggested that it is glycosylated within five minutes of its synthesis while still in the cytoplasm (Davis & Blobel 1987). Using chemically deglycosylated or hexosaminidase-treated human erythrocyte 65 kDa protein as a substrate, we have developed an assay for the UDP-GlcNAc:polypeptide *O*-GlcNAc transferase that is dependent upon exogenous substrate, enzyme and time of incubation. Using this assay, we have found that the enzyme is not inhibited by GlcNAc-1-phosphate, GlcNAc or tunicamycin. However, it is inhibited by UDP, and has an absolute requirement for Mn^{2+}. Latency experiments with sealed microsomes show that the transferase's active site is cytosolic, but surprisingly the enzyme appears to be membrane-bound.

During studies on nuclear pore glycoproteins, we and others (Davis & Blobel 1987) have observed *O*-GlcNAc-bearing proteins within the nucleus. Several well-studied transcription factors for RNA polymerase II were found to be multiply glycosylated by *O*-GlcNAc (Jackson & Tjian 1988). As illustrated in Fig. 5, we have found an astonishing abundance of *O*-GlcNAc-bearing proteins associated with polytene chromosomes of *Drosophila* embryos. In addition to the intense labelling by galactosyltransferase (Fig. 5), which biochemical analyses show is

TABLE 4 Sequences surrounding sites of *O*-GlcNAc addition

Band 4.1
....Thr-Ala-Gln-Thr-Ile-**Ser*-Glu-Thr*-Pro-**Ser-Ser-Thr-Thr-Thr....
65 kDa protein
....**Asp-(Ser*)-Pro-**Val-Ser-Gln-Pro-Ser-Leu-Val-Gly-Ser-Lys....
Nuclear pore protein
....Ala-Gly-Gly-Pro-Ala-**Asp-Thr-Ser*-Asp-Pro-**Leu-Gln-Gln-Ile....

*Tentative sites of *O*-GlcNAc addition.
From D'Onofrio et al (1988).

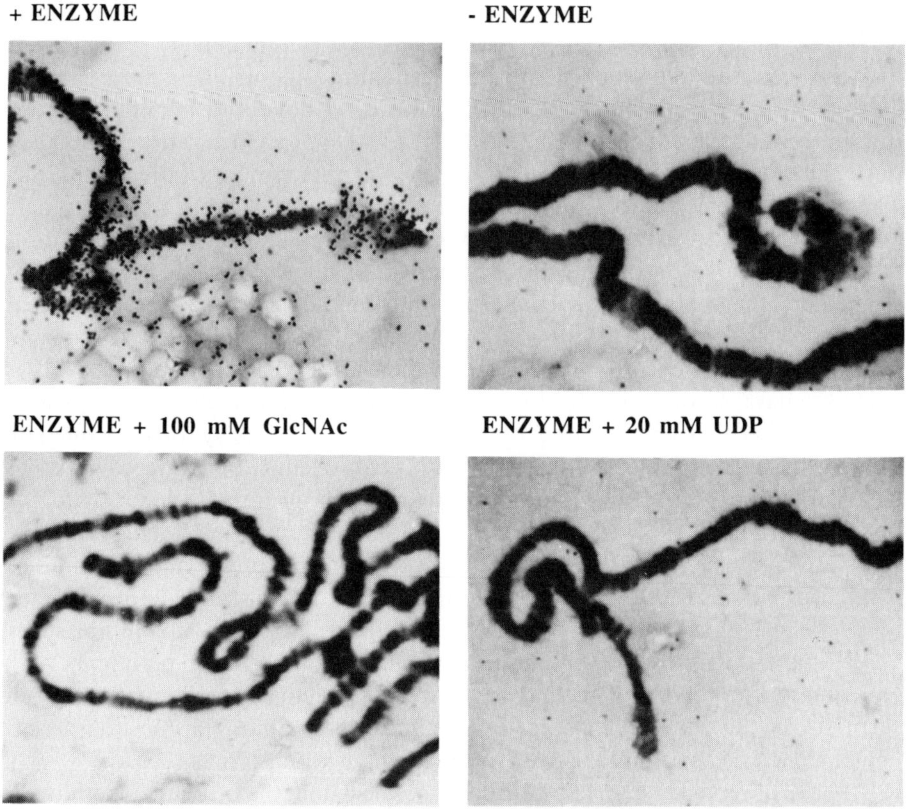

FIG. 5. *Drosophila* polytene chromosomes contain large amounts of *O*-GlcNAc-bearing proteins. Polytene chromosomes were prepared from third instar *Drosophila* larvae by standard procedures and galactosylated by UDP-[^{3}H]galactose and bovine milk galactosyltransferase. Note: Labelling is dependent upon enzyme, and is inhibited by GlcNAc and UDP. Biochemical analysis of chromatin indicates that most of the incorporated radioactivity is in *O*-GlcNAc-bearing proteins.

almost entirely on *O*-GlcNAc, fluorescent WGA also specifically and intensely stains polytene chromosomes. Thus, it appears that *O*-GlcNAc is a common modification of many chromatin-associated proteins.

The functions of *O*-GlcNAc remain largely unknown. However, there is clear evidence, as discussed above, that it plays some role in nucleocytoplasmic transport by nuclear pores. In addition, most of the proteins that it is found on are also phosphorylated. Perhaps it blocks phosphorylation sites and is removed by a regulated nuclear and/or cytoplasmic hexosaminidase. On many proteins, *O*-GlcNAc occurs in clusters. Perhaps it protects these regions of the polypeptide from proteolysis. Certain *O*-GlcNAc-bearing structures could also serve as a targeting signal to the nucleus (Schindler et al 1987), or *O*-GlcNAc might be involved in mediating or modulating important protein–protein interactions in the formation of multiprotein structures such as cytoskeletal networks (Holt et al 1987a) or the nuclear pore complex (Holt et al 1987b).

Acknowledgements

This work was supported by Grant HD 13563 awarded by the National Institutes of Child Health and Human Development, and by Grant CA 42486 awarded by the National Cancer Institute, Department of Health and Human Services, USA. It was also performed during the tenure of an American Heart Association Established Investigatorship (to G.W.H.). R.S.H. is supported by a postdoctoral fellowship from the Arthritis Foundation, and W.G.K. is a predoctoral fellow of the March of Dimes.

References

Abeijon C, Hirschberg CB 1988 Intrinsic membrane glycoproteins with cytosol-oriented sugars in the endoplasmic reticulum. Proc Natl Acad Sci USA 85:1010–1014

Baglia F, Maul GG 1983 Nuclear ribonucleoprotein release and nucleoside triphosphatase activity are inhibited by antibodies directed against one nuclear matrix glycoprotein. Proc Natl Acad Sci USA 80:2285–2289

Benko DM, Haltiwanger RS, Hart GW, Gibson W 1988 Virion basic phosphoprotein from human cytomegalovirus contains O-linked *N*-acetylglucosamine. Proc Natl Acad Sci USA 85:2573–2577

Bennett V 1985 The membrane skeleton of human erythrocytes and its implications for more complex cells. Annu Rev Biochem 54:273–304

Beyer TA, Sadler JE, Rearick JI, Paulson JC, Hill RL 1981 Glycosyltransferases and their use in assessing oligosaccharide structure and structure–function relationships. Adv Enzymol 52:23–175

D'Onofrio M, Starr CM, Park MK et al 1988 Partial cDNA sequence encoding a nuclear pore protein modified by O-linked *N*-acetylglucosamine. Proc Natl Acad Sci USA 85:9595–9600

Dabauvalle MC, Schulz B, Scheer U, Peters R 1988 Inhibition of nuclear accummulation of karyophilic proteins in living cells by microinjection of the lectin wheat germ agglutinin. Exp Cell Res 174:291–296

Davis LI, Blobel G 1986 Identification and characterization of a nuclear pore complex protein. Cell 45:699–709

Davis LI, Blobel G 1987 Nuclear pore complex contains a family of glycoproteins that includes p62: glycosylation through a previously unidentified cellular pathway. Proc Natl Acad Sci USA 84:7552–7556

Featherstone C, Darby MK, Gerace L 1988 A monoclonal antibody against the nuclear pore complex inhibits nucleocytoplasmic transport of protein and RNA in vivo. J Cell Biol 107:1289–1297

Finlay DR, Newmeyer DD, Price TM, Forbes DJ 1987 Inhibition of *in vitro* nuclear transport by a lectin that binds to nuclear pores. J Cell Biol 104:189–200

Hanover JA, Cohen CK, Willingham MC, Park MK 1987 O-linked *N*-acetylglucosamine is attached to proteins of the nuclear pore. Evidence for cytoplasmic and nucleoplasmic glycoproteins. J Biol Chem 262:9887–9894

Holt GD, Snow CM, Senior A, Haltiwanger RS, Gerace L, Hart GW 1987a Nuclear pore complex glycoproteins contain cytoplasmically disposed O-linked *N*-acetylglucosamine. J Cell Biol 104:1157–1164

Holt GD, Hart GW 1986 The subcellular distribution of terminal *N*-acetylglucosamine moieties. Localization of a novel protein–saccharide linkage, O-linked GlcNAc. J Biol Chem 261:8049–8057

Holt GD, Haltiwanger RS, Torres CR, Hart GW 1987b Erythrocytes contain cytoplasmic glycoproteins. O-linked GlcNAc on Band 4.1. J Biol Chem 262:14847–14850

Jackson SP, Tjian R 1988 O-Glycosylation of eukaryotic transcription factors: implications for mechanisms of transcriptional regulation. Cell 55:125–133

Newmeyer DD, Forbes DJ 1988 Nuclear import can be separated into distinct steps in vitro: nuclear pore binding and translocation. Cell 52:641–653

Nyame K, Cummings RD, Damian RT 1987 *Schistosoma mansoni* synthesizes glycoproteins containing terminal O-linked *N*-acetylglucosamine residues. J Biol Chem 262:7990–7995

Park MK, D'Onofrio M, Willingham MC, Hanover JA 1987 A monoclonal antibody against a family of nuclear pore proteins (nucleoporins): O-linked *N*-acetylglucosamine is part of the immunodeterminant. Proc Natl Acad Sci USA 84:6462–6466

Schindler M, Hogan M, Miller R, DeGaetano D 1987 A nuclear specific glycoprotein representative of a unique pattern of glycosylation. J Biol Chem 262:1254–1260

Snow CM, Senior A, Gerace L 1987 Monoclonal antibodies identify a group of nuclear pore complex glycoproteins. J Cell Biol 104:1143–1156

Torres CR, Hart GW 1984 Topography and polypeptide distribution of terminal *N*-acetylglucosamine residues on the surfaces of intact lymphocytes. Evidence for *O*-GlcNAc. J Biol Chem 259:3308–3317

Yoneda Y, Imamoto-Sonobe N, Yamaizumi M, Uchida T 1987 Reversible inhibition of protein import into the nucleus by wheat germ agglutinin injected into cultured cells. Exp Cell Res 173:586–595

DISCUSSION

Ruoslahti: It has been suggested that glycosylation is an alternative to phosphorylation of serine residues. Is there a resemblance between your glycosylation sequences and any of the preferred sequences phosphorylated by kinases?

Hart: These glycosylation sites look similar to phosphorylation sites in casein kinase II enzymes. We don't have any data yet, but one of our working hypotheses is that this type of glycosylation blocks phosphorylation sites. Perhaps there is a very active hexosaminidase that is also regulated. Figge et al (1988) made a computer search on several nuclear oncogenes and found that the sequence which seems to be involved in the transforming properties coincides with the apparent glycosylation sites, which may have something to do with phosphorylation in those proteins.

Several groups have shown that antibodies to *O*-GlcNAc or lectins that specifically bind *O*-GlcNAc (e.g. wheat germ agglutinin) block nuclear transport.

They do not appear to do this at the binding of proteins to the pore, but rather by blocking the active transport steps.

Czech: These glycosylation sequences (Table 4) are not classic consensus sequences for casein kinase II, in that optimal phosphorylation usually occurs when there are many acidic amino acids, especially at the third residue on the C-terminal side of the serine or threonine. Have you made peptides from the sequences and investigated whether they are good phosphorylation sites by using purified kinases?

Hart: We have several peptides, but we are mainly using them to go after the GlcNAc transferase. We have tried casein as a substrate for the O-linked GlcNAc transferase, but the one experiment we have done didn't work.

Stanley: Is the disposition of the GlcNAc transferase trypsin-sensitive in your microsomes?

Hart: We haven't done those experiments yet.

Stanley: Does the RL-2 monoclonal antibody (Holt et al 1987) recognize O-linked GlcNAc on cell surface glycoproteins as well?

Hart: I am no longer convinced that O-linked GlcNAc is on the cell surface. About 3% of the cells being leaky could easily account for all the O-linked GlcNAc that we saw. We are investigating that, because if O-linked GlcNAc is on the surface there are many implications, for example about topology and how it gets there.

Stanley: We have a Chinese hamster ovary (CHO) double glycosylation mutant called ldlD.Lec1 (Stanley 1989) which might be useful for selecting a mutant in the *O*-GlcNAc biosynthetic pathway.

Hughes: Your galactosyltransferase method is very good for picking up terminal GlcNAc residues of glycoproteins. Are some of the GlcNAc residues linked to hydroxy amino acids elongated with other sugars?

Hart: We are working on that. We metabolically label the glucosamine, do β-elimination and fractionate everything on a P4 column. Then we look for anything that has *N*-acetylglucosaminitol at the reducing end after the β-elimination and is larger than a monosaccharide. Those experiments were done in an R1 lymphoma cell line where only the cytosolic components, not the nuclear components, were released. We saw no evidence of elongation; more than 95% of the residues were monosaccharide. However, it is my feeling that there is elongation and that it will depend on the protein and the cell type, but we don't have any evidence. The simian virus 40 (SV40) large T antigen may have an elongated O-linked GlcNAc. It has an elongated base-labile linkage that is not typical—it contains glucosamine and galactose—but it hasn't been characterized more than that.

Hughes: Does it have a consensus sequence for O-glycosylation by a GlcNAc residue?

Hart: Yes; it was one of the proteins reported by Figge et al (1988) to contain these types of sequences. We have confirmed the presence of a putative consensus sequence for *O*-GlcNAc addition.

Rademacher: Does your glycosyltransferase transfer galactose to O-linked GalNAc?

Hart: No, it does not.

Wang: You indicated that you get neither wheat germ agglutinin labelling nor [^{3}H]UDP-Gal transfer at the chromosome puffs of the *Drosophila* polytene chromosomes. Perhaps these are sites of localization of the transcription factor which has already elongated the O-linked GlcNAc residue with galactose. If so, you would need the next-step enzyme to label these puffs. Have you tried using sialyltransferase to label these sites?

Hart: That's a good experiment. The only problem is that the specificities of the transferases are so great that one makes the assumption that it is going to be a Gal(β1–4) linkage. We have purified liver (α2–6)-sialyltransferase and we shall do that experiment.

Hughes: You could use galactose oxidase to pick up any terminal galactose and then see if a Gal(β1–4)GlcNAcitol is released by β-elimination.

Barondes: I am also interested in the possibility that the *O*-GlcNAc residues might be modified by additions of galactose, because this would give rise to galactosides that could bind the complementary lectins which appear to be present in the cytoplasm. Is there evidence for a galactosyltransferase in, or facing, the cytoplasm?

Hart: There was little O-linked GlcNAc in the Golgi compartment compared to other parts of the cell. Nancy and Joel Shaper have demonstrated that murine galactosyltranferase has two transcripts (Shaper et al 1988). One product is 13 amino acids longer than the other, because of alternative splicing. The positions of the cleavable signal sequence and membrane-spanning domain in the longer one suggest that it faces the cytosol. That is a minor component; the major galactosyltransferase faces the cisternae of the Golgi. This has not been proved at the protein level, but if it does occur, then perhaps the major galactosyltransferase is localized to the Golgi. If our *O*-GlcNAc transferase is localized, say, in the endoplasmic reticulum, and proteins go into the nucleus elsewhere, it is easy to see how the *O*-GlcNAc glycoproteins would not meet the galactosyltransferases; even though these transferases are not compartmentalized, they are segregated in that they are facing the wrong direction.

Barondes: But how would they ever meet?

Hart: Perhaps there are intracellular transport mechanisms that somehow bring specific proteins to the Golgi at the wrong face. I know that's a pretty bizarre idea.

Lindahl: There are other examples from glycosaminoglycan biosynthesis where enzymes and substrates seem to be in the same subcellular compartment and yet are functionally segregated until a brief period of reaction, after which they are again kept separate. For example, in heparin biosynthesis the C5-epimerase converts D-glucuronic into L-iduronic acid units but some of the potential substrates escape modification.

Feizi: In immunofluorescence experiments using antibodies to the soluble β-galactoside-binding lectin from bovine heart, cell nuclei were stained in sections of several bovine tissues (Childs & Feizi 1980). Dr Barondes made similar observations (Beyer & Barondes 1980). When we applied the soluble lectin to sections of the nuclei before applying the anti-lectin serum, the nuclear immunofluorescence was substantially enhanced. This enhancement could be inhibited by lactose, which is an inhibitor of the lectin. These results are consistent with the presence of Gal-GlcNAc units in the nuclei.

Hart: I would like to know more about those findings. In reports of compositional and other analyses of sugars in nuclei and nuclear membranes, the one sugar linkage that is notably absent is β-galactoside. Perhaps it has been missed because it is bound to something.

Fukuda: You said B cells contain many O-linked GlcNAcs. Are they in the nucleus or some other compartment?

Hart: We didn't investigate that. But B cells do not have much else containing terminal GlcNAc residues.

Fukuda: Can you use lectin affinity columns to isolate these glycoproteins containing *N*-acetylglucosamine, instead of going through the galactosylation?

Hart: Some of them will stick to wheat germ agglutinin-Sepharose. It depends on how many sugars they have and how close together they are. Most of the transcription factors that Steve Jackson and Bob Tjian have studied stick to these columns. We don't use lectin columns very much, but about half the 65 kDa erythrocyte protein sticks; we use that half.

Wassarman: The 62 kDa nuclear pore complex protein has ten glycosylation sites. Have you identified a consensus sequence?

Hart: The 62 kDa protein has at least two of our recognition signals, but next to them are long stretches of serines and threonines. Many of the proteins on which this type of sugar is found have these long stretches. We think the enzyme recognizes and glycosylates the signal site, but any nearby serines or threonines are also glycosylated. That is the only explanation for having so many sugars in the same area. A similar phenomenon is observed in transcription factor Sp-1.

Wassarman: Are all the GlcNAc residues localized in one area of the protein?

Hart: From the limit pronase fragments, we know that the sugars tend to be clustered together.

Wassarman: Do histones have O-linked GlcNAc?

Hart: Histones definitely do not have it. None of the major structural proteins in the nucleus has *O*-GlcNAc residues. There are about a hundred chromatin proteins that have these sugars and they are mostly proteins that are not visible on a Coomassie-stained gel.

Sharon: What is known about *O*-GlcNAc linkages in plants?

Hart: We haven't looked at plants. This linkage is found in yeast, but apparently not in *E. coli*, which is the only prokaryote we have studied.

Finne: It's striking that you have *N*-acetylglucosamine available almost only in the intercellular compartment. Our experience in trying to label various cells with galactosyltransferase is that there are almost no available *N*-acetylglucosamine residues (Viitala & Finne 1984, Spillmann & Finne 1987). Perhaps cells avoid having *N*-acetylglucosamine on the cell surface at the same time as they have it inside the cell. There are also very few glycolipids which end in *N*-acetylglucosamine.

Ruoslahti: That's probably why it is immunogenic.

Hart: A B cell-activating lymphokine has been studied by Katoh et al (1986). It recognizes GlcNAc residues and stimulates B cell proliferation. Therefore, exposed *O*-GlcNAc residues may be involved in regulating the immune system.

Kobata: There are many glycoproteins which end with *N*-acetylglucosamine in plasma membranes. For example, γ-glutamyltranspeptidase contains complex-type sugar chains enriched in non-reducing terminal *N*-acetylglucosamine residues (Kobata & Yamashita 1984).

Finne: Perhaps galactosyltransferase does not see those proteins.

Kobata: The sugar chains of γ-glutamyltranspeptidase can be galactosylated. But we haven't tried to galactosylate intact glycoproteins. There might be some mechanism to protect the oligosaccharides of glycoproteins in a membrane.

Hart: One advantage of this approach using transferases is that you are looking on the outside of the cell at what is seen by other cells or macromolecules. It is a very selective way of investigating what is happening.

Wang: Larry Gerace's monoclonal antibody recognizes O-linked GlcNAc. In the electron microscopy study, is the staining or labelling uniform throughout the nuclear envelope or is it only concentrated at the pores?

Hart: Using the RL-2 antibody, it is only at the pores. When you do electron or light microscopy you can play with the conditions a lot. I wouldn't want to say that there isn't O-linked GlcNAc elsewhere in the nuclear envelope, but the data so far indicate it only at the pores.

Gahmberg: Have you found any glycoproteins which are N- or O-linked externally at the plasma membrane and have this O-linked GlcNAc on the cytoplasmic aspect?

Hart: No. That is an important question. When I hear about a transmembrane receptor, I always look at the C-terminal portion to see if it has our sequences. So far, several have, but that doesn't necessarily mean they have O-linked GlcNAc.

Wassarman: When does O-linked glycosylation with GlcNAc take place? Does it occur co-translationally?

Hart: The only information on that comes from Davis & Blobel (1987). They were able to do a pulse-chase experiment with methionine looking at p62 because they had an antibody that did not recognize O-GlcNAc. They found that the sugar was added within 5 min, before association with the pore complex in the cytoplasm. So addition is either co-translational, or very early, for the most part.

Wassarman: Where does glycosylation take place?

Hart: In the cytoplasmic compartment. We have no idea where the enzyme is localized; we haven't done any subcellular fractionations.

Hindsgaul: Have you isolated intact GlcNAc O-linked to either serine or threonine and obtained a structure?

Hart: We haven't done that yet. We have shown that the amino acid linkage is to serine and, to a small extent, to threonine, by alkali-induced β-elimination using sodium borotritide. Sequence analysis has shown the sugar to be GlcNAc and not GalNAc.

Hindsgaul: There is no doubt that your β-elimination leads to GlcNAc, but you have not excluded unknown intermediates. It would be preferable to see the actual structure, not just circumstantial evidence.

Hart: I agree. That is why we are going to do some mass spectrometry of glycopeptides. But we have also done careful kinetic analysis of the β-elimination reaction and there are no detectable intermediates.

Rademacher: You do β-eliminations at 37 °C which is not the classical condition. Are O-linked sugars on mucins β-eliminated under your conditions? In our experience, we would be waiting weeks to get some mucins to release at 37 °C. Are you using selective conditions to release your O-linked GlcNAc and therefore not seeing the other O-linked structures?

Hart: We have compared the rates of β-elimination of the O-linked GlcNAc and O-linked mucin and it's clear that O-linked GlcNAc eliminates more rapidly. But we use the conditions reported by Robert Spiro (1972)—37 °C, 0.1 M base, 1 M sodium borohydride. As far as I can tell, those work very well for mucins.

Sharon: Have you examined the nuclear magnetic resonance spectrum of glycopeptides which contain one or two *N*-acetylglucosamine residues?

Hart: We haven't had enough material to do that. Our only hope at the moment is to do either tandem or fast atom bombardment mass spectrometry.

Watkins: When you remove the GlcNAc with β-hexosaminidase no sugar remains. Surely that indicates that GlcNAc is the only sugar linked to your peptide?

Hart: Yes, one could make that assumption. We use very pure glycosidases—we purify our own from *Diplococcus*. But that doesn't exclude other possibilities. Perhaps it is linked by some unknown structure.

Hindsgaul: Do you investigate whether a non-substituted serine is released when you do that experiment?

Hart: We have never shown that in hexosaminidase-treated proteins the serine no longer β-eliminates.

Sharon: Did you sequence the hexosaminidase-released compounds?

Hart: No.

Fukuda: You can combine Edman degradation with fast atom bombardment mass spectrometry. When you remove the amino acid residues to the point where you have GlcNAc, you get a very nice spectrum. Tandem mass spectrometry requires more material than these methods.

References

Beyer EC, Barondes SH 1980 Chicken tissue binding sites for a purified chicken lectin. J Supramol Struct 13:219–227

Childs RA, Feizi T 1980 β-galactoside-binding lectin of human and bovine tissues. Cell Biol Int Rep 4:775

Davis LI, Blobel G 1987 Nuclear pore complex contains a family of glycoproteins that includes p62: glycosylation through a previously unidentified cellular pathway. Proc Natl Acad Sci USA 84:7552–7556

Figge J, Webster T, Smith TF, Paucha E 1988 Prediction of similar transforming regions in simian virus 40 large T, adenovirus E1A, and *myc* oncoproteins. J Virol 62:1814–1818

Holt GD, Snow CM, Senior A, Haltiwanger RS, Gerace L, Hart GW 1987 Nuclear pore complex glycoproteins contain cytoplasmically disposed O-linked *N*-acetylglucosamine. J Cell Biol 104:1157–1164

Katoh Y, Ono S, Takahama Y, Miyake K, Hamaoka T 1986 Polyclonal B cell activation by a B cell differentiation factor, B151-TRF2. J Immunol 137:2871–2877

Kobata A, Yamashita K 1984 The sugar chains of γ-glutamyltransferase transpeptidase. Pure Appl Chem 56:821–832

Shaper NL, Hollis GF, Douglas JG, Kirsch IR, Shaper JH 1988 Characterization of the full length cDNA for murine β-1,4-galactosyltransferase. J Biol Chem 263:10420–10428

Spillman D, Finne J 1987 Poly-N-acetyllactosamine glycans of cellular glycoproteins: predominance of linear chains in mouse neuroblastoma and rat pheochromocytoma cell lines. J Neurochem 49:874–883

Spiro RG 1972 Study of the carbohydrates of glycoproteins. Methods Enzymol 28:3–42

Stanley P 1989 Chinese hamster ovary mutants with multiple glycosylation defects for production of glycoproteins with minimal carbohydrate heterogeneity. Mol Cell Biol 9:377–393

Viitala J, Finne J 1984 Specific cell-surface labeling of polyglycosyl chains in human erythrocytes and HL-60 cells using endo-β-galactosidase and galactosyltransferase. Eur J Biochem 138:393–397

Bioactive ganglioside-mediated carbohydrate recognition in coupling with ecto-protein phosphorylation

Yoshitaka Nagai*‡ and Shuichi Tsuji*

Department of Biochemistry, Faculty of Medicine, University of Tokyo, Hongo 7-3-1, Bunkyo-ku, Tokyo-113, Japan and *‡Department of Neurobiology, Institute of Brain Research, Niigata University, Niigata-951, Japan*

Abstract. Recent studies, including ours, on bioactive gangliosides revealed that certain gangliosides have an interesting ability to modulate a variety of cell functions. For instance, we demonstrated that a tetrasialoganglioside, GQ_{1b}, promotes neurite outgrowth when added in nanomolar concentrations to cells from two human neuroblastoma cell lines. Also, phosphorylation of several cell surface proteins was observed on addition of ATP. Several lines of evidence indicated that this phosphorylation is probably catalysed by a novel cell surface membrane-bound protein kinase which is specifically activated by a particular ganglioside (Gg). Because of its location on the cell surface we proposed calling this type of kinase(s) ecto-Gg kinase. A procedure to inhibit the phosphorylation of the cell surface protein resulted in suppression of the GQ_{1b}-dependent promotion of neuritogenesis, strongly suggesting that these two cellular events are intricately coupled. Other evidence also indicated that the GQ_{1b}-dependent neuritogenesis is mediated through a receptor-coupled process of the cell surface membrane. Thus, it is likely that this represents a new type of biosignal transduction that is mediated through cell surface carbohydrate recognition (ecto biosignal transduction system).

1989 Carbohydrate recognition in cellular function. Wiley, Chichester (Ciba Foundation Symposium 145) p 119–134

Gangliosides are a family of glycosphingolipids. They contain sialic acid and are a major and essential cell membrane constituent of vertebrates. Their contribution to cell biology as useful markers for cell surfaces is well known, but the function of these markers remains unclear. On the other hand, several biological activities of gangliosides have now been reported with reference to a variety of biological functions including: cell growth, differentiation, and cycle; receptor function; ion and water transport; immunological function; and neural function. Most of these observations were made by adding gangliosides to cultured cells. We have named these unique gangliosides 'bioactive gangliosides' (Nagai 1985).

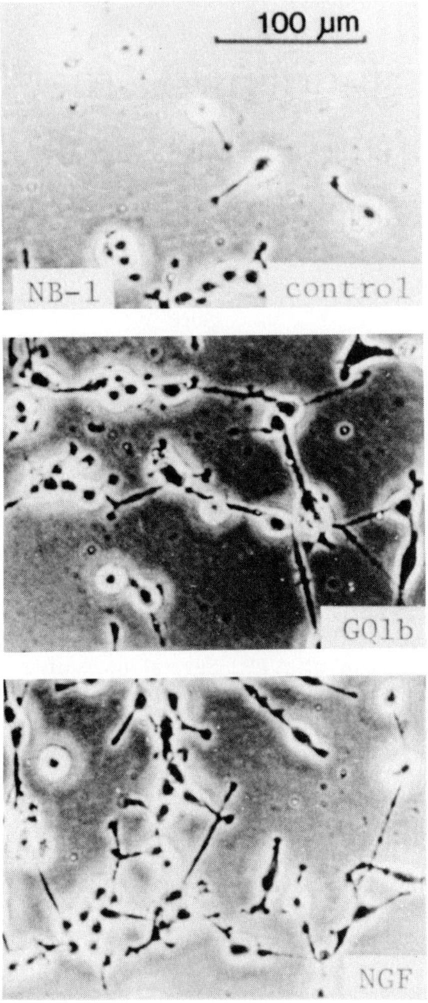

FIG. 1. Cultures of neuroblastoma cells (NB-1) grown with or without purified GQ_{1b} or NGF. Cells were grown exponentially to 2 to 6×10^4 cells per 35 mm polystyrene plate at 37 °C in a humidified atmosphere of 5% CO_2/95% air in medium containing serum for one or two days. After removal of the medium, fresh serum-free medium (2 ml), with or without GQ_{1b} (5 nM) or NGF (100 ng/ml), was added. Photomicrographs were taken 24 h later.

Our finding of neurite outgrowth-promoting (neuritogenic) activity of a tetrasialoganglioside, GQ_{1b}, seems particularly interesting because it may represent a novel carbohydrate signal transduction system. In this system a message contained in the carbohydrate structure is entrapped by a cell surface membrane receptor (receptor lectin). Coupling with cell surface protein phosphorylation catalysed by ecto-protein kinase results in the expression of the biological activity.

Ganglioside-dependent ncuritogenesis and carbohydrate recognition

We reported (Tsuji et al 1983) that the tetrasialoganglioside GQ_{1b} (1)

```
      Gal(β1-3)GalNAc(β1-4)Gal(β1-4)Glc(β1-1)ceramide
       |                    |
NeuAc(α2-3)          NeuAc(α2-3)
       |                    |
NeuAc(α2-8)          NeuAc(α2-8)
                               (1)
```

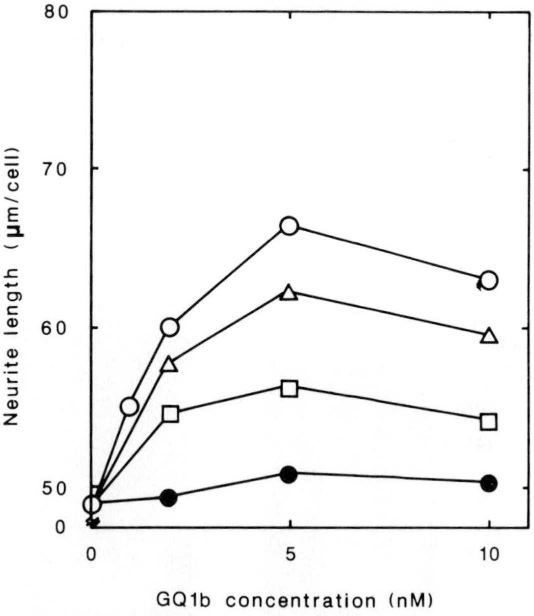

FIG. 2. Effects of the ganglioside GQ_{1b} and GQ_{1b} plus the oligosaccharide oligo-GQ_{1b} on the neuroblastoma cell line, GOTO. Cells were cultured in the presence of varying concentrations of GQ_{1b} with or without oligo-GQ_{1b} at various concentrations. Photographs were taken as described in the legend to Fig. 1, and the total neurite lengths for more than 10 000 cells were determined in several randomly chosen fields. ○, without oligo-GQ_{1b}; △, 0.1 μM oligo-GQ_{1b}; □, 0.2 μM oligo-GQ_{1b}; ●, 1 μM oligo-GQ_{1b}.

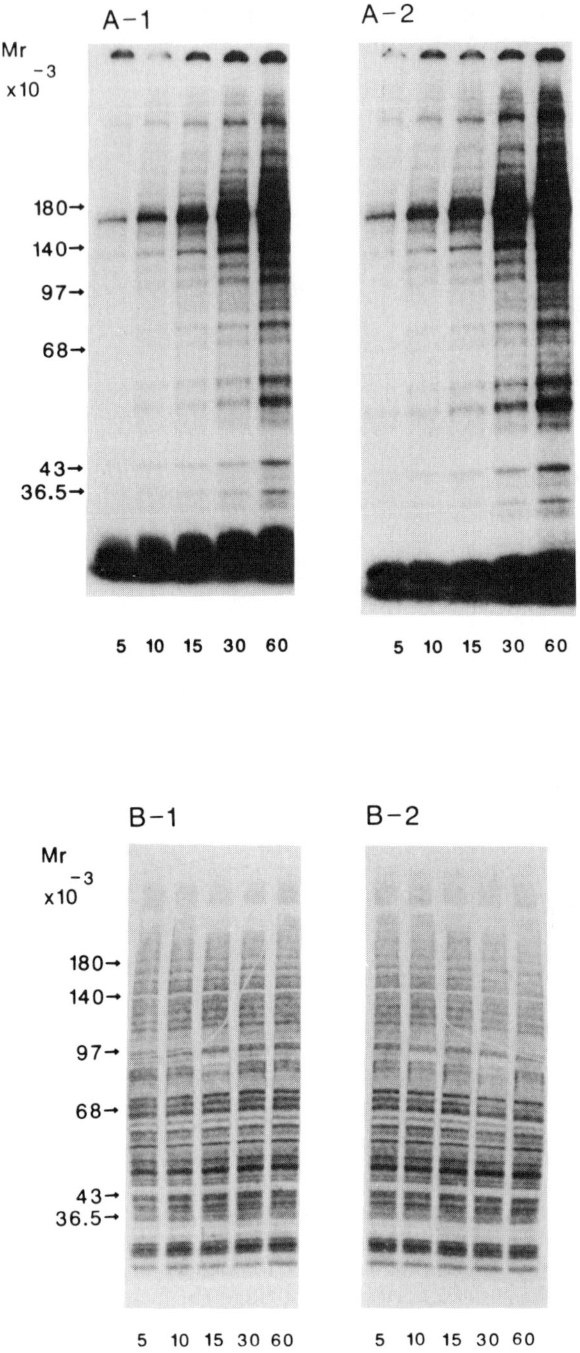
A-1
A-2
Mr
x10^-3
180→
140→
97→
68→
43→
36.5→
5 10 15 30 60
5 10 15 30 60
B-1
B-2
Mr
x10^-3
180→
140→
97→
68→
43→
36.5→
5 10 15 30 60
5 10 15 30 60

exhibits remarkable nerve growth factor (NGF)-like activity in two human neuroblastoma cell lines, GOTO and NB-1 in the absence of serum in the culture medium (Fig. 1). Later it was found that the ganglioside tends to promote extension of single neurites from each cell body, whereas NGF promotes multiple generation of neurites, from the cells (Tsuji et al 1988a). The biological effects of exogenous gangliosides on cultured cells or subcellular fractions have been observed so far in the range of micromolar concentrations. The GQ_{1b}-dependent neuritogenesis, however, was observed at concentrations as low as a few nanomolar in these two human neuroblastoma cell lines. To distinguish between these different responses to exogenous gangliosides, which presumably reflect different molecular mechanisms, we have named the former 'M-type' responses and the latter 'N-type' responses (Nagai & Tsuji 1988a, Tsuji et al 1988a). Thus, the M-type response is characterized by the relatively high dose of ganglioside required, in the μM range, and relatively broad specificity, whereas the N-type response is characterized by low effective doses, in the nM range, and high specificity regarding molecular species. Moreover, simultaneous addition of oligo-GQ_{1b} (the tetrasialosaccharide oligomer prepared from GQ_{1b}) distinctly inhibited the neuritogenic effect of GQ_{1b} in a concentration-dependent manner at concentrations of more than twentyfold those of GQ_{1b} (Nakajima et al 1986) (Fig. 2). This strongly suggested the involvement of a cell surface receptor that recognizes oligo-GQ_{1b} in the development of GQ_{1b}-dependent activity (Nakajima et al 1986). The carbohydrate recognition of this receptor must be quite strict, because structure–function analysis revealed that GQ_{1b} activity depends solely upon the tetrasialosyl structure of GQ_{1b}: the loss of any of the four sialic acid residues results in total loss of the GQ_{1b} activity (Nakajima et al 1986).

FIG. 3. Endogenous substrates for the ecto-type protein kinase activity in intact GOTO cells. The ecto-protein kinase assay was performed by slightly modifying the method of Ehrlich et al (1986). Neuroblastoma cells in the exponential growth phase were distributed into 96-well plates at the inoculum size of 2×10^4 cells. The cells were then incubated for two days. Before the experiment, the cells were rinsed three times with a fetal calf serum (FCS)-free medium and then incubated in serum-free fresh medium containing ganglioside GQ_{1b} (5 nM) for one hour at 37 °C. The reaction was started by adding 1 μM [γ-^{32}P]ATP (10 μCi/well) to the attached cells and terminated after one hour incubation with gentle shaking at 30 °C by adding the termination buffer. Samples were resolved on 4–20% polyacrylamide gel gradients and then autoradiographed for one or two weeks. A. Protein phosphorylation in intact GOTO cells with radiolabelled ATP. A-1, without GQ_{1b} treatment; A-2, preincubated with GQ_{1b} (5 nM) for one hour. *, 64 kDa; **, 60 kDa; ***, 54 kDa proteins. B. The pattern for protein stained with silver, which appeared to be the same as in A. B-1, without GQ_{1b} treatment; B-2, preincubated with GQ_{1b} (5 nM) for one hour. The figures below the lanes in A and B indicate the reaction times in minutes.

GQ_{1b}-dependent cell surface protein phosphorylation

During our investigation of the molecular mechanism of GQ_{1b}-dependent neuritogenesis, we found that GQ_{1b} stimulated phosphorylation of cell surface proteins in the presence of extracellular ATP (Nagai & Tsuji 1988a,b, Tsuji et al 1988b) (Fig. 3). The experiment was simple. Human neuroblastoma cells, GOTO, in the exponential growth phase were placed in 96-well plates, each inoculum containing 2×10^4 cells. The cells were incubated for two days. Before the experiment, the cells were rinsed three times with a fetal calf serum (FCS)-free medium (RPMI (Roswell Park Memorial Institute medium): MEM (minimal essential medium Eagle), 1:1) and then incubated in serum-free fresh medium (50 µl) containing ganglioside GQ_{1b} (5 nM) for one hour at 37 °C. The reaction was started by adding 1 µM [γ-^{32}P]ATP (10 µCi/well) to the attached cells. After one hour incubation with gentle shaking at 30 °C, the reaction was terminated by adding the termination buffer (50 µl, 0.5 M Tris-HCl, 10% SDS, 6.0 M urea, 1% 2-mercaptoethanol, pH 6.8). Samples were resolved on 4–20% polyacrylamide gel gradients and then autoradiographed for one or two weeks.

In the presence of [γ-^{32}P]ATP alone, about 28 cell surface proteins were radiolabelled. Among these 28 proteins, however, phosphorylation of at least three cell surface-associated proteins of M_r 64 000, 60 000 and 54 000 was specifically stimulated (Fig. 3). Phosphorylation was directed toward threonine and serine residues, but not toward tyrosine. Time-course studies revealed that protein phosphorylation proceeds very quickly, attaining a plateau after 30 min (Fig. 4). GQ_{1b}, at 5 nM, was the most potent of the gangliosides tested (GQ_{1b}, GT_{1b}, GD_{1a}, GM_1, GA_1 and oligo-GQ_{1b}; see Tables 1 and 2), indicating a highly specific effect on phosphorylation. GQ_{1b} was more effective when added to cells before the [γ-^{32}P]ATP was added, which suggests a direct insertion of GQ_{1b} molecules into the cell membrane bilayer. It should be emphasized that the concentrations required for stimulation of phosphorylation remain within the range required for the GQ_{1b}-dependent neuritogenesis described earlier. Again, the simultaneous addition of an excess amount of oligo-GQ_{1b} inhibited the GQ_{1b}-stimulated phosphorylation, indicating the necessity for the sialosaccharide moiety (Table 1). Preliminary results showed that the ecto-type phosphorylation was not activated by kinase activators, such as cAMP (10 µM), cGMP (10 µM), calmodulin (1 µM), or phosphatidylserine (50 µg/ml) (S. Tsuji et al, unpublished work 1988). Moreover, known protein kinases (cAMP-dependent kinase, cGMP-dependent kinase and Ca^{2+}/calmodulin-activated kinase) seemed to be suppressed rather than stimulated by GQ_{1b}, particularly in the case of Ca^{2+}/phospholipid-activated kinase (protein kinase C) (Tsuji et al 1985). Thus this extracellular, ecto-type system must be different from the well known intracellular, endo-type protein kinase system. The evidence is summarized in Table 3.

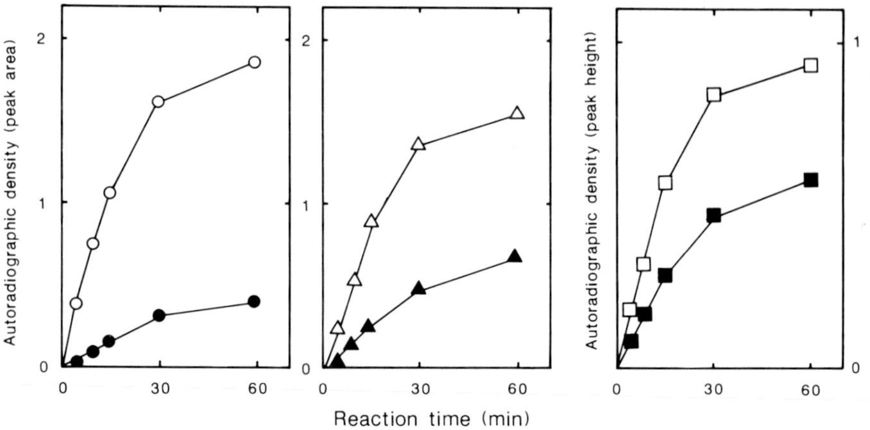

FIG. 4. Time course of protein phosphorylation in intact GOTO cells with [γ-^{32}P]ATP. Cells were grown and maintained in the medium for a few days and then assayed as described for Fig. 3. After one hour preincubation with ganglioside GQ_{1b}, the reaction was started as described and stopped at the indicated time points, followed by resolution by SDS-PAGE. After autoradiography, the autoradiogram was traced with a densitometer. The peak areas for the 64 and 60 kDa proteins, and the peak height for the 54 kDa protein were estimated. Open symbols, $+GQ_{1b}$ (5 nM); filled symbols, $-GQ_{1b}$. ○, ●, phosphorylated 64 kDa; △, ▲, phosphorylated 60 kDa; and □, ■, phosphorylated 54 kDa proteins.

TABLE 1 The inhibitory effect of oligo-GQ_{1b} on GQ_{1b}-stimulated phosphorylation

	Stimulation rate[a]			
Conditions	*64 kDa*	*60 kDa*	*54 kDa*	*75 kDa*[b]
GQ_{1b} (5 nM)	4.7 ± 1.8	2.3 ± 0.6	1.6 ± 0.3	1.1 ± 0.2
GQ_{1b} (5 nM) + oligo-GQ_{1b} (25 nM)	3.8 ± 1.3	2.1 ± 0.7	1.6 ± 0.4	1.0 ± 0.3
GQ_{1b} (5 nM) + oligo-GQ_{1b} (1 μM)	1.9 ± 0.5	1.4 ± 0.4	1.2 ± 0.3	1.2 ± 0.4
oligo-GQ_{1b} (1 nM)	1.2 ± 0.3	1.1 ± 0.3	1.1 ± 0.3	1.1 ± 0.2
oligo-GQ_{1b} (25 nM)	0.9 ± 0.4	1.2 ± 0.4	0.8 ± 0.4	0.9 ± 0.3
oligo-GQ_{1b} (1 μM)	0.9 ± 0.3	0.9 ± 0.2	1.1 ± 0.3	0.8 ± 0.4

[a]The stimulation rate is defined as the rate of phosphorylation in the presence of GQ_{1b} with or without oligo-GQ_{1b}, or in the presence of oligo-GQ_{1b} alone, divided by the rate of phosphorylation in the absence of both GQ_{1b} and oligo-GQ_{1b}.
[b]A typical example of a protein whose phosphorylation is insensitive to GQ_{1b} stimulation.

TABLE 2 Glycolipid specificity of the stimulation of ecto-type protein (64 kDa) phosphorylation in GOTO cells.

Glycolipids	*Optimum concentration (nM)*	*Relative activity*[a]
GQ_{1b}	5	4.7
GT_{1b}	50	2.0
GD_{1a}	–	–
GM_1	–	–
GA_1	–	–
Control	–	1.0

–, Insignificant change from control (no added glycolipid) values.

TABLE 3 Evidence for the existence of ecto-type protein kinase(s)

(1) ATP does not appear to permeate through cell surface membranes into cells, although there are a few exceptions to this.

(2) Phosphorylation did not occur with ortho-[^{32}P]phosphoric acid (10 μCi/well) for half an hour, whereas with [γ-^{32}P]ATP (1 μM; 10 μCi/well) phosphorylation had already occurred by this time. On longer incubation after the addition of ortho-[^{32}P]phosphoric acid to the medium, phosphorylation did occur but with a different pattern from those of the short term phosphorylations. The added ortho-[^{32}P]-phosphate may be incorporated intracellularly into ATP. The newly synthesized ATP will be used in the later phosphorylations, including intracellular phosphorylation.

(3) Exogenous GQ_{1b} and [γ-^{32}P]ATP stimulated ^{32}P-labelling of three proteins (64, 60 and 54 kDa) specifically. The phosphorylation occurred at specific amino acids, Ser and Thr.

(4) On treatment of the intact cells with trypsin the three phosphorylated proteins (64, 60 and 54 kDa bands) disappeared almost immediately; after 5 min, almost all the 64 and 60 kDa proteins had already disappeared and the 54 kDa band remained only faintly. During brief trypsin (0.01%) treatment, the total amount of SDS-solubilized protein remained almost unchanged and the treated cells remained attached to the plates. No changes in the cell shape were observed.

(5) AMP-PCP (0.1 mM) or 5′-*p*-fluoro-sulphonyl-benzoyl-adenosine (1 mM) inhibited almost all the ecto-type phosphorylations. Protein kinase inhibitors (K-252a, H-7 and others) also inhibited these phosphorylations. Moreover, photoaffinity labelling of the cell surface with 8-azido-ATP led to complete inhibition of the ecto-type phosphorylations.

(6) GQ_{1b}-dependent phosphorylation in intact cells was inhibited with GQ_{1b}-oligosaccharide in a concentration-dependent manner.

TABLE 4 Evidence for coupling of GQ_{1b}-dependent ecto-type protein phosphorylation with GQ_{1b}-dependent neuritogenesis

(1) Exogenous GQ_{1b}-oligosaccharide inhibited GQ_{1b}-dependent neuritogenesis in a concentration-dependent manner.
(2) Photoaffinity labelling of the cell surface with 8-azido-ATP led to complete inhibition of the ecto-type phosphorylation. Under these conditions, exogenous GQ_{1b} could not stimulate neuritogenesis.

This novel ganglioside-dependent cell surface phosphorylation (ecto-type phosphorylation) is probably catalysed by ecto-type protein kinase(s). We have named this ganglioside-dependent ecto-protein kinase 'ecto-Gg kinase' (Nagai & Tsuji 1988a,b).

Biological implication of a novel carbohydrate signal-mediated cell surface protein phosphorylation

We have presented data for two types of phenomena: GQ_{1b}-dependent neuritogenesis which suggests the involvement of a highly carbohydrate-specific receptor on the cell surface; and carbohydrate-specific ecto-protein phosphorylation catalysed by ecto-type protein kinase(s). It is likely that these two events are closely coupled. In support of the existence of such coupling (Table 4), photoaffinity labelling of the cell surface with 8-azido-ATP led to complete inhibition of the ecto-type phosphorylation. Under these conditions, exogenous GQ_{1b} could not stimulate neuritogenesis.

Based on receptor-mediated carbohydrate signal transduction coupled with ecto-Gg kinase, three models might be postulated for carbohydrate recognition and carbohydrate message transfer (Fig. 5). In the *cis* recognition model, the ganglioside (Gg), the receptor lectin coupled with the ecto-Gg kinase (R-K) and the phosphate acceptor protein (A) are all located at the same cell surface membrane. In the second model (*trans* recognition), R-K and A are located in a different cell which interacts with a Gg-bearing cell. In the sandwich model, the phosphate acceptors (A) are intercellular molecules which interact with Gg- and R-K-bearing cells. Subsequently, phosphorylated A affects other adjacent cells. The *trans* recognition and sandwich models seem to be particularly favoured for understanding the interaction and message transfer between cells.

At present we do not know about any real source of extracellularly acting ATP (exogenous ATP). With respect to neuronal cells, it is well known that synaptic vesicles contain very high concentrations of ATP, and that it is secreted

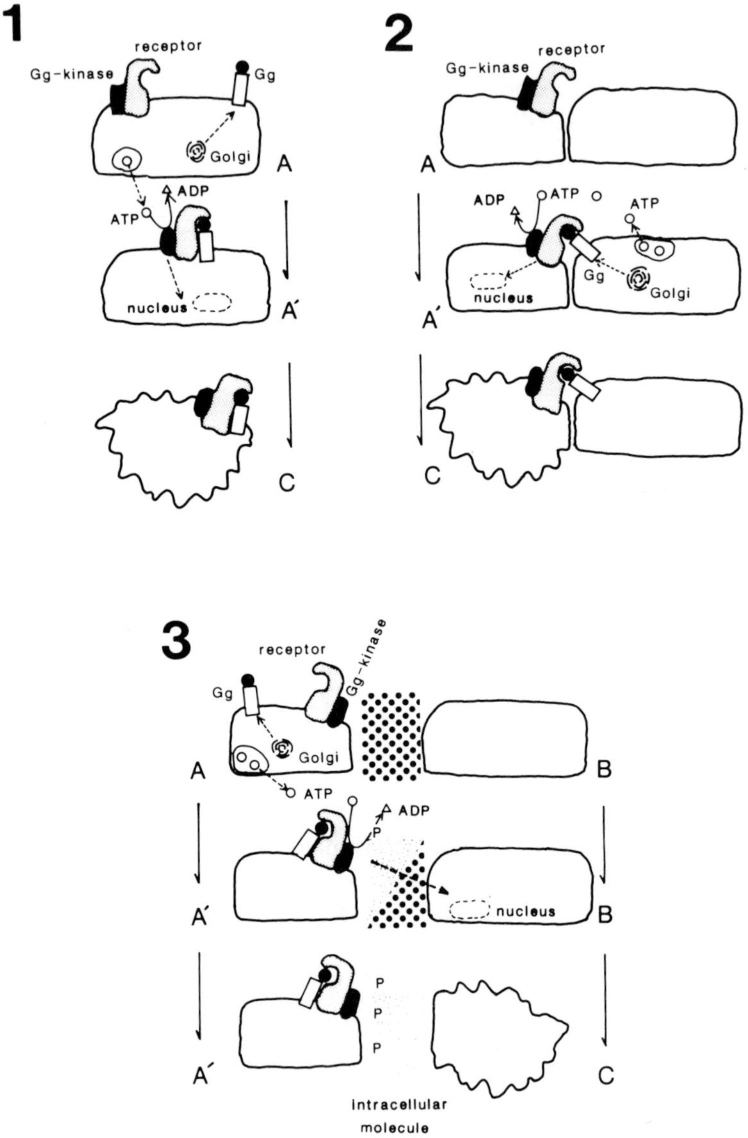

FIG. 5. Three models for an ecto signal transduction system involving gangliosides (Gg) and ecto-protein kinase. 1. *Cis* recognition model. 2. *Trans* recognition model. 3. Sandwich model.

from neurons in association with neurotransmitter at certain synapses and neuromuscular junctions. ATP at synapses may function as a neurotransmitter or as a ligand for the adenosine receptor (purinergic receptor). However, the extracellular phosphorylation of cell surface proteins by such secreted ATP may play an important role in cell–cell interactions, cell differentiation and other carbohydrate-associated cell surface phenomena. Ehrlich et al (1986), using a hybrid cell line of neuronal origin, clone NG108-15, observed ecto-type phosphorylation and suggested that one substrate of the ecto-protein kinase might be neuronal cell adhesion molecules (N-CAM). They proposed that phosphorylation of N-CAM might modulate cell–cell interactions. Some of the ecto-protein kinase may be localized around the synaptic junctions and may play a role in the regulation of synaptic transmission and other synapse-associated neuronal functions. Purification of ecto-Gg kinase and the elucidation of the function of the phosphorylated cell surface protein will shed light on the role of this novel, carbohydrate-mediated, ecto signal transduction system.

Summary

We have described a tetrasialoganglioside GQ_{1b} that has a strong and highly specific ability to promote neuritogenesis of human neuroblastoma cell lines. This function might be mediated by a carbohydrate-recognizing cell surface receptor (receptor lectin). GQ_{1b} also specifically enhances ATP-dependent phosphorylation of cell surface proteins, particularly three proteins with molecular mass of 54, 60 and 64 kDa, at serine and threonine residues, catalysed by ecto-Gg kinase. It is possible that these two responses, one biological and the other biochemical, are closely coupled through receptor-mediated carbohydrate recognition on the cell surface.

Acknowledgements

This work was supported in part by a Grant-in-Aid for Scientific Research and a Grant-in-Aid for Cancer Research, both from the Ministry of Education, Science and Culture of Japan. Support by a Grant-in-Aid from the Foundation for Metabolic Diseases is also gratefully acknowledged. We also thank the Haraguchi Memorial Cancer Foundation for support.

References

Ehrlich YH, Davis TB, Bock E, Kornecki E, Lenox RH 1986 Ecto-protein kinase activity in the external surface of neural cells. Nature (Lond) 320:67–70

Nagai Y 1985 Bioactive gangliosides: a proposal of the ganglioside-mediated bimodal modulation of cellular activity through cell membrane microdomain and the gangliosides cascade. In: Davidson EA et al (eds) Glycoconjugates. Prager Scientific, New York (Proc 8th Internal Symp Houston, 1985), p 525–526

Nagai Y, Tsuji S 1988a Cell biological significance of gangliosides in neural differentiation and development: critique and proposals. In: Ledeen RW et al (eds) New trends in ganglioside research. Neurochemical and neurogenerative aspects. Fidia Research Series vol 14:329–350

Nagai Y, Tsuji S 1988b Gangliosides in recognitive cell interactions and carbohydrate-mediated biosignalling. In: Gabius HJ, Nagel GA (eds) Lectins and glycoconjugates in oncology. Springer-Verlag, Berlin, p 59–66

Nakajima J, Tsuji S, Nagai Y 1986 Bioactive gangliosides: analysis of functional structures of the tetrasialoganglioside GQ_{1b} which promotes neurite outgrowth. Biochim Biophys Acta 876:65–71

Tsuji S, Arita M, Nagai Y 1983 GQ_{1b}, a bioactive ganglioside that exhibits novel nerve growth factor (NGF)-like activities in the two neuroblastoma cell lines. J Biochem (Tokyo) 94:303–306

Tsuji S, Nakajima J, Sasaki T, Nagai Y 1985 Bioactive gangliosides. IV. Ganglioside GQ_{1b}/Ca^{2+} dependent protein kinase activity exists in the plasma membrane fraction of neuroblastoma membrane fraction of neuroblastoma cell line, GOTO. J Biochem (Tokyo) 97:969–972

Tsuji S, Yamashita T, Tanaka M, Nagai Y 1988a Synthetic sialyl compounds as well as natural gangliosides induce neuritogenesis in a mouse neuroblastoma cell line (Neuro2a). J Neurochem 50:414–423

Tsuji S, Yamashita T, Nagai Y 1988b A novel, carbohydrate signal-mediated cell surface protein phosphorylation: ganglioside GQ_{1b} stimulates ecto-protein kinase activity on the cell surface of a human neuroblastoma cell line, GOTO. J Biochem (Tokyo) 104:498–503

DISCUSSION

Hart: Have you measured the effect of GQ_{1b} on the permeability of the intact cells to ATP?

Nagai: We tried to do that, but we failed to incorporate ATP into neuroblastoma cells. If we incubate cells with the ganglioside for 20 min, calcium uptake occurs and there may be some permeability change, either with or without delay. At the moment we have no idea how to interpret such data.

Hart: A nice control experiment would be to take a non-hydrolysable analogue of ATP and look at the effect of the glycolipid on its uptake.

Nagai: Yes; some analogues specifically blocked the phosphorylation and also inhibited neurite outgrowth.

Feizi: In your models, the ectokinase is depicted as a bifunctional protein with a lectin-like domain. Have you tried to enrich for your kinase activity on the immobilized ganglioside?

Nagai: GQ_{1b} is a minor component and is not available in sufficient quantities. It is difficult to synthesize such a complicated structure. We obtain the oligo-GQ_{1b} from GQ_{1b} by ozonolysis and purifying on a gel filtration column. We are trying to purify the kinase so that we can determine its localization.

Jessell: Antibodies against GQ_{1b} reveal that most neural cells express very high levels of this ganglioside. Therefore I don't understand why the endogenous ganglioside isn't maximally activating the kinase. Why do you see an additional effect on adding more GQ_{1b}?

Nagai: That is a good question. We have obtained a monoclonal antibody to GQ_{1b}, and by TLC immunostaining we did not find any GQ_{1b} in the human neuroblastoma cells used for the *in vitro* assay. Nagata et al (1987) used rat nodose ganglia *in vitro*, which do not contain GM_1. The addition of 5 nM GM_1 accelerated amino acid transport in the cells by a different mechanism from that of the action of nerve growth factor. These results suggest that if a cell is deficient in a particular ganglioside it may develop some new function if that ganglioside is added.

Jessell: If you study primary neurons, which do express GQ_{1b}, with your neurite outgrowth assay, do you see any effect on axon extension?

Nagai: We have not looked at primary neurons yet. However, there are several experiments which show the necessity of GQ_{1b} for neural development. Firstly, the concentration of GQ_{1b} is elevated prior to synaptogenesis and axonal and dendritic arborization (Rösner & Rahmann 1987). Seyfried (1987) examined a T-locus mutant mouse embryo, t^{w1}/t^{w1}, which dies from failed neural differentiation in the ventral portion of the neural tube. At Embryonic Day 11 (E-11) there was a marked deficiency of GQ_{1b}, but GT_{1b}, the immediate precursor of GQ_{1b}, was present. On the other hand, in mice homozygous for the dominant T mutation, T/T, which die from failed mesodermal differentiation in the notocord, no major ganglioside abnormalities were found at E-10. The importance of the T-locus or T/t complex in mouse chromosome 17 has already been pointed out by many investigators with reference to cellular differentiation and embryonic development, and cell–cell recognition in particular.

Fukuda: Does GT_1 bind to some of the extracellular matrix components?

Nagai: We use serum-free media for such assays. If serum is present, the serum components preferentially bind ganglioside so that no activity can be observed.

Ruoslahti: The ectokinase hypothesis is attractive, but have you considered the possibility that the molecule affected is the laminin receptor? Laminin is known to be a very efficient promoter of neuritogenesis. On the other hand, it is known that gangliosides are associated with the functions of integrins.

Nagai: Some extracellular materials bind particular glycolipids. For example, Roberts (1986) reported that laminin specifically binds a sulphate-containing glycolipid (galactocerebroside-3-sulphate). But the promotion of neurite outgrowth in our experimental system requires the highly specific oligosaccharide structure of GQ_{1b}.

Ruoslahti: I am not thinking of binding of the ganglioside to an extracellular matrix protein, but to a receptor for an extracellular matrix protein, specifically that for laminin.

Jessell: Isn't the idea there that you are interfering with the calcium binding and that it would be a charge effect, in which case you might not see the specificity that is observed?

Ruoslahti: I don't know whether that is the case. David Cheresh reconstituted an integrin function having first shown that if he removed gangliosides and divalent cations there was no binding of the integrin to its ligand. When he restored the divalent cation he recovered most of the activity. On addition of the ganglioside he got more activity back. This doesn't agree with the idea that the ganglioside would interfere with cation binding: the ganglioside seems to be necessary in addition to the cations.

Schachner: The Ca^{2+}-independent adhesion molecule, L1, is implicated in neurite outgrowth. Have you specifically looked for adhesion molecules?

Nagai: No.

Czech: Might the ganglioside allow kinase activity to leak out of the cells rather than increase ATP permeability? Have you investigated whether the media of these cells at different times have kinase activity? Does the putative ectokinase phosphorylate exogenous substrates added to the cells and to the media?

Nagai: Phosphorylation seemed to occur within a few seconds of addition of ATP.

Czech: Have you been able to obtain phosphorylation of exogenous proteins or peptides?

Nagai: No, we have not tried to do that.

Hughes: Now that you are showing that gangliosides have these biologically important activities, what happens in insects? They don't have gangliosides.

Nagai: Even if a particular glycoconjugate is definitely involved in some function of certain cells, not all cells with this function contain the same glycoconjugate molecules. Thus one cellular function cannot always be related to one particular glycoconjugate. Insects which have no gangliosides in the nervous tissues have instead other acidic glycolipids, which presumably might fulfil the same role.

Jessell: Ron Schnarr's group have shown that gangliosides will support the adhesion of neural retinal cells (Blackburn et al 1986). Similar assays using GQ_{1b} to see whether not only neurite outgrowth but also the adhesive properties of these cells are dependent on this ganglioside might allow you to identify the cell surface proteins to which these gangliosides bind.

Nagai: Perhaps the added GQ_{1b} is inserted into the bimolecular layer of the plasma membrane. It is known that polysialogangliosides at such low concentrations are dispersed monomolecularly in aqueous media and can easily and quickly be inserted into plasma membranes, compared with the micelle-type dispersion of gangliosides. We have evidence that preincubation of the cells with GQ_{1b} for 30 min is sufficient for starting protein phosphorylation with exogenous ATP.

Ruoslahti: After that, did you look at the distribution of the molecule at the cell surface? That might be informative, because if GQ_{1b} does bind to integrins you might see that the adhesion plaques have now become positive.

Nagai: We didn't do that. But the mode of insertion of the ganglioside into the plasma membrane is disputed. There is no reliable direct method to prove insertion of the lipid molecules. As far as I know, only electron spin resonance spectroscopy can provide evidence for this at present.

Schachner: Might the quantity of adhesion molecules in the surface membrane be altered by changing the ganglioside composition?

Nagai: That is an interesting idea. It is possible, but the vigorous requirement for the GQ_{1b} oligosaccharide structure rather suggested the involvement of a receptor-type recognition function at the cell surface.

Barondes: Has it been possible to examine the effects of oligosaccharides separated from the lipid component of the gangliosides? That might be useful in clarifying the mechanism of the cellular response to the gangliosides.

Nagai: We did an assay using very high concentrations of the oligo-GQ_{1b}. Neuritogenesis was promoted, but we could not exclude the possibility that a tiny amount of intact GQ_{1b} remained as a contaminant in such large quantities of the oligosaccharide preparation. As I said, a hormonal dose of GQ_{1b} is sufficient for the development of the activity.

Feizi: But if the stimulation of kinase activity is due to its interaction with a membrane-associated ganglioside, one might predict that this effect would be inhibited by free oligosaccharide. Have you performed inhibition experiments?

Nagai: It is possible that cross-linking occurs through the oligosaccharide, in a manner similar to your model of protein–saccharide linking at cell surfaces (Feizi, this volume). But concentration-dependent inhibition was observed with both ecto-phosphorylation and neuritogenesis.

References

Blackburn CC, Swank-Hill P, Schnarr RL 1986 Gangliosides support neural retina cell adhesion. J Biol Chem 261:2873–2881

Feizi T 1989 Glycoprotein oligosaccharides as recognition structures. In: Carbohydrate recognition in cellular function. Wiley, Chichester (Ciba Found Symp 145) p 62–79

Nagata Y, Ando M, Iwata M, Hara A, Taketomi T 1987 Effect of exogenous gangliosides on amino acid uptake and Na^+,K^+-ATPase activity in superior cervical and nodose ganglia of rats. J Neurochem 49:201–207
Roberts DD 1986 Sulfatide-binding proteins. Chem Phys Lipids 42:173–183
Rösner H, Rahmann H 1987 Ontogeny of vertebrate brain gangliosides. In: Rahmann H (ed) Gangliosides and modulation of neuronal functions. Springer Verlag, Berlin, New York (NATO ASI Series H: Cell Biology) vol 7:373–390
Seyfried TH 1987 Ganglioside abnormalities associated with failed neural differentiation in a T-locus mutant mouse embryo. Develop Biol 123:286–291

Role of carbohydrates in receptor-mediated fertilization in mammals

Paul M. Wassarman

Department of Cell and Developmental Biology, Roche Institute of Molecular Biology, Roche Research Center, Nutley, New Jersey 07110, USA

Abstract. The mouse sperm receptor, called ZP3, is a glycoprotein (83 000 M_r) that consists of a 44 000 M_r polypeptide chain (402 amino acids), three or four N-linked oligosaccharides, and an undetermined number of O-linked oligosaccharides. There are more than 10^9 copies of ZP3 present throughout the mouse egg extracellular coat, or zona pellucida. As a prelude to fertilization, each acrosome-intact sperm binds in a relatively species-specific manner to tens-of-thousands of copies of ZP3 at the surface of the zona pellucida. Binding to ZP3 induces sperm to undergo the acrosome reaction (membrane fusion) and, consequently, enables them to penetrate through the zona pellucida and to reach, and then fuse with, egg plasma membrane (fertilization). Purified ZP3, as well as a specific class of ZP3-derived O-linked oligosaccharides (3900 M_r), exhibit sperm receptor activity *in vitro*. The oligosaccharides, which represent a relatively low percentage of total ZP3 O-linked oligosaccharides, account for the glycoprotein's sperm receptor activity *in vitro* (i.e., recognition and binding). Furthermore, either enzymic removal or modification of certain sugars that constitute these oligosaccharides results in destruction of sperm receptor activity. These and other findings strongly suggest that during mammalian fertilization carbohydrates play a fundamental role in species-specific sperm–egg interactions.

1989 Carbohydrate recognition in cellular function. Wiley, Chichester (Ciba Foundation Symposium 145) p 135–155

For both plants and animals, the ability of male and female gametes of the same species to recognize one another is a ubiquitous and essential feature of the fertilization process. Whether sperm and eggs encounter one another in water, on land, or, in the case of mammals, in the confines of the female reproductive tract, there are mechanisms to support species-specific fertilization and, conversely, prevent interspecific fertilization. For reviews of various aspects of species-specific fertilization in mammals see Gwatkin (1977), Wassarman et al (1985), Wassarman (1987a,b, 1988a) and Yanagimachi (1988).

For some time it has been known that, among mammals, it is the egg extracellular coat, or zona pellucida (ZP), that confers a degree of species-specificity on fertilization. When eggs from one mammal are mixed with sperm from a different mammal *in vitro*, fusion of the two cells rarely occurs. On the other hand, when the ZP is first removed from eggs, fertilization frequently does take place, although development aborts shortly thereafter. These observations support the idea that the ZP of unfertilized eggs contains receptors that sperm recognize and bind to in a relatively species-specific manner. Furthermore, because sperm from a particular mammal bind to the ZP of unfertilized eggs, but not to the ZP of either fertilized eggs or embryos from the same mammal, it would appear that sperm receptors are rendered ineffective following fertilization. Thus, mammalian sperm receptors regulate interactions between sperm and eggs both before and after fertilization. In the former case, the aim is to ensure that the combination of gametes is compatible with normal development, and, in the latter, to prevent polyspermy, a condition that is incompatible with normal development.

Research carried out in my laboratory during the past ten years or so has led to the identification and isolation of the mouse egg sperm receptor, which we call ZP3. Biochemical characterization of ZP3, one of three ZP glycoproteins (ZP1, ZP2, ZP3), carried out in conjunction with functional studies *in vitro*, has revealed that a specific class of ZP3 O-linked oligosaccharides is responsible for the glycoprotein's ability to serve as a sperm receptor. Thus, as in both plants and non-mammalian animal species, apparently carbohydrates play an essential role in gamete recognition, gamete binding, and other aspects of fertilization in mammals.

The fertilization pathway in mice

The pathway leading to fertilization of mouse eggs consists of several steps that occur in a compulsory order (Wassarman 1987a,b). This pathway is derived from studies carried out by many investigators during the past century. The following briefly describes certain steps of this pathway and introduces some relevant terminology.

Sperm first associate with ovulated eggs at the surface of the egg extracellular coat, the ZP (Fig. 1). This relatively loose association, called 'attachment', is not species-specific. Attached sperm can then form a relatively tenacious, species-specific association with the egg ZP; a state of adhesion called 'binding'. Sperm bind to the ZP through the plasma membrane overlying the anterior region of the sperm head. Binding is mediated by sperm receptors present in the ZP and complementary egg-binding proteins present in sperm plasma membrane. Bound sperm then undergo the acrosome reaction. The acrosome is a lysosome-like organelle located in the sperm head, overlying the nucleus. The acrosome reaction involves fusion of outer acrosomal membrane and sperm plasma membrane at many sites. The small, hybrid membrane vesicles that form are

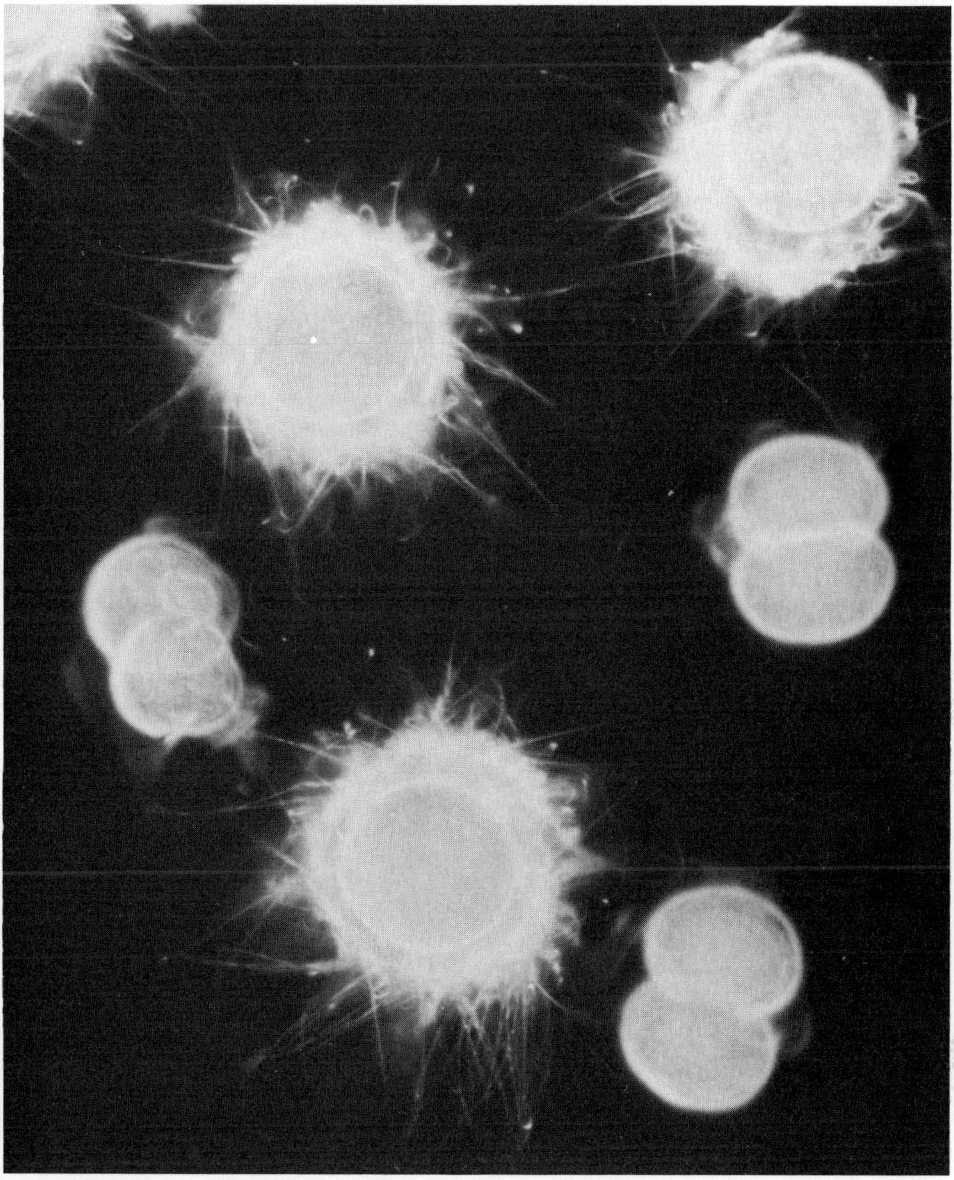

FIG. 2. Photomicrograph illustrating the binding of mouse sperm to unfertilized mouse eggs, but not to two-cell mouse embryos *in vitro* (Wassarman 1987a). Although the ZP of both eggs and embryos contains ZP3, the sperm receptor, only ZP3 in the egg ZP is functional for binding of sperm and induction of the acrosome reaction.

MET ALA SER SER TYR PHE LEU PHE LEU CYS LEU LEU LEU CYS GLY GLY PRO GLU LEU CYS
1 *20*
ASN SER GLN THR LEU TRP LEU LEU PRO GLY GLY THR PRO THR PRO VAL GLY SER SER SER
21 40
PRO VAL LYS VAL GLU CYS LEU GLU ALA GLU LEU VAL VAL THR VAL SER ARG ASP LEU PHE
41 60
GLY THR GLY LYS LEU VAL GLN PRO GLY ASP LEU THR LEU GLY SER GLU GLY CYS GLN PRO
61 80
ARG VAL SER VAL ASP THR ASP VAL VAL ARG PHE ASN ALA GLN LEU HIS GLU CYS SER SER
81 100
ARG VAL GLN MET THR LYS ASP ALA LEU VAL TYR SER THR PHE LEU LEU HIS ASP PRO ARG
101 120
PRO VAL SER GLY LEU SER ILE LEU ARG THR ASN ARG VAL GLU VAL PRO ILE GLU CYS ARG
121 140
TYR PRO ARG GLN GLY ASN VAL SER SER HIS PRO ILE GLU PRO THR TRP VAL PRO PHE ARG
141 160
ALA THR VAL SER SER GLU GLU LYS LEU ALA PHE SER LEU ARG LEU MET GLU GLU ASN TRP
161 180
ASN THR GLU LYS SER ALA PRO THR PHE HIS LEU GLY GLU VAL ALA HIS LEU GLN ALA GLU
181 200
VAL GLN THR GLY SER HIS LEU PRO LEU GLN LEU PHE VAL ASP HIS CYS VAL ALA THR PRO
201 220
SER PRO LEU PRO ASP PRO ASN SER SER PRO TYR HIS PHE ILE VAL ASP PHE HIS GLY CYS
221 240
LEU VAL ASP GLY LEU SER GLU SER PHE SER ALA PHE GLU VAL PRO ARG PRO ARG PRO GLU
241 260
THR LEU GLN PHE THR VAL ASP VAL PHE HIS PHE ALA ASN SER SER ARG ASN THR LEU TYR
261 280
ILE THR CYS HIS LEU LYS VAL ALA PRO ALA ASN GLN ILE PRO ASP LYS LEU ASN LYS ALA
281 300
CYS SER PHE ASN LYS THR SER GLN SER TRP LEU PRO VAL GLU GLY ASP ALA ASP ILE CYS
301 320
ASP CYS CYS SER HIS GLY ASN CYS SER ASN SER SER SER SER GLN PHE GLN ILE HIS GLY
321 340
PRO ARG GLN TRP SER LYS LEU VAL SER ARG ASN ARG ARG HIS VAL THR ASP GLU ALA ASP
341 360
VAL THR VAL GLY PRO LEU ILE PHE LEU GLY LYS ALA ASN ASP GLN THR VAL GLU GLY TRP
361 380
THR ALA SER ALA GLN THR SER VAL ALA LEU GLY LEU GLY LEU ALA THR VAL ALA PHE LEU
381 400
THR LEU ALA ALA ILE VAL LEU ALA VAL THR ARG LYS CYS HIS SER SER SER TYR LEU VAL
401 420
SER LEU PRO GLN
421 424

FIG. 3. Primary structure of the ZP3 polypeptide chain as predicted from exon mapping and DNA sequencing (Kinloch et al 1988). The entire sequence of the 424 amino acid polypeptide chain, including the predicted signal sequence (amino acids 1–22; italicized), is shown. Locations of six potential N-linked glycosylation sites (consensus sequence, Asn-X-Ser/Thr; underlined) are also shown (only three or four sites are used; Salzmann et al 1983).

1985, Wassarman 1987a,b). Each acrosome-intact sperm recognizes and binds to tens-of-thousands of copies of ZP3 at the surface of the egg ZP. ZP3 is located every 15 nm or so along the ZP filaments that constitute the extracellular coat, placing tens-of-millions of copies of ZP3 at the ZP surface available for sperm binding. ZP3 also serves as the acrosome reaction inducer following sperm binding, thereby enabling bound sperm to penetrate the extracellular coat. Thus, egg ZP3 performs three different functions: (a) It plays a structural role during assembly of ZP filaments; (b) It serves as a receptor for sperm during species-specific binding of sperm to eggs; and (c) It induces bound sperm to undergo the acrosome reaction. After fertilization, ZP3 is inactivated as a sperm receptor, making the embryo ZP refractory to sperm.

Biochemical and molecular aspects of ZP3: the sperm receptor

ZP3 has an apparent M_r of about 83 000, is quite acidic, having an isoelectric point of about 4.7, and exhibits considerable heterogeneity on SDS-PAGE (polyacrylamide gel electrophoresis) (Wassarman 1988b). The low isoelectric point and considerable heterogeneity of ZP3 are due entirely to the glycoprotein's oligosaccharides. ZP3 is synthesized and secreted exclusively by oocytes during their growth phase, the period of oogenesis when the ZP first appears (Bleil & Wassarman 1980b, Shimizu et al 1983, Salzmann et al 1983, Philpott et al 1987, Wassarman 1988b). The mature ZP3 polypeptide chain consists of 402 amino acids (43 900 M_r) and has an isoelectric point of about 6.7. It is encoded by a single copy gene, about 10 kilobases in length, that contains 8 exons (ZP3 poly(A)$^+$ mRNA is about 1.5 kilobases in length) (Ringuette et al 1986, 1988, Kinloch et al 1988). Nascent ZP3, which consists of 424 amino acids, has a 22 amino acid 'signal sequence' at its N-terminus that is cleaved during intracellular processing of the glycoprotein by growing oocytes (Fig. 3). The polypeptide chain is especially rich in serine, threonine, and proline residues, contains little, if any, alpha-helix, and, overall, is not particularly hydrophobic or hydrophilic. Three or four high mannose-type (oligomannosidic), N-linked oligosaccharides are added co-translationally to the ZP3 polypeptide chain (giving rise to 53 000 and 56 000 M_r precursors, respectively), and these are subsequently processed to complex-type oligosaccharides in the Golgi. In addition, an undetermined number of O-linked oligosaccharides are added to ZP3 prior to secretion from oocytes (Salzmann et al 1983, Roller & Wassarman 1983). The addition of N- and O-linked oligosaccharides to the ZP3 polypeptide chain results in the mature, 83 000 M_r glycoprotein that is assembled into ZP filaments.

Role of ZP3 oligosaccharides in sperm receptor function

Several lines of evidence strongly suggest that the sperm receptor function of ZP3 is completely dependent on the glycoprotein's oligosaccharides, not its polypeptide chain. Such a situation is consistent with the unusual stability of ZP3 as a sperm receptor. For example, exposure of ZP3 to high temperatures (100 °C), denaturants (e.g. urea), or detergents (e.g. SDS) does not abolish ZP3's sperm receptor activity (or its acrosome reaction-inducing activity) *in vitro*. In this context, numerous investigators have reported that various lectins, monosaccharides, and glycoconjugates prevent binding of sperm to mammalian eggs. Furthermore, it has been proposed that the egg-binding protein on sperm, a macromolecule that recognizes and binds to sperm receptors in the ZP, is a glycosyltransferase (specifically, a galactosyltransferase) (Macek & Shur 1988). These and other findings suggest that carbohydrates play a role in mammalian sperm receptor function.

To evaluate the role of ZP3's polypeptide chain and oligosaccharides in sperm receptor function, we subjected the purified glycoprotein to both chemical and enzymic dissection (Florman et al 1984, Florman & Wassarman 1985, Wassarman 1987a,b):

(a) To assess the role of polypeptide chain, ZP3 was digested extensively with insolubilized pronase, producing a mixture of small glycopeptides (M_r 1500 to 6000), and these were tested for sperm receptor activity *in vitro*. The glycopeptides were found to be virtually as effective as intact ZP3 in inhibiting binding of sperm to eggs (i.e. receptor activity *present*).

(b) To assess the role of oligosaccharides, ZP3 was treated with trifluoromethanesulphonic acid, a reagent that removes both N- and O-linked oligosaccharides from glycoproteins. Such treatment converted ZP3 to a species with a M_r approximating that of the glycoprotein's polypeptide chain (44 000). The denuded polypeptide chain had no effect on binding of sperm to eggs *in vitro* (i.e. receptor activity *absent*).

The above results suggest that ZP3's oligosaccharides, rather than polypeptide chain, play a role in the glycoprotein's sperm receptor function.

(c) To assess the role of N-linked oligosaccharides, ZP3 was subjected to mild alkaline hydrolysis (β-elimination), resulting in specific removal of O-linked oligosaccharides from ZP3. The N-glycosylated polypeptide chain that remained had no effect on binding of sperm to eggs *in vitro* (i.e. receptor activity *absent*).

(d) To assess the role of O-linked oligosaccharides, ZP3 was digested with endoglycosidase F, an enzyme that selectively removes N-linked oligosaccharides from glycoproteins. The O-glycosylated polypeptide chain that remained inhibited binding of sperm to eggs *in vitro* (i.e. receptor activity *present*).

The results of experiments (c) and (d) suggest that ZP3's O-linked oligosaccharides, rather than N-linked oligosaccharides, play a role in the glycoprotein's sperm receptor function.

(e) To strengthen the case for involvement of O-linked oligosaccharides in sperm receptor function, ZP3 was subjected to β-elimination in the presence of sodium borohydride. Such treatment resulted in the release of intact O-linked oligosaccharides having *N*-acetyl-D-galactosaminitol at their reducing termini. These O-linked oligosaccharides inhibited binding of sperm to eggs *in vitro* (i.e. receptor activity *present*). In this context, it was noted that O-linked oligosaccharides released by alkaline-borohydride hydrolysis from either ZP2 or human chorionic gonadotropin had no effect on binding of sperm to eggs *in vitro*. (It is of interest to note that neither small ZP3 glycopeptides nor ZP3

O-linked oligosaccharides induced sperm to undergo the acrosome reaction *in vitro*.)

Collectively, the results of these five experiments strongly suggest that ZP3's O-linked oligosaccharides account for the glycoprotein's ability to serve as a receptor for sperm.

Role of specific ZP3 O-linked oligosaccharides in sperm receptor function

Although the number of O-linked oligosaccharides on ZP3 has not been determined accurately, it is clear that there are a relatively large number. Therefore, we asked whether or not *all* ZP3 O-linked oligosaccharides possessed sperm receptor activity *in vitro* (Florman & Wassarman 1985). O-Linked oligosaccharides released from ZP3 by alkaline-borohydride hydrolysis (done in the presence of sodium borotritide to radiolabel oligosaccharides) were subjected to gel-filtration on Bio-Gel P-2 (1800 M_r cut-off). Sperm receptor activity was found associated only with the void volume fractions, not with the included volume. Accordingly, void volume fractions recovered from Bio-Gel P-2 columns were pooled and applied to Bio-Gel P-6 (6000 M_r cut-off) in order to fractionate the oligosaccharides according to size. Such experiments revealed that only a specific size-class of ZP3 O-linked oligosaccharides, having an apparent M_r between 3400 and 4600, possessed sperm receptor activity *in vitro* (Fig. 4). This fraction of oligosaccharide contained 10% or less of the radiolabel incorporated into ZP3 O-linked oligosaccharides. Thus, it appears that a relatively small percentage of ZP3 O-linked oligosaccharides is responsible for the glycoprotein's sperm receptor function.

The results presented here suggest that sperm recognize and bind to a specific size-class of ZP3 O-linked oligosaccharides. To pursue this possibility, we incubated sperm with O-linked oligosaccharides released from ZP3 in the manner described above, recovered the sperm by centrifugation through dibutyl phthalate into sucrose-Triton X-100, and fractionated the sperm-associated, radiolabelled oligosaccharides on Bio-Gel P-2 and P-6 (as described above). As anticipated, the profile of radiolabel following Bio-Gel P-6 fractionation displayed a tremendous enrichment of material in the region 3400 to 4600 M_r (as compared to profiles of ZP3 O-linked oligosaccharides not incubated with sperm) and only material in this region exhibited sperm receptor activity *in vitro* (Fig. 4). (It is interesting that analogous experiments using mouse adipocytes, rather than sperm, did not demonstrate any selective binding of oligosaccharides to these cells.) Therefore, sperm recognize and bind to certain ZP3 O-linked oligosaccharides and ignore others.

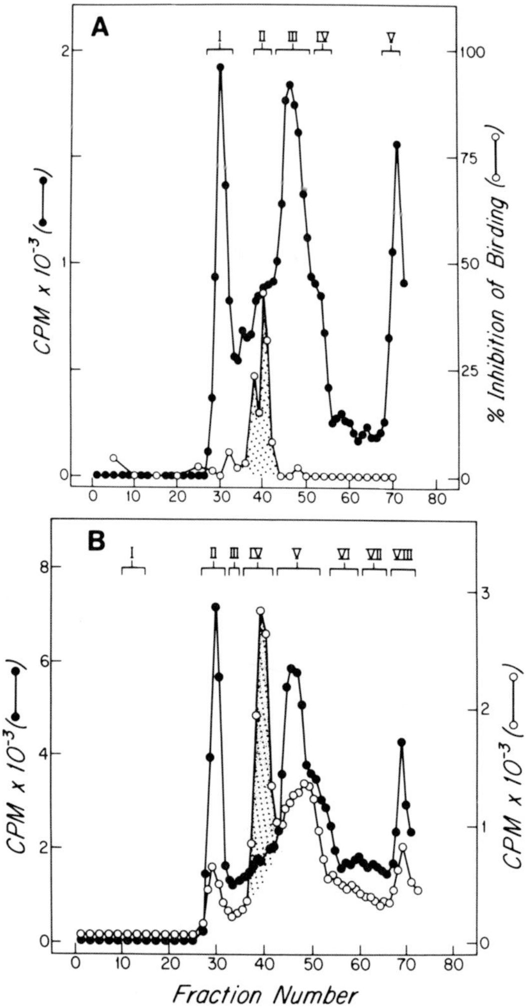

FIG. 4. Size-fractionation and competition binding analyses of O-linked oligosaccharides released from purified egg ZP3 by alkaline-borohydride hydrolysis (Florman & Wassarman 1985). A. Bio-Gel P-6 elution profile ($CPM \times 10^{-3}$) for radiolabelled ZP3-derived oligosaccharides (filled circles) and the sperm receptor activity profile (% inhibition of binding) for the released oligosaccharides (open circles). The region of the elution profile displaying sperm receptor activity (region II) is stippled. B. Bio-Gel P-6 elution profile ($CPM \times 10^{-3}$) and sperm receptor activity of ZP3-derived O-linked oligosaccharides bound to sperm. Shown are the profiles for radiolabelled oligosaccharides not incubated with sperm (as in panel A; filled circles) and those associated with sperm after incubation for one hour (open circles). The region of the elution profile that was significantly enriched following incubation of oligosaccharides with sperm (region IV; equivalent to region II in A) is stippled. Only region IV exhibited sperm receptor activity, reducing sperm binding by more than 50% as compared with controls.

ZP3 carbohydrate determinants responsible for sperm receptor function

We have initiated studies aimed at identifying individual sugars on ZP3 O-linked oligosaccharides that are required for sperm binding (i.e. receptor activity). In one experimental approach, purified ZP3 and purified ZP3-derived O-linked oligosaccharides (Fig. 5) were treated with a variety of exoglycosidases (Bleil & Wassarman 1988). Following such treatment, both substrates were tested for retention of sperm receptor activity, and any released monosaccharides were dansylated and analysed by reverse-phase HPLC. These studies have revealed the highly specific nature of interactions between sperm and ZP3 oligosaccharides.

Of six exoglycosidases tested for their ability to destroy the sperm receptor activity of purified ZP3 and purified ZP3-derived O-linked oligosaccharides, only α-galactosidase and α-fucosidase had a significant effect. Consistent with this finding, much lower levels of sperm binding were observed with formaldehyde-fixed eggs exposed to α-galactosidase and α-fucosidase than with either untreated eggs or eggs exposed to the four other exoglycosidases [β-galactosidase, β-glucuronidase, β-*N*-acetylglucosaminidase (containing β-*N*-acetylgalactosaminidase), or neuraminidase]. These results suggest that removal of either terminal galactose or fucose from ZP3 O-linked oligosaccharides destroys the glycoprotein's ability to serve as a sperm receptor. However, it is important to note that, while we were able to detect the release of galactose from ZP3 oligosaccharides treated with α-galactosidase, we did *not* detect release of fucose (or any other sugar) from α-fucosidase-treated samples. Consequently, we can only make a strong case for involvement of terminal galactose in sperm receptor activity.

Additional evidence supporting a vital role for galactose in sperm receptor function comes from our experiments using galactose oxidase (Bleil & Wassarman 1988). This enzyme converts the C-6 position alcohol of terminal galactose and *N*-acetylgalactosamine residues of oligosaccharides to an aldehyde. We found that exposure of purified ZP3 and purified ZP3-derived O-linked oligosaccharides to galactose oxidase abolished their sperm receptor activity *in vitro*. However, sperm receptor activity was restored to galactose oxidase-treated samples by regeneration of a C-6 position alcohol on galactose by reduction with sodium borohydride. Thus, conversion of a ZP3 sugar alcohol to an aldehyde is sufficient to preclude binding of sperm to the glycoprotein. These results are not unexpected, because interactions between carbohydrate-binding proteins and sugar ligands are stabilized by hydrogen bonds and van der Waals contacts, with the former providing the major contribution to binding (Quiocho 1986). Sugar hydroxyls participate extensively in hydrogen bonding, probably accounting for the inhibitory effect of galactose oxidase treatment on carbohydrate-mediated sperm–egg interaction.

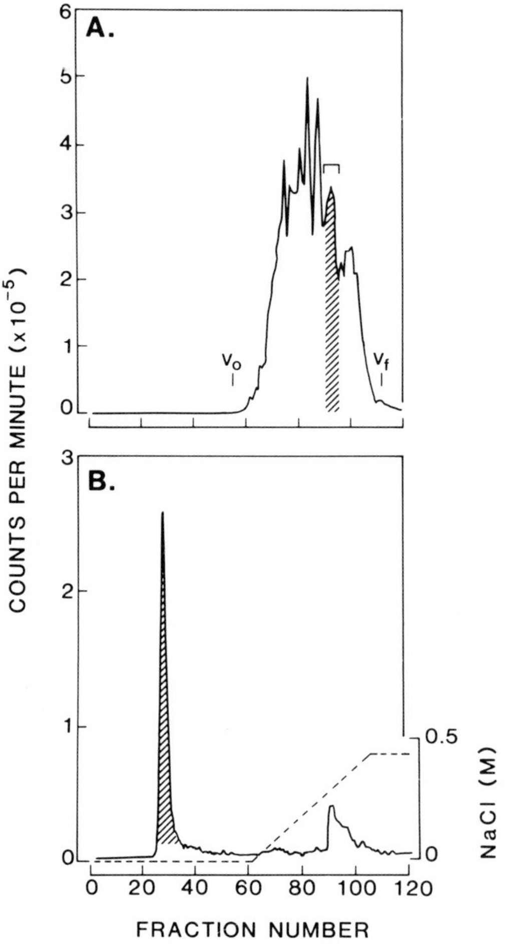

FIG. 5. Purification of ZP3-derived O-linked oligosaccharides possessing sperm receptor activity *in vitro* (Bleil & Wassarman 1988). ZP3 purified from approximately 80 000 ZP was subjected to alkaline-borohydride hydrolysis and the radiolabelled O-linked oligosaccharides were fractionated according to size by HPLC on an SW-2000 column (A). Fractions were assayed for radioactivity and tested for sperm receptor activity, and active fractions were pooled (A; hatched region, fractions 89–95), desalted, lyophilized, resuspended, and subjected to ion-exchange chromatography by HPLC on a DEAE-5PW column (B). Fractions were assayed for radioactivity and tested for sperm receptor activity, and active fractions were pooled (B; hatched region, fractions 27–31) and stored frozen prior to use. V_o, void volume; V_f, final volume.

Summary and final comments

In this paper, I have summarized research done in my laboratory which has led us to conclude that, as a prelude to fertilization, mouse sperm recognize and interact with a specific class of O-linked oligosaccharides on the mouse egg ZP glycoprotein, ZP3 (the sperm receptor). Each acrosome-intact sperm binds to tens-of-thousands of copies of ZP3, probably through interactions with multiple O-linked oligosaccharide chains on each ZP3 molecule. As a consequence of binding to ZP3, sperm undergo the acrosome reaction (membrane fusion) which enables them to penetrate the ZP and fuse with the egg plasma membrane. After sperm–egg fusion, ZP3 is inactivated both as a sperm receptor and an acrosome reaction inducer. Whether ZP3 and ZP3-derived O-linked oligosaccharides exhibit any species specificity has not yet been evaluated adequately. Certainly the great diversity of known oligosaccharide structures is compatible with generation of species specificity for mammalian sperm receptors. The variety of compositions, sequences, branching patterns, conformations, and other features of oligosaccharides associated with glycoproteins provide for a staggering number of combinatorial possibilities (Ginsburg & Robbins 1984).

Our results suggest that galactose, located in α-linkage at the non-reducing terminus of ZP3 O-linked oligosaccharides (3900 M_r), is at least one of the carbohydrate determinants responsible for binding of sperm to the ZP. In this connection, it is interesting that oligosaccharides with repeating *N*-acetyllactosamine units (i.e. lactosaminoglycans) are usually associated with developmentally regulated glycoconjugates (Pink 1980, Feizi et al 1981). In certain instances, such sequences are often terminated by a galactose in α-linkage with the galactose of the *N*-acetyllactosamine unit (Feizi et al 1981, Dodd & Jessell 1985, Jessell & Dodd 1985). In fact, preliminary results obtained in my laboratory suggest that the carbohydrate epitope recognized by the monoclonal antibody designated LA4 (Dodd & Jessell 1985, Jessell & Dodd 1985), an epitope containing a terminal galactose in α-linkage with a penultimate galactose (structure based on the type-2 [galactose(β1–4)-*N*-acetylglucosamine] lactoseries), is present on ZP3, but not on ZP1 or ZP2 (R. Shalgi and P. Wassarman, unpublished work 1987). (In view of our results obtained with α-fucosidase, it is of interest to note that the epitope for LA4 can include a fucose in α-linkage with the penultimate galactose (Dodd & Jessell 1985).) Additional experiments are necessary to examine the potential role of lactosaminoglycans in ZP3's sperm receptor activity.

Finally, it is tempting to speculate that inactivation of ZP3 as a sperm receptor and acrosome reaction inducer, which occurs following fertilization, is due to removal or modification of one or more sugars of ZP3 O-linked oligosaccharides. Our results indicate that removal or modification of a single sugar that serves as a sperm-binding determinant leads to loss of ZP3's receptor activity. It seems likely that cortical granule enzyme(s), released into the ZP following fertilization, could be responsible for inactivation of mammalian sperm receptors as part of the secondary block to polyspermy (Wassarman 1987a,b, 1988a). Preliminary results obtained in my laboratory suggest that O-linked oligosaccharides responsible for ZP3's sperm receptor activity are, indeed, modified following either fertilization or artificial activation of mouse eggs (J. Bleil & P. Wassarman, unpublished work 1988).

Acknowledgements

I am delighted to acknowledge the important contributions of my colleagues, Drs Jeffrey Bleil and Harvey Florman, who carried out much of the research on ZP3 oligosaccharides that is described here. We benefited from discussions with and advice given generously by Dr Jacques Baenziger. Some of this research was supported in part by the National Institute of Child Health and Human Development and the Rockefeller Foundation.

References

Bleil JD, Wassarman PM 1980a Mammalian sperm–egg interaction: identification of a glycoprotein in mouse egg zonae pellucidae possessing receptor activity for sperm. Cell 20:873–882

Bleil JD, Wassarman PM 1980b Synthesis of zona pellucida proteins by denuded and follicle-enclosed mouse oocytes during culture *in vitro*. Proc Natl Acad Sci USA 77:1029–1033

Bleil JD, Wassarman PM 1983 Sperm–egg interactions in the mouse: sequence of events and induction of the acrosome reaction by a zona pellucida glycoprotein. Dev Biol 95:317–324

Bleil JD, Wassarman PM 1986 Autoradiographic visualization of the mouse egg's sperm receptor bound to sperm. J Cell Biol 102:1363–1371

Bleil JD, Wassarman PM 1988 Galactose at the nonreducing terminus of O-linked oligosaccharides of mouse egg zona pellucida glycoprotein ZP3 is essential for the glycoprotein's sperm receptor activity. Proc Natl Acad Sci USA 85:6778–6782

Dodd J, Jessell TM 1985 Lactoseries carbohydrates specify subsets of dorsal root ganglion neurons projecting to the superficial dorsal horn of rat spinal cord. J Neurosci 5:3278–3294

Feizi T, Kapadia A, Gooi HC, Evans MJ 1981 Human monoclonal autoantibodies detect changes in expression and polarization of the Ii antigens during cell differentiation in early mouse embryos and teratocarcinomas. In: Muramatsu T, Ikawa Y (eds) Teratocarcinoma and cell surface. North-Holland Biomedical Press, Amsterdam, p 167–181

Florman HM, Bechtol KB, Wassarman PM 1984 Enzymatic dissection of the functions of the mouse egg's receptor for sperm. Dev Biol 106:243–255

Florman HM, Wassarman PM 1985 O-Linked oligosaccharides of mouse egg ZP3 account for its sperm receptor activity. Cell 41:313–324

Ginsburg V, Robbins PW (eds) 1984 Biology of carbohydrates. John Wiley & Sons Inc, New York, vol 2

Gwatkin RBL 1977 Fertilization mechanisms in man and mammals. Plenum Press, New York

Jessell TM, Dodd J 1985 Structure and expression of differentiation antigens on functional subclasses of primary sensory neurons. Philos Trans R Soc Lond B Biol Sci 308:271–281

Kinloch RA, Roller RJ, Fimiani CM, Wassarman DA, Wassarman PM 1988 Primary structure of the mouse sperm receptor polypeptide determined by genomic cloning. Proc Natl Acad Sci USA 85:6409–6413

Macek MB, Shur BD 1988 Protein–carbohydrate complementarity in mammalian gamete recognition. Gamete Res 20:93–110

Philpott CC, Ringuette MR, Dean J 1987 Oocyte-specific expression and developmental regulation of ZP3, the sperm receptor of the mouse zona pellucida. Dev Biol 121:568–575

Pink JRL 1980 Changes in T-lymphocyte glycoprotein structures associated with differentiation. Contemp Top Mol Immunol 9:89–113

Quiocho FA 1986 Carbohydrate-binding proteins: tertiary structures and protein–sugar interactions. Annu Rev Biochem 55:287–315

Ringuette MJ, Sobieski DA, Chamow SM, Dean J 1986 Oocyte-specific gene expression: molecular characterization of a cDNA for ZP3, the sperm receptor of the mouse zona pellucida. Proc Natl Acad Sci USA 83:4341–4345

Ringuette MJ, Chamberlin ME, Baur AW, Sobieski DA, Dean J 1988 Molecular analysis of cDNA coding for ZP3, a sperm binding protein of the mouse zona pellucida. Dev Biol 127:287–295

Roller RJ, Wassarman PM 1983 Role of asparagine-linked oligosaccharides in secretion of glycoproteins of the mouse egg's extracellular coat. J Biol Chem 258:13243–13249

Salzmann GS, Greve JM, Roller RJ, Wassarman PM 1983 Biosynthesis of the sperm receptor during oogenesis in the mouse. EMBO (Eur Mol Biol Organ) J 2:1451–1456

Shimizu S, Tsuji M, Dean J 1983 In vitro biosynthesis of three sulfated glycoproteins of murine zonae pellucidae by oocytes grown in culture. J Biol Chem 258:5858–5863

Wassarman PM 1987a The biology and chemistry of fertilization. Science (Wash DC) 235:553–560

Wassarman PM 1987b Early events in mammalian fertilization. Annu Rev Cell Biol 3:109–142

Wassarman PM 1988a Fertilization in mammals. Sci Am 255 (December):78–84

Wassarman PM 1988b Zona pellucida glycoproteins. Annu Rev Biochem 57:415–442

Wassarman PM, Florman HM, Greve JM 1985 Receptor mediated sperm–egg interactions in mammals. In: Metz CB, Monroy A (eds) Biology of fertilization. Academic Press, New York, vol 2:341–360

Yanagimachi R 1988 Mammalian fertilization. In: Kobil E, Neill JD (eds) The physiology of reproduction. Raven Press, New York, vol 1:135–185

DISCUSSION

Thomas-Oates: Do you know any more about the structure of ZP3?

Wassarman: Not a great deal more. Tom Jessell provided a library of monoclonal antibodies directed against carbohydrate epitopes (Dodd & Jessell 1985). We screened them, and found that one, LA4, which is directed against an *N*-acetyllactosamine group with a terminal galactose, reacted with ZP3, but not with ZP1 and ZP2 (R. Shalgi & P. Wassarman, unpublished results). The epitope can also have a fucose located at the penultimate galactose. In addition, some antibodies against blood group B determinants gave a mild reaction on a Western blot (R. Shalgi & P. Wassarman, unpublished results). We tried to use LA4 to block fertilization, but were unsuccessful. I think this was because LA4 is a relatively weak antibody.

Gahmberg: It is difficult to get enough material. Have you looked at mucins or other glycoproteins for this blocking activity?

Wassarman: We have considered doing that, but haven't as yet. I suspect that the ZP3 oligosaccharide, or a portion of it, is present on mucins and, perhaps, elsewhere.

Hart: How do your results fit with Barry Shur's proposal that galactosyltransferase on the head of the sperm is the receptor that binds a poly-*N*-acetyllactosamine glycan (Shur & Neely 1988)?

Wassarman: Barry Shur reported that galactosyltransferase on the plasma membrane of sperm recognizes sperm receptors in the egg zona pellucida and binds to them (Shur & Hall 1982). Our finding of a terminal galactose on the receptor oligosaccharide is not completely consistent with that proposal. I think the jury is still out on the question of whether or not galactosyltransferase is an egg-binding protein. We shall have to wait and see.

Hart: Do you have evidence that ZP3 has poly-*N*-acetyllactosamine glycans?

Wassarman: We lack direct evidence for mouse ZP3. However, Yurewicz et al (1987) have shown that the analogous glycoprotein from pig zona pellucida does have poly-*N*-acetyllactosamine glycans.

Kobata: What are the sources of your α-fucosidase and α-galactosidase?

Wassarman: The α-fucosidase came from beef kidney and the α-galactosidase from coffee bean (Bleil & Wassarman 1988).

Gahmberg: Is there sialic acid on the ZP3 molecule?

Wassarman: Yes. The original oligosaccharides that we isolated and used to block sperm binding were a mixture of molecules with and without sialic acid. However, apparently only oligosaccharides lacking sialic acid have sperm receptor activity (J. Bleil, J. Baenziger & P. Wassarman, unpublished results).

Kobata: You say that sialic acid is not present on the sperm receptor oligosaccharide. What about sulphate?

Wassarman: Sulphate may be present, and that could prove to be important.

Kobata: Is the active oligosaccharide charged?

Wassarman: All we can say is that the oligosaccharide elutes from DEAE-Sephadex before the salt gradient (Bleil & Wassarman 1988).

Jessell: Given what you know about the receptor, it is difficult to understand how you get species specificity. Does it arise from different carbohydrate modifications? Is a completely different system used in each mammalian species?

Wassarman: I think differences in carbohydrate structure could give rise to the relative degree of species specificity seen in mammalian fertilization. Certain carbohydrate epitopes may be shared by different mammals, however, and that may account for the small amount of heterospecific sperm–egg interaction observed both *in vitro* and *in vivo*.

Feizi: What are the relative activities of intact ZP3 and the oligosaccharide fraction you isolated by HPLC? One molecule of a large glycoprotein may contain many sugar chains. Do you know how active the oligosaccharide is on a molar basis?

Wassarman: We have found the intact glycoprotein to be more effective than small glycopeptides and small glycopeptides to be more effective than the purified oligosaccharide. I can't provide quantitative data, but the activities are in the nanomolar range in all three cases.

Feizi: Do you have direct evidence that the labelled material is terminated by galactosaminitol?

Wassarman: Yes, that has been demonstrated (Florman & Wassarman 1985).

Feizi: All the antibodies to human milk galactosyltransferase that we have received from Eric Berger (Childs et al 1986, Feizi et al 1987), even those raised against the extensively deglycosylated enzyme (unpublished work) contained antibodies to oligosaccharide structures, including those found on embryonic cells such as SSEA-1. So the labelling of the cell surface by antibodies to galactosyltransferase may not necessarily correspond exclusively to the distribution of the enzyme. In my opinion, the distribution of the enzyme needs to be re-evaluated by a more direct approach.

Sharon: In your fucosidase experiment, it is possible that there is contamination by other glycosidases. One could exclude that by doing what Winifred Watkins did many years ago—adding fucose as an inhibitor.

Wassarman: I wish to emphasize that not only did we not release fucose, we did not release any sugar.

Sharon: The α-linked galactose can usually be blocked by lectins such as that of *Griffonia simplicifolia*. Does that lectin inhibit your sperm–egg interaction?

Wassarman: We haven't done any lectin experiments as yet because until relatively recently we didn't know which lectins to use. But it has been known since the early 1970s that some lectins prevent binding of sperm to eggs and therefore block fertilization (Oikawa et al 1973). However, it has not been demonstrated that lectins do this by specifically modifying sperm receptor oligosaccharides.

Sharon: What is the difference between ZP3 of the unfertilized egg and of the fertilized two-cell embryo?

Wassarman: We looked for protein changes resulting from fertilization, and were never able to find any. From our most recent work, we feel that modification of sperm receptor oligosaccharides results in the inactivation of

ZP3 after fertilization. We have compared patterns of ZP3 oligosaccharides from two-cell embryos and unfertilized eggs. Preliminary results suggest that the sperm receptor oligosaccharide (3900 M_r) either disappears or is significantly depleted on fertilization (J. Bleil & P. Wassarman, unpublished results).

Sharon: This raises the possibility of involvement of glycosidases in fertilization.

Wassarman: It certainly does, and we are studying that.

Sharon: New endoglycosidases are being found. For example, one endoglycosidase cleaves any type of GalNAc-Ser linkage. Have you examined the effect of this enzyme on the activity of ZP3?

Muramatsu: This is an enzyme which even cleaves the GalNAc-Ser/Thr linkage in a long chain (Iwase et al 1988). I don't know whether the enzyme was tested against the 3900 M_r sperm receptor. Endo-β-galactosidases of *Escherichia freundii* may cleave the oligosaccharide and render it inactive. If so, perhaps the epitope is on poly-*N*-acetyllactosamine. There are also two endo-β-galactosidases that act on α-linked galactose determinants. One is blood group endo-β-galactosidase, which attacks the determinant on type-2 chains, and the other acts on Gal(α1–3)Gal linkages.

Barondes: In *Xenopus laevis* eggs a different type of carbohydrate–protein interaction has been implicated in the fertilization process; there is evidence that egg lectins may be released upon fertilization (Greve & Hedrick 1978, Roberson & Barondes 1983). It has been suggested that these lectins may bind glycoconjugates around the egg and play a role in blocking polyspermy. Have similar lectins been looked for in mouse eggs?

Wassarman: I don't think any comparable work has been published.

Barondes: There is also evidence for soluble carbohydrate-binding proteins in some types of sperm. Has that been examined in mice?

Wassarman: Little is known, but in sea urchins sperm binding is thought to be mediated by a lectin, called bindin, that recognizes sugars on receptors in the egg vitelline envelope. It is tempting to think that there is a plasma membrane lectin on mammalian sperm that recognizes ZP3. A number of proteins have been identified in different mammalian sperm which are reportedly lectins.

Barondes: Are these integral membrane proteins?

Wassarman: It is not always clear.

Barondes: ZP3 would be a wonderful reagent to use for identifying soluble sperm lectins. Unfortunately it is hard to get large quantities of it. However, other readily available glycoconjugates could be used to try to purify sperm lectins. These might not have as high affinity as a natural glycoconjugate receptor, but one might be able to obtain large amounts of these ligands and couple them at high density to affinity columns. With other lectins, highly substituted affinity columns made with low affinity ligands are routinely used for lectin purification.

Wassarman: Your point is well taken.

Fukuda: The poly-*N*-acetyllactosamine oligosaccharides that we isolated from red blood cells are also excluded from DEAE-Sephadex, despite the fact they have sialic acid (Fukuda et al 1984). Poly-*N*-acetyllactosamine extensions are found on O-linked oligosaccharide (Fukuda et al 1986). It is likely that you are seeing some O-linked oligosaccharides which contain poly-*N*-acetyllactosamine repeats and do not have a high total charge. α-Galactose is absent in humans. Have you looked at ZP3 in human eggs?

Wassarman: We are beginning experiments along those lines, but it is difficult to do biochemistry in the human system because of the limited amounts of material available. Cloning is relatively easy because you can use mouse probes to pull out the analogous human gene, but work on human sperm receptor carbohydrates is much more difficult.

Ruoslahti: Have you expressed ZP3 in a mammalian expression system? It would be interesting to see whether you get an active or inactive product in a mouse system.

Wassarman: We are trying to do that. We suspect we shall get plenty of protein but not the correct O-linked glycosylation.

Ruoslahti: If you don't get the right glycosylation and the product is inactive, that would tend to prove that the carbohydrate is important.

Wassarman: Yes, it would provide further support for our current interpretation.

Finne: Can you exclude fucosidase binding to the α-galactosides which are present? You could get inhibition by fucosidase if fucosidase binds to α-galactose.

Wassarman: That is possible.

Finne: ZP3 did not totally inhibit the binding; you always have 20% remaining. Perhaps carbohydrates are not the whole story. Additional factors, such as protein–protein interactions, may be involved.

Wassarman: I can't exclude that. We rarely see more than 10% binding remaining with high doses of ZP3, but I don't know the reason for the residual binding.

Rademacher: I don't think that something not being released by endo-glycosidase F but being released by β-elimination defines it as an O-linked glycan. Many people have found that β-elimination releases N-linked glycans.

Wassarman: We have analysed the amino acids that are modified during reductive β-elimination (serine and threonine), and also analysed the sugar at the reducing terminus (*N*-acetyl-D-galactosaminitol).

Rademacher: Could the bioactive compound be the 8–10% of the N-links that are released on β-elimination? We have seen up to 25% depending on the protein we use.

Hart: β-Elimination seems to be controversial. To dismiss it as a useful technique because people use it carelessly is a mistake. If you do careful kinetics and product characterization whenever possible, β-elimination is a very good method for distinguishing N- and O-linked oligosaccharides. Most problems arise from destruction of the polypeptide structure and the production of glycopeptides that can't be distinguished from oligosaccharides.

Feizi: Dr Wassarman, you suggest ZP3 oligosaccharides are modified during fertilization. Have you put your radiolabelled active oligosaccharide into a fertilization mixture and looked for evidence of modification?

Wassarman: We have taken active oligosaccharide, incubated it with unfertilized eggs, and activated the eggs. The idea was to modify both the endogenous zona pellucida around the eggs and the added radiolabelled oligosaccharide. Our results have not been very promising. But technically this is a very difficult experiment.

Fukuda: Are the oligosaccharides from ZP1 and ZP2 active?

Wassarman: The oligosaccharides from egg ZP1 and ZP2, and the oligosaccharides from ZP3 from embryos, have all been tested, and are all inactive as sperm receptors.

Carver: When you talk about the post-fertilization oligosaccharides, what are you following? You no longer have a receptor-binding to follow.

Wassarman: We haven't done these kinds of experiments until recently because we didn't know what to look for. However, now we have a pattern of radiolabelled material that we see under standard conditions for the oligosaccharides from unfertilized egg zonae, and we compare it with fertilized egg zonae.

Carver: What is the time-course for that inactivation?

Wassarman: It was formerly thought that inactivation took 20 to 30 minutes. Our data suggest that it is occurring in five minutes or less.

Lee: Presumably this interaction of lectin and oligosaccharide won't occur until the sperm reaches the surface of the egg. Is there a process, such as chemotaxis, to cause orientation of sperm in that direction?

Wassarman: That's an interesting question. I don't know the answer.

Lee: When sperm reach the surface of the egg, they are wagging their tails, which generates force. So the adhesive force must overcome the locomotive force. Is there any information on the thermodynamics?

Wassarman: No.

Sharon: Since you are so limited in the amount of material, you could collect a large amount of information by examining the behaviour of the oligosaccharide on mini-columns of lectins.

Feizi: Have you been able to determine whether the active oligosaccharide peak is a single species, or whether there are isomeric structures?

Wassarman: The material we currently use appears to be homogeneous, but we certainly can't dismiss the possibility of isomeric structures.

References

Bleil JD, Wassarman PM 1988 Galactose at the non-reducing terminus of O-linked oligosaccharides of mouse egg zona pellucida glycoprotein ZP3 is essential for the glycoprotein sperm receptor activity. Proc Natl Acad Sci USA 85:6778–6782

Childs RA, Berger EG, Thorpe SJ, Aegerter E, Feizi T 1986 Blood-group-related carbohydrate antigens are expressed on human milk galactosyltransferase and are immunogenic in rabbits. Biochem J 238:605–611

Dodd J, Jessell TM 1985 Lactoseries carbohydrates specify subsets of dorsal root ganglion neurons projecting to the superficial dorsal horn of rat spinal cord. J Neurosci 5:3278–3294

Feizi T, Thorpe SJ, Childs RA 1987 Blood group genetic markers on human milk galactosyltransferase: relevance to the immunohistochemical approach to enzyme localization. Biochem Soc Trans 15:614–617

Florman HM, Wassarman PM 1985 O-linked oligosaccharides of mouse egg ZP3 account for its sperm receptor activity. Cell 41:313–324

Fukuda M, Dell A, Fukuda MN 1984 Structure of fetal lactosaminoglycan, the carbohydrate moiety of Band 3 isolated from umbilical cord erythrocytes. J Biol Chem 259:4782–4791

Fukuda M, Carlsson SR, Klock JC, Dell A 1986 Structures of O-linked oligosaccharides isolated from normal granulocytes, chronic myelogenous leukemia cells and acute myelogenous leukemia cells. J Biol Chem 261:12796–12806

Greve LC, Hedrick JL 1978 An immunocytochemical localization of the cortical granule lectin in fertilized and unfertilized eggs of *Xenopus laevis*. Gamete Res 1:13–18

Iwase H, Ishii I, Ishihara K, Tanaka Y, Omura S, Hotta K 1988 Release of oligosaccharides possessing reducing end N-acetylgalactosamine from mucus glycoprotein in streptomyces SP OH-11242 culture medium through action of endo type glycosidase. Biochem Biophys Res Commun 151:422–428

Oikawa T, Yanagimachi R, Nicolson GL 1973 Wheat germ agglutinin blocks mammalian fertilization. Nature (Lond) 241:256–259

Roberson MM, Barondes SH 1983 *Xenopus laevis* lectin is localized at several sites in *Xenopus* oocytes, eggs, and embryos. J Cell Biol 97:1875–1881

Shur BD, Hall NG 1982 A role for mouse sperm surface galactosyltransferase in sperm binding to the egg zona pellucida. J Cell Biol 95:574–579

Shur BD, Neely CA 1988 Plasma membrane association, purification, and partial characterization of mouse sperm β1,4-galactosyltransferase. J Biol Chem 263:17706–17714

Yurewicz EC, Sacco AG, Subramanian MG 1987 Structural characterization of the $M_r = 55,000$ antigen (ZP3) of porcine oocyte zona pellucida. J Biol Chem 262:564–571

Families of neural adhesion molecules

Melitta Schachner

Department of Neurobiology, University of Heidelberg, Heidelberg, Federal Republic of Germany

Abstract. The neural cell adhesion molecules L1 and N-CAM share a common carbohydrate epitope that is recognized by the monoclonal antibodies L2 and HNK-1. The L2/HNK-1 epitope is also present on the myelin-associated glycoprotein (MAG) and secreted J1 glycoprotein, both of which have been identified as cell adhesion molecules. Each of the four adhesion molecules is differentially expressed during development on distinct cell types. Expression of the L2/HNK-1 epitope is regulated independently of the protein backbone, is phylogenetically conserved, and plays a role in cell–cell and, particularly, cell–substrate interactions. Another set of glycoproteins shares a common carbohydrate epitope designated L3. This epitope is present on the novel adhesion molecule on glia (AMOG), L1 and MAG, but not on J1 and N-CAM. As in the L2/HNK-1 family, the number of glycoproteins expressing this epitope is not yet known. It is therefore possible that heterogeneities in carbohydrate structures are associated with different sets of adhesion molecules and may have functional implications.

1989 Carbohydrate recognition in cellular function. Wiley, Chichester (Ciba Foundation Symposium 145) p 156–172

It is becoming evident that diverse cell surface and extracellular matrix glycoproteins, operationally termed adhesion molecules, play important roles in the specification of cell interactions in the developing nervous system. These adhesion molecules subserve distinct functions involving different cells at different developmental stages, but may cooperate when coexpressed. Despite their structural and functional diversity, cell adhesion molecules share common traits at the carbohydrate and protein levels. Adhesion molecules can be grouped into distinct families according to the presence of functionally important carbohydrate domains. Some of these molecules belong to more than one family. At the protein level, three Ca^{2+}-independent adhesion molecules have been recognized as belonging to the immunoglobulin superfamily.

For building a complex organ, such as the nervous system, a number of cellular and molecular mechanisms would seem necessary. In terms of cell surface interactions the questions are: how many adhesion molecules are needed for nervous system development; how do they function at the cellular and molecular levels; are they dependent on each other in function; and are there structural and functional similarities among adhesion molecules? One cue that guided our search was the monoclonal antibody, designated L2, that recognizes a carbohydrate epitope common to several adhesion molecules (Fig. 1). At the time the L2 monoclonal antibody was discovered, only L1 and N-CAM had been described as *bona fide* adhesion molecules (Kruse et al 1984). The myelin-associated glycoprotein (MAG), which is one of an unknown number of L2 carbohydrate-positive glycoproteins, had been hypothesized, but not demonstrated, to be involved in axon-myelinating cell interactions (for reference, see Poltorak et al 1987). From the observation that both functionally characterized L2 epitope-positive molecules were indeed adhesion molecules, we suggested that other glycoproteins expressing this carbohydrate structure might also be involved in adhesion. To verify this, we have functionally characterized other L2 carbohydrate-carrying glycoproteins.

MAG, J1 and other members of the L2 carbohydrate family are adhesion molecules

That MAG is an adhesion molecule was shown by using antibodies against it to interfere with cell adhesion *in vitro* (Poltorak et al 1987). MAG could

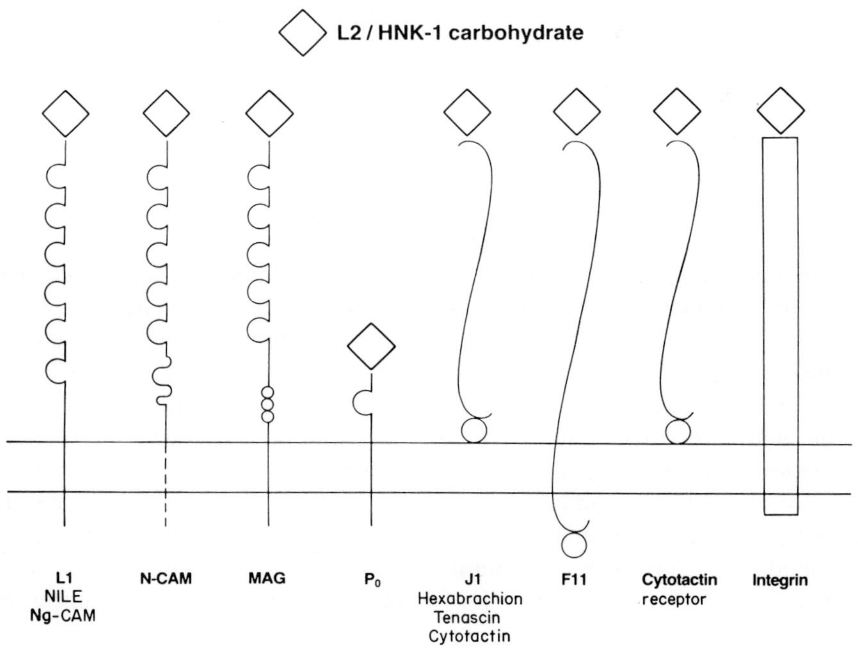

FIG. 1. Schematic presentation of the members of the L2/HNK-1 carbohydrate-based family of adhesion molecules in the nervous system.

thus be implicated in oligodendrocyte–neuron and oligodendrocyte–oligodendrocyte, but not oligodendrocyte–astrocyte interactions. MAG could also be incorporated into liposomes and shown to bind to neurites and neurite bundles in cultures of neurons known to engage in myelination, such as spinal cord and dorsal root ganglion neurons. Cerebellar granule cell neurons that are not myelinated were poor binding partners for these MAG-containing liposomes. *In situ*, MAG is localized at the interface between axon and myelinating cell and on opposing surfaces of the turning loops of myelin-forming cells, lending support to the notion that MAG is involved in many types of cell interactions (Martini & Schachner 1986). MAG became detectable on myelinating Schwann cells once they had turned approximately 1.5 to 2 loops around the axon, signifying a precise timing in the onset of expression. MAG not only acts as a ligand between cell surfaces but also binds to extracellular matrix molecules, such as different types of collagens and heparin (Fahrig et al 1987). The association of MAG with the extracellular matrix in the peripheral, but not central, nervous system may thus have functional implications (Martini & Schachner 1986, 1988).

Another adhesion molecule belonging to the L2 family is the J1 complex of glycoproteins (Kruse et al 1985) that appears to be immunochemically related to tenascin. The J1 glycoproteins are secreted molecules and extracellularly localized, also binding to several types of collagens and heparin. During development, J1 immunoreactivity shifts from a group of glycoproteins with molecular masses above 200 kDa to glycoproteins with masses between 190 and 160 kDa in the adult nervous system. J1-immunoreactive material is found on astrocytes, oligodendrocytes and fibroblast-like cells, but not on neurons. J1 mediates neuron–astrocyte and neuron–Schwann cell interactions, but not astrocyte–astrocyte or neuron–neuron adhesion.

Inferences about functional roles of J1 glycoproteins can be derived from immunohistological observations in the optic nerve and at the neuromuscular junction. In the adult rat optic nerve, nodes of Ranvier were preferentially labelled by J1 and L2 antibodies (ffrench-Constant et al 1986). J1 immunoreactivity appeared concentrated at the site of interdigitation into a nodal region, where the axon is not covered by myelin and lies free to be contacted by astrocytes. Interestingly, other parts of these astrocytes did not show much J1 immunoreactivity, indicating a high degree of topographically selective deposition of J1. Because J1 is a secreted molecule that is also expressed by oligodendrocytes, conclusions about its mode of synthesis and deposition cannot be made yet. However, at least its receptor(s) should show this specialized accumulation at the interfaces between oligodendrocyte, astrocyte and axon. The observation that J1 is concentrated at the node of Ranvier may have important functional implications in view of the fact that J1 is involved in neuron–glia adhesion *in vitro*. These *in vivo* and *in vitro* observations raise the possibility that J1 plays an important part in neuron–glia interactions which are involved in assembly and/or maintenance of the exquisite cytoarchitecture at the node of Ranvier.

J1 also appears to play a role in neuron–muscle interactions. During development, J1 was detectable in small, discrete deposits occupying spaces between myotubes (Sanes et al 1986). Upon innervation of the muscle these deposits became sparser and they disappeared completely after birth. After denervation of adult muscle, J1 was re-expressed around the original synaptic site within two days of the denervation. Re-expression of J1 was also seen after paralysis of nerve activity by tetrodotoxin. When muscles were re-innervated, J1 disappeared again. Thus, axonal regeneration reverses the denervation-induced accumulation of J1. The accumulation of J1 at original synaptic sites is intriguing in view of the capability of regenerating axons to preferentially re-innervate the muscle at these original synaptic sites. Thus, J1 may be one of the cues that guide the re-growing axon to its target.

Besides MAG and J1, other cell surface glycoproteins have now been recognized as belonging to the L2 family. These are the major glycoprotein of peripheral nervous system, myelin P_0 (Bollensen & Schachner 1987), the F11 adhesion molecule involved in fasciculative neurite outgrowth (Rathjen et al 1987), the receptor for cytotactin (Hoffman & Edelman 1987) and two molecules belonging to the integrin family that are cell surface receptors for extracellular matrix constituents, CSAT (Pesheva et al 1987) and fibronectin receptor. Other glycoproteins also carry the L2 epitope, but their functions have not yet been identified.

The adhesion molecule L1

The observation that a cell can express more than one adhesion molecule at a particular developmental stage and a topographically distinct site on the cell surface (e.g. axon versus dendrite) begs the question why a cell permits itself this luxury. One could argue that with variation in the steady-state levels of a particular adhesion molecule and by its localization at topographically distinct sites, one adhesion molecule would suffice to specify cell surface contacts. I shall present evidence that L1 and N-CAM may not act independently of each other within the surface membrane, but first I shall summarize present knowledge about L1.

L1, (which is immunochemically identical to NILE and Ng-CAM; Bock et al 1985, Friedlander et al 1986), is, like N-CAM, involved in Ca^{2+}-independent adhesion among neural cells. During central nervous system development, L1 appears first and is coexpressed with N-CAM on postmitotic neurons. In contrast to N-CAM, L1 displays a previously unrecognized restricted expression by particular neurons; it is found, for example, in the cerebellum on granule and Purkinje cells, but not on stellate and basket cells (Persohn & Schachner 1987). Also, L1 is expressed only on axons, and not on dendrites or cell bodies of neurons. L1 is expressed on fasciculating axons in the central and peripheral nervous system. It is particularly intriguing that an axon may change L1 expression along its course. A set of axons in the embryonic spinal chord expresses L1 only when they fasciculate, but not when they are not fasciculated and are individually interdispersed among neuroepithelial cells (Holley & Silver 1987, J. A. Holley & M. Schachner, unpublished observations). Likewise, fasciculating axons in the sciatic and optic nerves express L1, but when they become myelinated more distally they lose L1 expression (Martini & Schachner 1986, 1988). Thus, an axon is able to express L1 together with N-CAM along its length when fasciculating; and when an axon is non-fasciculating or surrounded by myelin, L1 expression is not detectable and expression of N-CAM is reduced. Therefore, the superposition of expression of L1 onto that of N-CAM may be responsible for axon–axon interactions. Two other surface molecules have also been shown to be characteristic of fasciculating axons, F11 and neurofascin, suggesting either redundancy or interdependency in functional mechanisms. Antibodies to these molecules interfere with fasciculation and neurite outgrowth on other neurites.

However, L1 has also been observed on single axons in the inner ear and in Vater-Pacini corpuscles in an L1-free territory (Mbiene et al 1988, C. Nolte & R. Martini, unpublished observations), suggesting that it may not only interact with other L1-positive structures, such as axons or Schwann cells, but also engage in other interactions. It is pertinent that L1 is itself a ligand that promotes neurite outgrowth (Lagenaur & Lemmon 1987, Kadmon et al 1989).

The involvement of L1 in cerebellar granule neuron migration (Lindner et al 1983, 1986) was initially thought to result from an L1-mediated apposition of the migrating neuronal cell body and its leading and trailing processes with Bergmann glia. However, in contrast to Ng-CAM, L1 has so far not been observed to affect neuron–glia, but only neuron–neuron, adhesion *in vitro* Keilhauer et al 1985). In the telencephalic anlage, L1 was detectable on neurons only after cessation of neuron migration along radial glial processes (Fushiki & Schachner 1986). Furthermore, in the developing cerebellum L1 is always confined to contact sites between apposing neuronal surface membranes and excluded from contacts between L1-positive and L1-negative surface membranes (Persohn & Schachner 1987). Thus, the evidence so far suggests that L1 specifies neuron migration in allowing neuronal cell bodies to sort out at the onset and aggregate for the termination of migration.

Interestingly, L1 is not confined to nervous tissue, but is expressed on epithelial cells in the intestinal tract, where it is localized on the proliferating epithelial progenitor cells of crypts, but not in the more differentiated epithelial cells of villi (Thor et al 1987). These epithelial cells constitute another migratory system;

they are constantly generated in the depth of the crypts, from where they migrate to the villi to be shed from their tips into the intestinal lumen. Because L1 engages in crypt cell adhesion, its function may be to promote the tight apposition of migratory cells, disappearing from epithelial cells when they prepare for shedding.

The interdependence of L1 and N-CAM

When aggregation of cell bodies in single cell suspensions was measured, L1 and N-CAM appeared to act synergistically with each other (Faissner et al 1984). When adhesion between single cell suspensions and substrate-attached monolayer cells and migration of cerebellar granule cells were measured, L1 and N-CAM antibodies blocked each other; in the presence of the two antibodies, blocking levels were less than additive (Lindner et al 1986, Keilhauer et al 1985). It was suggested that this interdependence in their function results from interactions between the two molecules, either between neighbouring partner cells or within the plasma membrane of one cell. While no evidence has been obtained that L1 and N-CAM interact with each other in a specific manner, association of L1 and N-CAM has been demonstrated by antibody-induced re-distribution within the surface membrane of live cells (Thor et al 1986). These studies showed that L1 was specifically associated with the largest molecular form (180 kDa) of N-CAM (N-CAM 180), but not with the lower molecular mass form, N-CAM 140. A close association between the two molecules may underlie functional cooperativity, as has been seen with the components of the T cell receptor complex in the immune system.

The feature that distinguishes N-CAM 180 from the smaller N-CAMs is its long cytoplasmic domain. Its extracellular part is supposedly identical to those of N-CAM 140 and N-CAM 120. During development of the cerebellum, N-CAM 180 is the last component to appear, in that it is only detectable in granule cell bodies and axons once migration is finished (Persohn & Schachner 1987). Thus, N-CAM 180 only becomes apparent when neurons engage to form stable cell contacts. Also, when cultured neuroblastoma cells are promoted into morphological differentiation, they shift N-CAM expression towards a predominance of N-CAM 180 (Pollerberg et al 1985, 1986). L1 and N-CAM 180 were found accumulated together at contact sites between neighbouring cells, whereas N-CAM 140 was more uniformly distributed over the whole cell body (Pollerberg et al 1985, 1986, 1987). The cytoskeleton membrane linker protein brain spectrin (fodrin) and actin were also accumulated at contact sites, whereas other cytoskeleton-associated proteins, such as neurofilament, alpha-actinin, filamin, vinculin, beta-tubulin, ankyrin, Band 4.1 and synapsin I, did not show this accumulation. These observations led to the hypothesis that N-CAM 180 may be involved in stabilization of cell contacts by association with the cytoskeleton. Indeed, brain spectrin specifically binds N-CAM 180, but not N-CAM 140 or N-CAM 120. In further support of this idea, N-CAM 180 showed a reduced lateral mobility within the surface membrane. It is therefore noteworthy that growth cones express only N-CAM 180 when in contact with the target cell. Conversely, N-CAM 140 is more apparent on growth cones when they are not in contact and are free 'searching'. Thus, restriction of lateral mobility by association with the cytoskeleton is likely to induce accumulation of N-CAM 180 at contact sites to selectively stabilize cell–cell contacts.

Regeneration

In the mammalian nervous system neurons of both peripheral and central nervous system origin show the capacity for functional regrowth, but Schwann cells are unique in allowing regrowth and regeneration (Aguayo 1985). In contrast to the peripheral glia, central nervous system glial cells beyond a certain developmental stage do not support regrowth (Smith et al 1986). In lower vertebrates, however, axon regrowth in the central nervous system takes place even in adulthood. These observations led us to search for the cellular and molecular signals that regulate adhesion molecule expression on Schwann cells during development and regeneration.

Immunohistological observations showed that L1 and N-CAM are expressed on Schwann cells before the establishment of a 1:1 relationship with axons (Martini & Schachner 1986) and remain expressed in the non-myelinating state into adulthood (Martini & Schachner 1986, Bollensen & Schachner 1987). P_o then becomes expressed on Schwann cells when these establish a 1:1 ratio with axons. When Schwann cell processes have turned 1.5 to 2 loops around the axons, L1 and N-CAM are reduced in their expression on both axons and Schwann cells. Thereafter, neither axons nor Schwann cells express L1, whereas N-CAM is found periaxonally and, more weakly, in compact myelin, thus exemplifying again the principle of a broader range of expression. With the disappearance of L1, MAG becomes detectable periaxonally and on the turning loops of Schwann cells, to disappear again once compaction occurs. P_o, in contrast, remains detectable in compact myelin. Thus axon–Schwann cell interactions are characterized by a precisely timed appearance of adhesion molecules in various compartments of Schwann cells and axons. From this sequence it may be deduced that L1 and N-CAM are involved in the initial interaction between axon and Schwann cell, and P_o is involved in the segregation of axon fascicles for ensheathment by Schwann cells on a 1:1 basis, apposition of turning loops of Schwann cells and the maintenance of compact myelin. MAG, in conjunction with P_o, may be involved in the spiralling of Schwann cell processes.

After lesioning of peripheral nerves by cut or crush, re-expression of adhesion molecules is reminiscent of development. Within several days after transection, L1 and N-CAM appear again on all Schwann cells, even if they are still in association with degenerating myelin (Martini & Schachner 1988). The temporal sequence and spatial distribution of adhesion molecule expression on axons and Schwann cells is then the same as during development.

The functional significance of re-expression of L1 and N-CAM on Schwann cells at the time of axon regrowth requires validation. It is likely that expression of L1 and N-CAM is a prerequisite for successful regrowth of axons and that the capacity for regeneration in the peripheral nervous system may be attributed to the plasticity in re-expression of these adhesion molecules. It is, therefore, interesting that both L1 and N-CAM are involved in adhesion of dorsal root ganglion neurons and Schwann cells (Seilheimer & Schachner 1988). Only L1, however, appears to be involved in promoting neurite outgrowth, reminiscent of the involvement of L1 in axon growth on other L1-positive axons. These observations suggest that L1 and N-CAM may, but not necessarily, be interdependent in function.

Elucidation of the cellular and molecular signals that underlie re-expression of adhesion molecules on Schwann cells is necessary for understanding the remarkable plasticity of these cells to re-juvenate after incurrence of a lesion. As a first step in this direction, we have studied the influence of nerve growth factor (NGF) on adhesion molecule expression by Schwann cells in culture. These experiments were instigated by the observation that upon denervation NGF is synthesized by Schwann cells (Rush 1984) and that Schwann cells may express NGF receptors at early developmental stages (Rohrer 1985). Furthermore, we knew that PC12 pheochromocytoma cells are induced by NGF to synthesize increased levels of NILE/L1 (Lee et al 1981). It was gratifying to find that NGF also leads to an increased expression of L1 in pure cultures of Schwann cells (Seilheimer & Schachner 1987). Because antibodies to NGF reduce L1 expression in these cultures, NGF probably acts by an autocrine mechanism. These experiments show for the first time that a neuronotrophic factor directly influences a functionally meaningful parameter, namely adhesion molecule expression, in a non-neuronal cell, the Schwann cell. Thus, NGF may play a dual role in regeneration: increase in L1 expression by both axons and Schwann cells to enhance successful neurite outgrowth.

Family traits

The three functionally characterized Ca^{2+}-independent adhesion molecules, N-CAM, MAG and L1, all carry five or six immunoglobulin homologous domains (Fig. 1) (Arquint et al 1987). The functional role of these domains, which are characteristic of all members of the immunoglobulin superfamily (Williams

1987), remains obscure. That members of the immunoglobulin superfamily in the immune system have been reported to interact with each other (not themselves) is an incentive to look at similar types of interactions in the nervous system.

Expression of the L2 carbohydrate epitope is another trait found in common among several Ca^{2+}-independent adhesion molecules (Fig. 1). One could argue that because this carbohydrate is shared by functionally important molecules, it itself may be functionally important. The carbohydrate epitope is not confined to the nervous system, but is also present on subpopulations of lymphoid cells, among them natural killer cells, giving it the name HNK-1 (Abo & Balch 1981). On neural adhesion molecules, the carbohydrate is regulated during development independently of the protein backbone (Martini & Schachner 1986). Also, in the adult state, N-CAM, L1 and MAG are heterogeneous with respect to the expression of the L2/HNK-1 epitope in that only subpopulations of each adhesion molecule express the carbohydrate (Faissner 1987, Kruse et al 1984, Poltorak et al 1986). The carbohydrate is carried by the 65 kDa N-terminal fragment of N-CAM, but is not present in the N-terminal 24 kDa region that contains the heparin-binding domain (Cole & Schachner 1987). Sera from patients with gammopathy and peripheral polyneuropathy also react with the carbohydate (for references, see Bollensen et al 1988). On glycolipids the L2/HNK-1 carbohydrate epitope contains a 3′-sulphated glucuronic acid (Chou et al 1986).

Indications that the L2/HNK-1 domain is involved in cell interactions came from investigations on the effect of L2 antibodies on neural cell adhesion (Faissner 1987, Keilhauer et al 1985). However, because antibodies do not only cover the epitope that they are directed against, but may sterically block the function of neighbouring molecular domains, a more direct demonstration of the importance of the L2/HNK-1 carbohydrate for cell interactions appeared necessary. We therefore took advantage of the possibility to assay the effects of the L2 glycolipid and the tetrasaccharide derived from it on cell–cell interactions in culture. The isolated carbohydrate structures interfered not only with cell-to-cell interactions, but, even more strikingly, cell-to-substrate interactions (Künemund et al 1988). Because the inhibitory effects observed in adhesion assays with the glycolipid or tetrasaccharide were qualitatively and quantitatively very similar to those observed with the L2 antibodies, the most straightforward interpretation of our findings is that the L2/HNK-1 carbohydrate is itself involved as a ligand in cell interactions. However, the involvement of carbohydrate structures in adhesion and recognition may be more complex and subtle, so that direct inferences about the molecular mechanisms of the carbohydrate's function may be premature. At the moment, the possibility that the L2/HNK-1 carbohydrate structure shares certain structural and functional properties with glycosaminoglycans would make it a likely candidate for binding to extracellular matrix molecules (Bronner-Fraser 1987, Künemund et al 1988). Furthermore, it is likely that the epitope also modulates protein–protein interactions, possibly by fine-tuning the extent and quality of affinities between adhesion molecules. Indeed, subsets of axons in the sciatic nerve express different carbohydrate epitopes that may correlate with distinct modalities conveyed by these axons (Martini et al 1988).

The adhesion molecule on glia (AMOG) and the L3 carbohydrate family

AMOG is a neural cell adhesion molecule that mediates neuron–astrocyte interaction *in vitro* (Antonicek et al 1987, Antonicek & Schachner 1988). It is expressed by astrocytes in the cerebellum at critical developmental stages of granule cell neuron migration. AMOG is not expressed on Bergmann glial cells before the onset of migration and disappears from these cells after migration has ceased. At the end of the migratory period, AMOG becomes detectable on astrocytic processes in the internal granular layer and remains there in adulthood. Granule neuron migration is strongly inhibited by monoclonal antibody raised against AMOG, probably by disturbing neuron–glia adhesion. Thus, AMOG is yet another cell adhesion molecule involved in granule cell migration.

AMOG does not express the L2/HNK-1 carbohydrate epitope, but expresses another carbohydrate structure recognized by the monoclonal antibody L3. The L3 carbohydrate is expressed by several adhesion molecules from mouse brain, including L1 and MAG, but is not detected in J1 and N-CAM (Kücherer et al 1987). The occurrence of this epitope thus makes AMOG the founding member of another family of cell adhesion molecules based on a carbohydrate structure which is also shared by some, but not all members of the L2 family. The number of members in the L3 family is not yet known, but the family appears to consist of at least nine glycoproteins in adult mouse brain.

Several other features of the L3 carbohydrate epitope are reminiscent of those of the L2/HNK-1 epitope. Like L2/HNK-1, the L3 carbohydrate domain appears to be involved in cell interactions, because L3 antibodies have been found to inhibit cell adhesion and cellular outgrowth patterns in explant cultures (D. Weber & M. Schachner, unpublished observations). Consistent with its functional importance is its phylogenetic conservation and occurrence in invertebrates (Dennis et al 1988). It is N-glycosidically linked and its expression is regulated independently of the protein backbone; not all molecules of each member of this family express the epitope. These findings are supported by the observation that only 10 to 20% of all MAG-positive oligodendrocytes express the L3 epitope. Similarly, in cultures of early postnatal mouse cerebellum, AMOG-positive astrocytes do not express the L3 epitope. Because the L3 carbohydrate epitope is expressed on both the L2-negative and L2-positive L1 molecules, a simple relationship between the L2 and L3 carbohydrate epitope-carrying variants of adhesion molecules appears unlikely.

These observations point to questions that need to be answered in the future. Is the next, functionally characterized member of the L3 family also an adhesion molecule? What is the structure and function of the L3 carbohydrate? How many families of adhesion molecules are there that are associated by a common, distinct carbohydrate structure? Is it possible that adhesion molecules are 'presenters' of functionally important carbohydrate structures with immense combinatorial possibilities? And, how do protein backbones and carbohydrates interact with each other? Elucidation of the functional and structural complexities in the rapidly expanding list of adhesion molecules will yield important insights into the morphogenetic mechanisms that underlie formation, maintenance and plasticity of the nervous system.

Acknowledgements

The author is grateful to Gemeinnützige Hertie-Stiftung, Hermann und Lilly Schilling-Stiftung, National Multiple Sclerosis Society, Bundesministerium für Forschung und Technologie and Deutsche Forschungsgemeinschaft for support. A similar review has appeared in *Neurology and Neurobiology* 1988 vol 48 (English) and in *Jahrbuch der Deutschen Naturforscher und Ärzte* 1989 (German).

References

Abo T, Balch CM 1981 A differentiation antigen of human NK and K cells identified by a monoclonal antibody (HNK-1). J Immunol 127:1024–1029

Aguayo AJ 1985 Axonal regeneration from injured neurons in the adult mammalian central nervous system. In: Cotman CW (ed) Synaptic plasticity. The Guilford Press, p 457–483

Antonicek H, Schachner M 1988 The adhesion molecule on glia (AMOG) incorporated into lipid vesicles binds to subpopulations of neurons. J Neurosci 8:2961–2966

Antonicek H, Persohn E, Schachner M 1987 Biochemical and functional characterization of a novel neuron-glia adhesion molecule that is involved in neuronal migration. J Cell Biol 104:1587–1595

Arquint M, Roder J, Chia LS et al 1987 Molecular cloning and primary structure of myelin-associated glycoprotein. Proc Natl Acad Sci USA 84:600–604

Bock E, Richter-Landsberg C, Faissner A, Schachner M 1985 Demonstration of immunochemical identity between the nerve growth factor-inducible large external (NILE) glycoprotein and the cell adhesion molecule L1. EMBO J 4:2765–2768

Bollensen E, Schachner M 1987 The peripheral myelin glycoprotein P_0 expresses the L2/HNK-1 and L3 carbohydrate structures shared by neural adhesion molecules. Neurosci Lett 82:77–82

Bollensen E, Steck A, Schachner M 1988 Reactivity with the peripheral myelin glycoprotein P_0 in sera from patients with monoclonal IgM gammopathy and polyneuropathy. Neurology 38:1266–1270

Bronner-Fraser M 1987 Perturbation of cranial neural crest migration by the HNK-1 antibody. Dev Biol 123:321–331

Chou DKH, Ilyas AA, Evans JE, Costello C, Quarles RH, Jungalwala FB 1986 Structure of sulfated glucuronyl glycolipids in the nervous system reacting with HNK-1 antibody and some IgM paraproteins in neuropathy. J Biol Chem 261:11717–11725

Cole GJ, Schachner M 1987 Localization of the L2 monoclonal antibody binding site on N-CAM and evidence for its role in N-CAM-mediated cell adhesion. Neurosci Lett 78:227–323

Dennis RD, Antonicek H, Yu RK, Wiegandt H, Schachner M 1988 Detection of the L2/HNK-1 carbohydrate epitope on glycoproteins and acidic glycolipids of the insect, Calliphora vicina. J Neurochem 51:1490–1496

Fahrig T, Landa C, Pesheva P, Kühn K, Schachner M 1987 Characterization of binding properties of the myelin-associated glycoprotein to extracellular matrix constituents. EMBO (Eur Mol Biol Organ) J 6:2875–2883

Faissner A 1987 Monoclonal antibody detects carbohydrate microheterogeneity on the murine cell adhesion molecule L1. Neurosci Lett 83:327–332

Faissner A, Kruse J, Goridis C, Bock E, Schachner M 1984 The neural cell adhesion molecule is distinct from the N-CAM related group of surface antigens BSP-2 and D2. EMBO (Eur Mol Biol Organ) J 3:733–737

ffrench-Constant C, Miller RH, Kruse J, Schachner M, Raff MC 1986 Molecular specialization of astrocyte processes at nodes of Ranvier in rat optic nerve. J Cell Biol 102:844–852

Friedlander DR, Grumet M, Edelman GM 1986 Nerve growth factor enhances expression of neuron-glia cell adhesion molecule in PC12 cells. J Cell Biol 102:413–419

Fushiki S, Schachner M 1986 Immunocytological localization of cell adhesion molecules L1 and N-CAM and the shared carbohydrate epitope L2 during development of the mouse neocortex. Dev Brain Res 289:153–167

Hoffman S, Edelman GM 1987 A proteoglycan with HNK-1 antigenic determinants is a neuron-associated ligand for cytotactin. Proc Natl Acad Sci USA 84:2523–2527

Holley JA, Silver J 1987 Growth pattern of pioneering chicks spinal cord axons. Dev Biol 123:375–388

Kadmon G, Kowitz A, Altevogt P, Schachner M 1989 The neural cell adhesion molecule N-CAM enhances L1-dependent cell–cell interactions. Submitted

Keilhauer G, Faissner A, Schachner M 1985 Differential inhibition of neurone–neurone, neurone–astrocyte and astrocyte–astrocyte adhesion by L1, L2 and N-CAM antibodies. Nature (Lond) 316:728–730

Kruse J, Mailhammer R, Wernecke H et al 1984 Neural cell adhesion molecules and myelin-associated glycoprotein share a common carbohydrate moiety recognized by monoclonal antibodies L2 and HNK-1. Nature (Lond) 311:153–155

Kruse J, Keilhauer G, Faissner A, Timpl R, Schachner M 1985 The J1 glycoprotein—a novel nervous system cell adhesion molecule of the L2/HNK-1 family. Nature (Lond) 316:146–148

Kücherer A, Faissner A, Schachner M 1987 The novel carbohydrate epitope is shared by some neural cell adhesion molecules. J Cell Biol 104:1597–1602

Künemund V, Jungalwala FB, Fischer G, Chou DKH, Keilhauer G, Schachner M 1988 The L2/HNK-1 carbohydrate of neural cell adhesion molecules is involved in cell interactions. J Cell Biol 106:213–223

Lagenaur C, Lemmon V 1987 An L1-like molecule, the 8D9 antigen, is a potent substrate for neurite extension. Proc Natl Acad Sci USA 84:7753–7757

Lee VM, Greene LA, Shelanski ML 1981 Identification of neural and adrenal medullary surface membrane glycoproteins recognized by antisera to cultured rat sympathetic neurons and PC12 pheochromocytoma cells. Neurosci 6:2773–2786

Lindner J, Rathjen FG, Schachner M 1983 L1 mono- and polyclonal antibodies modify cell migration in early postnatal mouse cerebellum. Nature (Lond) 305:427–430

Lindner J, Zinser G, Werz W, Goridis C, Bizzini B, Schachner M 1986 Experimental modification of postnatal cerebellar granule cell migration in vitro. Brain Res 377:298–304

Martini R, Schachner M 1986 Immunoelectron-microscopic localization of neural cell adhesion molecules (L1, N-CAM, MAG) and their shared carbohydrate epitope and myelin basic protein (MBP) in developing sciatic nerve. J Cell Biol 103: 2439–2448

Martini R, Schachner M 1988 Immunoelectron-microscopic localization of neural cell adhesion molecules (L1, N-CAM and myelin-associated glycoprotein) in regenerating adult mouse sciatic nerve. J Cell Biol 106:1735–1746

Martini R, Bollensen E, Schachner M 1988 Immunocytological localization of the major peripheral nervous system glycoprotein P_0 and its L2/HNK-1 and L3 carbohydrate structures in developing and adult mouse sciatic nerve. Devel Biol 129:330–338

Mbiene JP, Dechesne CJ, Schachner M, Sans A 1988 Immunocytological characterization of cell adhesion molecule L1 expression during early innervation of mouse otocysts. Cell Tissue Res 255:81–88

Persohn E, Schachner M 1987 Immunoelectron-microscopic localization of the neural cell adhesion molecules L1 and N-CAM during postnatal development of the mouse cerebellum. J Cell Biol 105:569–576

Pesheva P, Horwitz AF, Schachner M 1987 Integrin, the cell surface receptor for fibronectin and laminin, expresses the L2/HNK-1 and L3 carbohydrate structures shared by adhesion molecules. Neurosci Lett 83:303–306

Pollerberg E, Sadoul R, Goridis C, Schachner M 1985 Selective expression of the 180 kD component of the neural cell adhesion molecule N-CAM during development. J Cell Biol 101:1921–1929

Pollerberg E, Schachner M, Davoust J 1986 Differentiation-state dependent surface mobilities of two forms of the neural cell adhesion molecule. Nature (Lond) 324:462–465

Pollerberg E, Burridge K, Krebs K, Goodman S, Schachner M 1987 The 180 kD component of the neural cell adhesion molecule N-CAM is involved in cell–cell contacts and cytoskeleton–membrane interactions. Cell Tissue Res 250:227–236

Poltorak M, Steck AJ, Schachner M 1986 Reactivity with neural cell adhesion molecules of the L2/HNK-1 family in sera from patients with demyelinating diseases. Neurosci Lett 65:199–203

Poltorak M, Sadoul R, Keilhauer G, Landa C, Fahrig T, Schachner M 1987 The myelin-associated glycoprotein (MAG), a member of the L2/HNK-1 family of neural cell adhesion molecules, is involved in neuron–oligodendrocyte and oligodendrocyte–oligodendrocyte interaction. J Cell Biol 105:1893–1899

Rathjen FG, Wolff JM, Frank R, Bonhoeffer F, Rutishauser U, Schoeffski A 1987 Membrane glycoproteins involved in neurite fasciculation. J Cell Biol 104:343–353

Rohrer H 1985 Nonneuronal cells from chick sympathetic and dorsal root sensory ganglia express catecholamine uptake and receptors for nerve growth factor during development. Dev Biol 111:95–107

Rush RA 1984 Immunohistochemical localization of endogenous nerve growth factor. Nature (Lond) 312:364–367

Sanes JR, Schachner M, Covault J 1986 Distribution of several adhesive macromolecules in embryonic, adult, and denervated adult skeletal muscles. J Cell Biol 102:420–431

Seilheimer B, Schachner M 1987 Regulation of neural cell adhesion molecule expression on cultured mouse Schwann cells by nerve growth factor. EMBO J 6:1611–1616

Seilheimer B, Schachner M 1988 Studies of adhesion molecules mediating interactions between cells of peripheral nervous system indicate a major role for L1 in mediating sensory neuron growth on Schwann cells. J Cell Biol 107:341–351

Smith GM, Miller RH, Silver J 1986 Changing role of forebrain astrocytes during development, regenerative failure, and induced regeneration upon transplantation. J Comp Neurol 251:23–43

Thor G, Pollerberg E, Schachner M 1986 Molecular association of two neural cell adhesion molecules, L1 antigen and the 180 kD component of N-CAM, within the surface membrane of cultured neuroblastoma cells. Neurosci Lett 66:121–126

Thor G, Probstmeier R, Schachner M 1987 Characterization of the cell adhesion molecules L1, N-CAM and J1 in the mouse intestine. EMBO (Eur Mol Biol Organ) J 6:2581–2586

Williams AF 1987 A year in the life of the immunoglobulin superfamily. Immunol Today 8:298–303

DISCUSSION

Ruoslahti: Do brain glycoproteins that are not adhesion-related carry any of these epitopes? Is this brain-specific glycosylation or adhesion molecule-specific glycosylation?

Schachner: We have studied a number of identified and unidentified glycoproteins from brain. Na^+/K^+-ATPase, GABA receptor and acetylcholine receptor are L2/HNK-1-negative. There are several unidentified glycoproteins in brain that are void for the carbohydrate epitopes that I have described.

Ruoslahti: Does J1/tenascin from different tissues always have the L2/HNK-1 epitope?

Schachner: No; for example, some human fibroblast lines do not carry the L2/HNK-1 carbohydrate, but express J1/tenascin. J1 glycoproteins from brain express the L2/HNK-1 carbohydrate. However, as is the case for L1, N-CAM and myelin-associated glycoprotein, not all protein backbones of a particular adhesion molecule species may express the epitope.

Ruoslahti: So it is cell-type specific.

Schachner: This is difficult to say at the moment. All we can say is that if a molecule carries the L2/HNK-1 epitope we believe it to be an adhesion molecule inside or outside the nervous system. But there is differential expression of the carbohydrate independent of the protein backbone, during development, for example.

Barondes: Is anything known about the species distribution of these epitopes? If they are important for fundamental cellular or developmental processes they should be conserved over a wide range of species.

Schachner: They are phylogenetically conserved. For example, the L3 is in flies and leeches. The L4 and L5 haven't been studied carefully yet.

Ruoslahti: But if you take the same protein from different mammalian species, are they all L2-positive?

Schachner: That has not been done. There are some species differences; for example, the L2/HNK-1 epitope is detected on migrating neural crest cells in the chicken but not in the mouse. We don't know which molecules the epitope is on. But if the epitope is important in cell–cell interactions, why is there such species variability? The only answer is that there must be similar but, from an antigenic point of view, different epitopes.

Fukuda: I am intrigued by the observation that the L2 glycolipid and oligosaccharide both inhibit astrocyte–neuron cell interactions. How do you explain that?

Schachner: I don't see any conflict there. In one case, you have the oligosaccharide presented in a lipid micelle and, in the other, you have a monomeric oligosaccharide.

Fukuda: But glycolipids usually stick to other cells.

Schachner: You are saying that the glycolipids are multivalent in the micelle and could therefore cross-bridge. The same argument holds when you use antibodies in their bivalent state—IgGs rather than Fabs. If you add enough bivalent antibodies, you saturate the receptive molecule; there is a 1:1 relationship between the antibody and the structure that it recognizes. The same can be imagined for the glycolipids; you have fewer 'receptors' for the L2/HNK-1 epitope than L2 glycolipids in the micelle. The 'receptors' are covered so that there is no cross-bridging.

Gahmberg: Does N-CAM show a homotypic interaction?

Schachner: I don't think the evidence is tight enough yet.

Gahmberg: And what about L1?

Schachner: We have never been able to see a good self–self binding but Grumet and Edelman have shown some affinity. When we isolate the L1 molecules, they aggregate strongly (Sadoul et al 1988)—we can only separate them by SDS-PAGE—but it is audacious to call this homophilic binding, because we are dealing with membrane proteins out of their real environment.

Ruoslahti: But incoporation of the molecules into liposomes should abolish the non-specific interactions that arise from the hydrophobic domain.

Schachner: Yes, but you never know how much denaturing occurs during purification and how much that adds to non-specific binding that may even be specifically inhibited by antibody.

Ruoslahti: Do you have an alternative suggestion as to what they might be binding?

Schachner: No, we don't know the receptor molecules. There are receptors other than the known adhesion molecules.

Feizi: Are the other epitopes, as well as L2, susceptible to endo-β-galactosidase? Are they related to oligosaccharide epitopes that have already been characterized?

Schachner: The epitopes are removed by treatment with glycopeptidase or endoglycosidase F. We don't know what other monoclonal antibodies with defined carbohydrate reactivities recognize our adhesion molecules.

Jessell: From the work of Cole et al (1986) it seems that an immunoglobulin-based adhesion molecule similar to N-CAM has a heparin-binding domain. Is there any evidence that the sulphated glucuronic acid structure recognized by L2 interacts with that heparin-binding domain on N-CAM?

Schachner: The heparin-binding domain appears to be in a different domain than the L2 epitope.

Jessell: But perhaps adjacent molecules are interacting.

Schachner: You are wondering whether the L2 epitope goes into the heparin-binding site of N-CAM. We have studied competition between binding of heparin and L2 glycolipid to N-CAM. There is slight binding interference, but only at high concentrations of the L2 glycolipid.

Hughes: The inhibition by the L2 carbohydrate seemed to be much higher in the outgrowth assay than in the cell–cell adhesion assay. Is it possible that the sulphated oligosaccharide has more to do with cell–substratum adhesion and migration than cell–cell adhesion?

Schachner: The assay systems are not strictly comparable. Both use somewhat artificial conditions and what we measure within 30 min of adhesion between cells could be a kinetic, pseudo steady state that is not related to the situation in cells cultured for three days. But we are also thinking along those lines; the L2/HNK-1 carbohydrate epitope may have more to do with cell–substrate interactions, in fact with cell–laminin interactions, than with cell–cell interactions.

Hughes: Have you looked at a more biologically relevant substratum, like laminin, rather than polylysine?

Schachner: Using laminin, neurite outgrowth is nicely inhibited by the L2/HNK-1 carbohydrate. But using astrocytes as substrates we see no inhibition with the L2/HNK-1 carbohydrate.

Fukuda: If so many adhesion molecule share this epitope, are the cells confused by the common structure? How do you visualize specific interaction?

Schachner: That is a difficult question to answer. We like to consider the carbohydrate epitope in conjunction with the protein backbone or with other functionally important carbohydrate epitopes. It is interesting that in human chorionic gonadotropin there is cooperation between the protein backbone and the carbohydrate. Perhaps it is helpful to think along those lines.

Nagai: You suggested that sulphatide might be involved in cell adhesion. How do you explain this?

Schachner: Sulphatide and the L2 glycolipid have quite different structures, but have in common a sulphate in the 3′ position. This sulphate appears to be the important ingredient in the sulphatide- and the L2 glycolipid-mediated interaction, because galactocerebroside and the desulphated L2 glycolipid do not show inhibitory effects in our cell interaction assays (Künemund et al 1988).

Nagai: It is easy to synthesize galactocerebroside with sulphate attached at carbon 6 of the galactose. This positional isomer would provide a useful control. We have studied superoxide generation from polymorphonuclear leucocytes. Superoxide was generated when we used sulphatide, but not when we used the synthetic analogue sulphated at the C-6 position.

Schachner: We would love to have the sulphate at different positions.

Nagai: Cholesterol sulphate might also be useful as a control.

Jessell: Do cadherins have HNK-1/L2 carbohydrates? Jean Massoulie has evidence that acetylcholinesterase has HNK-1-like substitutions. Is there any evidence that they are involved in the adhesive properties of the globular form of the esterase?

Schachner: We have not been able to obtain cadherins to test. Acetylcholinesterase is present during development before acetylcholine is made and used as a synaptic transmitter. Paul Layer has speculated that the enzyme plays a role other than splitting acetylcholine. But acetylcholinesterase has not been tested for its adhesive functions. Another molecule that expresses the L2/HNK-1 carbohydrate epitope is ependymin, which has been implicated in the fish's ability to learn. Ependymin has been shown to be a good substrate for neurite outgrowth (Claudia Stürmer, personal communication).

References

Cole GJ, Loewy A, Cross NV, Akeson R, Glaser L 1986 Topographic localization of the heparin-binding domain of the neural cell adhesion molecule N-CAM. J Cell Biol 103:1739–1744

Künemund V, Jungalwala FB, Fischer G, Chou DKH, Keilhauer G, Schachner M 1988 The L2/HNK-1 carbohydrate of neural cell-adhesion molecules is involved in cell interactions. J Cell Biol 106:213–223

Sadoul K, Faissner A, Schachner M 1988 Biochemical characterization of different molecular forms of the neural cell adhesion molecule L1. J Neurochem 50:510–521

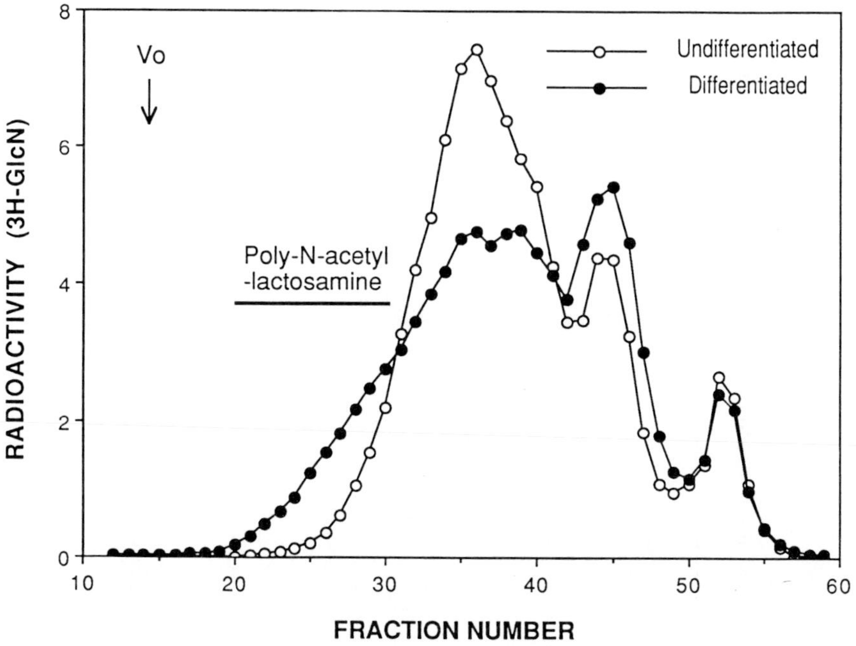

FIG. 2. Induction of poly-*N*-acetyllactosamine expression in differentiating mouse neuroblastoma cells. The N1E-115 neuroblastoma cells were grown in the undifferentiated form or induced to differentiate by serum starvation for ten days. The glycopeptides metabolically labelled with [^{3}H]glucosamine were prepared by exhaustive pronase digestion and fractionated by gel filtration on a column of Bio-Gel P-100. The poly-*N*-acetyllactosamine-containing glycopeptides were eluted in the fractions indicated under the bar. V_o, void volume (Spillmann & Finne 1987).

molecules to which they are bound. The carriers of several of these structures have been identified as molecules involved in cell adhesion (Table 1).

The only carrier of polysialic acid identified so far is the neural cell adhesion molecule N-CAM (see Finne et al 1987). That N-CAM is the major carrier of polysialic acid is indicated both by the amounts of polysialic acid isolated from whole brain and from purified N-CAM, and by immunoblotting data. In contrast, the poly-*N*-acetyllactosamine glycans were shown to occur on a selection of about six major molecular components in neural cell lines by a novel cell surface labelling method that we developed to specifically identify poly-*N*-acetyllactosamine-containing glycans (see Spillmann & Finne 1987). One carrier molecule is the nerve growth factor-inducible large external glycoprotein (NILE) that appears on rat pheochromocytoma cells after induced differentiation. This glycoprotein is immunologically similar to the L1 and the neuron-glia (Ng-CAM) cell adhesion molecules (see Spillman & Finne 1987).

TABLE 1 Properties of nervous tissue glycoprotein oligosaccharides: tissue specificity, developmental regulation and occurrence in adhesion molecules

Name	*Structure*	*Enrichment in neural cells*[a]	*Developmental regulation*[b]	*Carrier molecule*[c]	*Role in cell adhesion*	*References*
Polysialic acid	[NeuAc(α2–8)]$_n$	++	++	N-CAM	Modulator of adhesiveness	Finne (1985)
Poly-*N*-acetyl-lactosamine (linear form)	[Gal(β1–4)GlcNAc-(β1–3)]$_n$	+	++	NILE (L1/Ng-CAM), others		Spillmann & Finne (1987)
L2/HNK-1 antigen	SO_4^--3GlcUA(β1–3)-Gal(β1–4)GlcNAc(β1–	+		N-CAM, Ng-CAM, CSPG, others	Ligand in cell interactions?	Chou et al (1986)
O-Linked mannose	····Man-*O*-Ser(Thr)	++?	–	CSPG	CSPG binds laminin and tenascin	Finne et al (1979)
Sialyl X antigen	NeuAc(α2–3)Gal(β1–4)-[Fuc(α1–3)]-GlcNAc(β1–	+	–			Krusius & Finne (1978)
α-Galactosyl disaccharide	Gal(α1–3)GalNAc-(α1–*O*)-Ser(Thr)	++	–			Finne & Krusius (1976)

[a] ++, enrichment in nervous tissue with little, if any, so far revealed in other tissues; +, enrichment in nervous tissues but present also in some other tissues.

[b] Only structures known to undergo major changes during development and differentiation are indicated.

[c] The abbreviations are: N-CAM, neural cell adhesion molecule; Ng-CAM, neuron-glia cell adhesion molecule; NILE, nerve growth factor-inducible large external glycoprotein; CSPG, brain chondroitin sulphate proteoglycan.

The uronic acid- and sulphate-containing L2/HNK-1 antigen also occurs on a limited selection of glycoproteins (N-CAM, Ng-CAM, myelin-associated glycoprotein, brain chondroitin sulphate proteoglycan and others) that have all been proposed to be adhesion molecules (Künemund et al 1988). Still another new carbohydrate antigen (L3) of unknown structure also occurs on a selection of molecules and appears to be involved in cell adhesion (Kücherer et al 1987).

The brain chondroitin sulphate proteoglycan is a carrier of the O-linked mannose glycans (Finne et al 1979). It is so far not known whether other molecules also contain these glycans. The proteoglycan has been implicated in mediating contacts between neural and glial elements in nervous tissue by its ability to specifically bind the adhesion molecules laminin and cytotactin/tenascin (Hoffman & Edelman 1987). There is, as yet, no clear indication of a specific role for the carbohydrates of the proteoglycan in these interactions, but it is of interest that the proteoglycan contains the L2/HNK-1 determinant.

Modes of molecular interactions

At the molecular level, there are at least two basically different mechanisms by which the carbohydrates may participate in interactions with other molecules. The classical theory assumes that carbohydrate structures may function as specific receptors for the binding of other molecules, usually proteins. An example of this category is the recognition of the terminal saccharides of glycoproteins by the hepatic carbohydrate-specific receptors (see Ashwell & Harford 1982). The L2 carbohydrate determinant may be an additional example of this category, possibly by acting as a ligand in neuron–astrocyte and astrocyte–astrocyte interactions. This is indicated by the specific inhibition of these interactions by antibodies to L2 and by the L2 carbohydrate determinant itself (Künemund et al 1988).

The second way the protein-bound carbohydrates may influence molecular interactions is by acting as regulators of interactions specified by other molecules. The polysialic acid moieties of N-CAM are thought to function as negative modulators of N-CAM to N-CAM adhesion that appears to be based on the binding properties of the peptide portion of the molecule. This suggestion is based on the lesser adhesiveness of the embryonal, polysialylated, form of the molecule as compared to the adult form, and the increased adhesiveness of the embryonal form after sialic acid removal (Hoffman & Edelman 1983, Sadoul et al 1983, Rutishauser et al 1985). On the other hand, the polysialic acid moieties could also decrease the interactions of the cells by forming a negatively charged physical barrier on the cell surface (Acheson & Rutishauser 1988). The role of the second class of large units, poly-*N*-acetyllactosamine glycans, in neural cells is not known, but it is of interest that these glycans have been shown to act as negative modulators of the binding of fibronectin to gelatin (Zhu & Laine 1985).

Other biological roles

The biological interactions of the carbohydrates do not only involve molecules and cells intrinsic to the organism but also interactions with the outside world. One important question concerns the role of carbohydrates in disease processes such as bacterial sepsis and meningitis. Carbohydrates on both the host and the bacterial cells, as well as molecules involved in their recognition and binding, take part in these processes.

An important aspect of this is the establishment of the tissue specificity of the infections, a process that may involve cellular recognition processes analogous to those occurring during normal development. Our studies on *Escherichia coli* strains causing meningitis in newborn infants have revealed a new adhesion specificity that is based on the recognition of sialic acid-containing structures on the cell surface by a bacterial adhesin (Parkkinen et al 1986). On erythrocytes the receptors appear to be O-linked oligosaccharides of glycophorin, similar to those occurring abundantly, for example in the brain, but the structure of the receptors in meningeal tissue is not yet known. Sialic acid recognition, and also novel galactose-related recognition activities, occur in strains of *streptococci* capable of producing meningitis in piglets (Kurl et al 1989). These examples may be added to the list of molecular interactions involving recognition of specific carbohydrate structures.

Other aspects of the interplay between the host and bacterial carbohydrate and their binding proteins involve the molecular mimicry of polysialic acid by the capsular polysaccharides surrounding pathogenic bacteria. The capsular polysaccharides of group B *meningococci* and *E. coli* K1, both important agents causing meningitis in humans, and *Pasteurella haemolytica* serotype A2, an important pathogen causing septicaemia in sheep, have exactly the same poly(α2–8)-*N*-acetylneuraminyl structure as the polysialic acid of nervous tissue (see Finne 1985, 1988). The pathogenicity of these bacteria could in part be explained by their escape from immunological defence mechanisms. On the other hand, the explanation of the tissue specificity of the meningitis-causing bacteria is so far unknown, and possible interactions with, for example, polysialic acid-binding molecules have also to be considered.

A further consideration evolving from the study of the polysialic acid-expressing bacteria concerns the specific conformation of polysialic acid in its molecular interactions. Studies on the substrate specificity of endosialidases from bacteriophages cleaving polysialic acid have indicated that a long segment of the polysaccharide, consisting of eight sialic acid residues, is the minimum substrate size for efficient cleavage (Finne & Mäkelä 1985). Similarly, a minimum chain size of eight to ten sialyl residues is required for the binding of the monoclonal group B meningococcal antibody or its Fab fragment (Finne et al 1987). These specificities may be explained by the conformation of the polysialic acid chain that is required for its molecular interactions. Whether such a conformation is also a prerequisite for the interactions of polysialic acid in nervous tissue remains to be seen.

Finally, the structural similarity of the bacterial polysaccharides with tissue polysialic acid raises a question of the safety of proposed vaccines based on polysialic acid, or immune therapy employing anti-polysialic acid antibodies (Finne et al 1983). Due to the structural resemblance, the possibility of triggering an autoimmune reaction may have to be considered. This possibility is also supported by the involvement of other carbohydrate determinants, such as the L2/HNK-1 antigen as well as some rare gangliosides, as antigens in autoimmune conditions associated with neuropathy (Ilyas et al 1988).

Conclusions

Despite many similarities to glycoproteins of other tissues, it now seems evident that nervous tissue glycoproteins contain a number of structural features that are characteristic for this tissue. Many of the structures were in fact found for the first time in nervous tissue, and although some structures have later been found in some other tissues as well, nervous tissue remains where they are encountered in highest concentration. It is conceivable that these structures contribute to the functional properties of nervous tissue. In addition, the developmental regulation of the two large glycan types suggests a specific role for them in the cellular differentiation processes.

It is striking that many of the novel structures found in nervous tissue glycoproteins have turned out to be carried by molecules involved in cell adhesion. This suggests that some of the predictions on the role of carbohydrates in cell interactions of the nervous tissue (see Hughes 1976) may turn out to be correct. However, at the molecular level the structures may not only serve as mediators of interactions by serving as specific recognition ligands as previously thought, but may also serve as modulators of molecular interactions by fine-tuning the strength of the interactions.

The participation of carbohydrates in various ways in the pathogenic processes of microbial infections and autoimmunity provides additional examples of the types of biological interactions in which these complex carbohydrate structures may be involved. Thus a detailed knowledge of the structural properties of the carbohydrates is needed to enhance understanding of the roles of these molecules.

Acknowledgements

This work has been supported by grants from the Sigrid Jusélius Foundation, the Swiss Nationalfonds and the Academy of Finland.

References

Acheson A, Rutishauser U 1988 Neural cell adhesion molecule regulates cell contact-mediated changes in choline acetyltransferase activity of embryonal chick sympathetic neurons. J Cell Biol 106:479–486

Ashwell G, Harford J 1982 Carbohydrate-specific receptors of the liver. Annu Rev Biochem 51:531–554

Chou DKH, Ilyas AA, Evans JE, Costello C, Quarles RH, Jungalwala FB 1986 Structure of sulfated glucuronyl glycolipids in the nervous system reacting with HNK-1 antibody and some IgM paraproteins in neuropathy. J Biol Chem 261:11717–11725

Edelman GM 1985 Cell adhesion and the molecular processes of morphogenesis. Annu Rev Biochem 54:135–169

Finne J 1982 Occurrence of unique polysialosyl carbohydrate units in glycoproteins of developing brain. J Biol Chem 257:11966–11970

Finne J 1985 Polysialic acid—a glycoprotein carbohydrate involved in neural adhesion and bacterial meningitis. Trends Biochem Sci 10:129–132

Finne J 1988 The carbohydrate units of nervous tissue glycoproteins: structural properties and role in cell interactions. In: Edelman G et al (eds) Morphogenic molecules. The Neurosciences Institute, New York, in press

Finne J, Krusius T 1976 *O*-Glycosidic carbohydrate units from glycoproteins of different tissues: demonstration of a brain-specific disaccharide, α-galactosyl-(1→3)-*N*-acetylgalactosamine. FEBS (Fed Eur Biochem Soc) Lett 66:94–97

Finne J, Krusius T 1982 Preparation and fractionation of glycopeptides. Methods Enzymol 83:269–277

Finne J, Mäkelä PH 1985 Cleavage of the polysialosyl units of brain glycoproteins by a bacteriophage endosialidase. Involvement of a long oligosaccharide segment in molecular interactions of polysialic acid. J Biol Chem 260:1265–1270

Finne J, Krusius T, Margolis RK, Margolis RU 1979 Novel mannitol-containing oligosaccharides obtained by alkaline borohydride treatment of a chondroitin sulfate proteoglycan from brain. J Biol Chem 254:10295–10300

Finne J, Leinonen M, Mäkelä PH 1983 Antigenic similarities between brain components and bacteria causing meningitis. Implications for vaccine development and pathogenesis. Lancet II:355–357

Finne J, Bitter-Suermann D, Goridis C, Finne U 1987 An IgG monoclonal antibody to group B meningococci cross-reacts with developmentally regulated polysialic acid units of glycoproteins in neural and extraneural tissues. J Immunol 138:4402–4407

Hoffman S, Edelman GM 1983 Kinetics of homophilic binding by embryonic and adult forms of the neural cell adhesion molecule. Proc Natl Acad Sci USA 80:5762–5766

Hoffman S, Edelman GM 1987 A proteoglycan with HNK-1 antigenic determinants is a neuron-associated ligand for cytotactin. Proc Natl Acad Sci USA 84:2523–2527

Hughes RC 1976 Membrane glycoproteins. Butterworths, London

Ilyas AA, Li SC, Chou DKH, Li YT, Jungalwala FB, Dalakas MC, Quarles RH 1988 Gangliosides G_{M2}, $IV^4GalNAcG_{M1b}$, and $IV^4GalNAcG_{D1a}$ as antigens for monoclonal immunoglobulin M in neuropathy associated with gammopathy. J Biol Chem 263:4369–4373

Kiang WL, Krusius T, Finne J, Margolis RU, Margolis RK 1982 Glycoproteins and proteoglycans of the chromaffin granule matrix. J Biol Chem 257:1651–1659

Krusius T, Finne J 1978 Characterization of a novel sugar sequence from rat-brain glycoproteins containing fucose and sialic acid. Eur J Biochem 84:395–403

Krusius T, Finne J, Kärkkäinen J, Järnefelt J 1974 Neutral and acidic glycopeptides in adult and developing rat brain. Biochim Biophys Acta 365:80–92

Krusius T, Finne J, Margolis RK, Margolis RU 1986 Identification of an *O*-glycosidic mannose-linked sialylated tetrasaccharide and keratan sulfate oligosaccharides in the chondroitin sulfate proteoglycan of brain. J Biol Chem 261:8237–8242

Kücherer A, Faissner A, Schachner M 1987 The novel carbohydrate epitope L3 is shared by some neural cell adhesion molecules. J Cell Biol 104:1597–1602

Künemund V, Jungalwâla FB, Fischer G, Chou DKH, Keilhauer G, Schachner M 1988 The L2/HNK-1 carbohydrate of neural cell adhesion molecules is involved in cell interactions. J Cell Biol 106:213–233

Kurl DN, Haataja S, Finne J 1989 Hemagglutination activities of Group B, C, D and G streptococci: demonstration of novel sugar-specific cell-binding activities in *Streptococcus suis*. Infect Immun 57:384–389

Leppänen A, Korvuo A, Puro K, Renkonen O 1986 Glycoproteins of human teratocarcinoma cells (PA1) carry both anomers of *O*-glycosyl-linked D-galactopyranosyl-(1→3)-2-acetamido-2-deoxy-α-D-galactopyranosyl group. Carbohydr Res 153:87–95

Parkkinen J, Rogers GN, Korhonen T, Dahr W, Finne J 1986 Identification of the O-linked sialyloligosaccharides of glycophorin A as the erythrocyte receptors for S-fimbriated *Escherichia coli*. Infect Immun 54:37–42

Roth J, Taatjes D, Bitter-Suermann D, Finne J 1987 Polysialic acid units are spatially and temporally expressed in developing postnatal rat kidney. Proc Natl Acad Sci USA 84:1969–1973

Rutishauser U, Watanabe M, Silver J, Troy FA, Vimr ER 1985 Specific alteration of NCAM-mediated cell adhesion by an endoneuraminidase. J Cell Biol 101:1842–1849

Sadoul R, Hirn M, Deagostini-Bazin H, Rougon G, Goridis C 1983 Adult and embryonic mouse neural cell adhesion molecules have different binding properties. Nature (Lond) 304:347–349

Spillmann D, Finne J 1987 Poly-N-acetyllactosamine glycans of cellular glycoproteins: predominance of linear chains in mouse neuroblastoma and rat pheochromocytoma cells. J Neurochem 49:874–883

Zhu BLR, Laine RL 1985 Polylactosamine glycosylation on human placental fibronectin weakens binding affinity of fibronectin to gelatin. J Biol Chem 260:4041–4045

DISCUSSION

Ruoslahti: A transgenic mouse that expresses in its brain the phage enzyme that cleaves polysialic acid would be unable to add polysialic acid to N-CAM and other molecules. Have you started to produce one?

Finne: No, not yet. I was primarily interested to see whether there is an endogenous endosialidase in the brain. That would make sense in terms of the function of the polysialic acid moiety on N-CAM. The expression of polysialic acid might be selectively regulated on certain parts of the cell.

Schachner: Are the poly-*N*-acetyllactosamine residues present on L1 at different developmental stages? Were the pheochromocytoma cells in a morphologically differentiated or undifferentiated state?

Finne: The poly-*N*-acetyllactosamine glycans were not present on L1 in the undifferentiated state.

Schachner: And during normal brain development?

Finne: I haven't looked at that. In brain there is much less poly-*N*-acetyllactosamine than on cultured cells. Ganglia cells have poly-*N*-acetyllactosamine glycans but brain tissue has little, at least in the adult. Its expression could be related to the developmental stage.

Jessell: Given that the N-CAM is the predominant glycoprotein that has this long (α2–8)-sialylated structure, is there anything strange about the core or the backbone structure? How do you account for the specificity by which the polymerization occurs on N-CAM and not on any other brain glycoprotein?

Finne: McCoy & Troy (1987) have shown that fetuin and colominic acid do not function as acceptors for this type of addition, whereas N-CAM does. Therefore, something in the polypeptide appears to determine polysialylation.

Jessell: But nobody knows at a structural level. N-CAM has a number of potential N-link glycosylation sites. Is this (α2–8)-sialyl polymer a long chain on one of the potential sites or do all five potential glycosylation sites have these polysialyl chains?

Finne: That has been studied by Edelman's (1985) group. I understand that they think there are several polysialylated glycans, but the evidence is not entirely clear.

Stanley: What advantages does your method of cell surface labelling have compared with the galactose oxidase procedure, which should achieve the same result?

Finne: We don't get the same result because galactose oxidase labels all galactose-terminated structures. Our method just labels poly-*N*-acetyllactosamine. We also have very little background—with galactose oxidase there is a noticeable amount of background. We use commercial reagents, and the procedure is simple.

Feizi: Is polysialic acid entirely confined to the nervous system, as determined by histochemical experiments?

Finne: There is a small amount in the kidney (Finne et al 1987), where expression is also developmentally regulated.

Feizi: How did you determine that the minimum size for the polysaccharide to be antigenic is an octamer?

Finne: We measured direct binding of labelled oligomers to the antibodies and we only started to see binding from the octamer upwards. The labelling increases up to oligomers having about 14 sialic acid units. The highest relative increase is around 10 residues.

Feizi: So this is direct binding of the oligosaccharide to whole immunoglobulin. You are therefore looking at a high affinity reaction. With monovalent oligosaccharides one frequently cannot demonstrate binding. Is your oligosaccharide bi- or multivalent with respect to the epitopes?

Finne: You are asking whether the requirement for a long segment is due to multivalent binding. We investigated that by using Fab fragments, which are monomers and therefore exclude multivalency. The Fab binding requirement is the same as for the whole antibody (Häyrinen et al 1989).

Fukuda: The eight-unit polymer of sialic acid probably has some helical structure, and may have some super-structure. Has anybody made a nuclear magnetic resonance (NMR) spectroscopy study of this structure?

Finne: Michon et al (1987) have done NMR analysis of sialic acid oligomers. They found that the two terminal residues are differentially located at the ends of the polymer. In the middle portion, all the negative charges are on one side of the molecule and the N-acetyl groups are on the other side. The six residues in the middle might form the binding site for the antibody. That would correspond to the dimensions of a classical binding site. However, I do not understand why binding continues to increase when the polymer has 10 residues.

Sharon: Perhaps this is similar to peptides where a helix does not form below a certain size.

Carver: It's going to be an equilibrium, surely, between a helical form and a partially unwound form. As you go to larger oligomers that equilibrium will favour the helical form.

Sharon: How do your antibodies relate to those studied by Kabat et al (1986), which are specific for polysialic acid but also cross-react with DNA?

Finne: They are totally different. The human myeloma antibodies described by Kabat et al apparently recognize a certain pattern of charges—they are specific not only for polysialic acid but also for denatured DNA. Our antibodies do not react with denatured DNA. The different antibodies bind to different sides of the polymer.

Sharon: Was the combining site in the Kabat antibodies the same size?

Finne: It is in the same range (Kabat et al 1988).

Lindahl: What is the reason for the selective occurrence of O-mannosyl-linked oligosaccharides on the chondroitin sulphate proteoglycan in brain? Are the recognition sequences for mannosyl transfer present only in brain, or is the relevant enzyme expressed only in brain and not elsewhere?

Finne: We are not sure. We don't know whether other chondroitin sulphate proteoglycans could also have mannosyl oligosaccharides, because we have only done preliminary experiments to look for these structures in other tissues. They have not been reported elsewhere.

Hart: Does the presence of the O-linked mannose correlate with the presence of proteoglycan in the cytoplasm? Drs R. U. and R. K. Margolis and colleagues (e.g. Aquino et al 1984) have data suggesting that in adult brain this proteoglycan is cytoplasmic. Perhaps there is a mannosyltransferase in cytoplasm. Is there O-linked mannose in embryonic brain?

Finne: We have not studied that because we have only studied the whole proteoglycan sample that we obtained from the Drs Margolis. We don't know whether it is intracellular or extracellular. It is curious that in developing brain this proteoglycan is first extracellular and later cytoplasmic.

Ruoslahti: In granules or free in the cytoplasm?

Finne: In the cytoplasm.

Ruoslahti: I wonder how it gets there.

Hart: I seem to recall that in bacteria the (α2–8)-sialyl polymer is made via lipid intermediates. Are lipid intermediates involved in the mammalian biosynthesis of this (α2–8)-linked polymer?

Finne: That's an attractive hypothesis but, as far as I know, nobody has shown that. There have been reports on the biosynthesis of the (α2–8) polymer in mammalian cells. In pulse-chase experiments the polysialic acid appears very suddenly. It has been suggested that this may be due to an *en bloc* addition of the polysialic acid moiety, but I don't think one can differentiate between *en bloc* transfer and the addition of single residues in such an experiment. It might be that rapid additions could occur sequentially, once the enzyme–substrate complex has formed.

Hughes: That there are mannose residues linked to Ser and Thr residues in yeast glycoproteins is well documented by Widemar Tanner and his colleagues in Regensburg. The donor is a dolichol mannose. Have you tried inhibitors of dolichol synthesis in any of your cells to see if they block synthesis of your mammalian glycoproteins containing that linkage?

Finne: That is a good idea, but we are not working on the biosynthesis—Drs R. U and R. K. Margolis are working on that.

Rademacher: Do you think the 3′-sulphated glucuronic acid is on the O-linked mannose, or is it a contaminant that adheres to the proteoglycan? Would you have seen it in your analysis?

Finne: Glucuronic acid is not detected by the classical sugar analysis with methanolysis. There are improved methods now, but we have not used them. In principle, N-linked glycans, O-linked *N*-acetylgalactosamine, and O-linked mannose-containing glycans are all possible.

Rademacher: Glycopeptidase F removes the L2/HNK-1 epitope. By inference, the determinant is on the N-linked glycans. Do you have any evidence for that?

Finne: No, we have not investigated that.

Feizi: The mention of the O-linked mannose prompts one to ask whether the granules could represent a low-grade yeast infection in the rats. Are the same granules and the O-linked chains detected in rats reared in other laboratories?

Finne: If I take total fresh rat brain, I can detect the mannose-linked disaccharide by gas–liquid chromatography. It is also present in rabbit brain. The oligosaccharides are also present in the material purified by Dr Margolis using a different approach.

Ruoslahti: Are there any examples of an infectious agent bringing in an enzyme that assembles sugars?

Feizi: Yes: bacteria and yeasts. An unusual galactan reported among the glycoproteins in human gastric juice (Clamp et al 1983) now appears to be a polysaccharide which is of bacterial origin and has the composition galactose:arabinose:glucuronic acid:rhamnose:glucose, 5:2:1:1:0.5 (J. R. Clamp, unpublished observations).

Ruoslahti: Are any of the mitochondrial enzymes glycosyltransferases?

Sharon: Most bacteria do not produce sialic acid or polysialic acid. Roland Schauer (1982) thinks that the few bacteria that synthesize sialic acids (e.g. *E. coli*) may have picked up the genes for the enzymes required to synthesize these compounds from the intestinal flora. The similarity in biosynthetic pathways leading to sialic acid in bacteria and in higher animals supports this idea. He also presumes that such a gene transfer is responsible for the acquisition of sialidase and *N*-acetylneuraminate lyase by some bacteria.

Nagai: The monoclonal antibody M2590 to the ganglioside GM_3 was originally produced using mouse melanoma cells, but it reacts with cell surface GM_3 of human melanoma cells. People have questioned whether this antibody is truly directed to GM_3 molecules in the transformed cells. It cross-reacts not only with GM_3 sialic acid but also with lactones formed from the sialic acid. Moreover, this antibody reacts only with melanoma cells, and not with untransformed cells bearing GM_3. Dr S. Hakomori discussed this problem and drew attention to the importance of the density of GM_3 molecules at the cell surface. I wonder whether the residues in your polysialic acid may form lactones.

Finne: Two sialic acid residues may readily react with each other and make a lactone. Thus the molecule loses charge and has a new structure. I can only answer in a negative way. The formation of 10% lactone in polysialic acid abolishes the immune reactivity. This fits with the requirement of about 10 sialic acid residues for binding. It is interesting that monosialic acid does not readily make lactones, whereas polysialic acid makes lactones spontaneously at low pH. I wonder if that has something to do with the physiological role of polysialic acid.

Kobata: In 1965, mannosyl-serine and mannosyl-threonine were found in human urine (F. Tominaga, unpublished). Similar amounts of mannosyl-serine and xylosyl-serine were present (Tominaga et al 1968). Your material might be turning over quite quickly because of the excretion of the linkage region in large amounts. Xylosyl-serine was later found to be the linkage region of proteoglycans. Perhaps your results explain the origin of urinary mannosyl-serine and threonine.

Finne: That has to be examined in detail.

References

Aquino DA, Margolis RU, Margolis RK 1984 Immunocytochemical localization of a chondroitin sulfate proteoglycan in nervous tissue. II. Studies in developing brain. J Cell Biol 99:1130–1139

Clamp JR, Cooper B, Creeth JM, Ene D, Barrett J, Gough M 1983 The presence of polysaccharide in normal human gastric mucus. Biochem J 215:421–423

Edelman GM 1985 Cell adhesion and the molecular processes of morphogenesis. Annu Rev Biochem 54:135–169

Finne J, Bitter-Suermann D, Goridis C, Finne U 1987 An IgG monoclonal antibody to group B meningococci crossreacts with developmentally regulated polysialic acid units of glycoproteins in neural and extraneural tissues. J Immunol 138:4402–4407

Häyrinen J, Bitter-Suermann D, Finne J 1989 Interaction of meningococcal group B monoclonal antibody and its Fab fragment with α2–8 linked sialic acid polymers: requirement of a long oligosaccharide segment for binding. Mol Immunol, in press

Kabat EA, Nickerson KG, Liao J et al 1986 A human monoclonal macroglobulin with specificity for α(2–8)-linked poly-*N*-acetyl neuraminic acid, the capsular polysaccharide of group B meningococci and *Escherichia coli* K1, which crossreacts with polynucleotides and with denatured DNA. J Exp Med 164:642–654

Kabat EA, Liao J, Osserman EF, Gamian A, Michon F, Jennings HJ 1988 The epitope associated with the binding of the capsular polysaccharide of the group B meningococci and of *Escherichia coli* K1 to a human monoclonal macroglobulin, IgM^{NOV}. J Exp Med 168:699–711

McCoy RD, Troy FA 1987 CMF-NeuNAc:poly-α-2,8-sialosyl sialyltransferase in neural cell membranes. Methods Enzymol 138:627–637

Michon F, Brisson J-R, Jennings HJ 1987 Conformational differences between linear α(2–8)-linked homosialooligosaccharides and the epitope of the group B meningococcal polysaccharide. Biochemistry 26:8399–8405

Schauer R 1982 Introduction. In: Schauer R (ed) Sialic acids, chemistry, metabolism and function. Springer-Verlag, New York, p 3

Tominaga F, Oka K, Yoshida H 1968 The isolation and identification of O-xylosyl-serine and S-methionylcysteine sulfoxide from human urine. J Biochem (Tokyo) 57:717–720

Carbohydrate recognition in neuronal development: structure and expression of surface oligosaccharides and β-galactoside-binding lectins

M. A. Hynes‡, L. B. Buck‡, M. Gitt*, S. Barondes*, J. Dodd and T. M. Jessell‡

Center for Neurobiology and Behavior and ‡*Howard Hughes Medical Institute, Columbia University, College of Physicians and Surgeons, New York, NY 10032, USA* and **Department of Psychiatry, University of California, San Francisco, CA 94142, USA*

Abstract. The differentiation and development of vertebrate neurons is controlled in part by interactions with cell surface and extracellular matrix molecules, many of which are glycoproteins that mediate their developmental actions by homophilic or heterophilic binding to other glycoproteins. In addition there is increasing evidence that cell recognition and adhesion in some embryonic cell types involve interactions between cell surface oligosaccharides and complementary carbohydrate-binding proteins. Although a role for carbohydrate recognition in neuronal development has been proposed, the precise function of complex carbohydrate structures on neural cells has not been defined. To approach this problem, we have examined the structure and expression of cell surface oligosaccharides and carbohydrate-binding proteins by primary sensory neurons in the rat dorsal root ganglion (DRG). There are several functionally distinct subsets of DRG neurons, each of which conveys a different sensory modality to distinct target domains in the spinal cord. Monoclonal antibodies against defined oligosaccharide structures identify each of the major subsets of DRG neurons on the basis of their expression of a distinct set of complex oligosaccharides, derived from lacto-, globo- and ganglioseries backbone structures. In particular, small diameter DRG neurons involved in pain processing express β-galactoside-based lactoseries oligosaccharides. DRG and spinal cord neurons also express two soluble β-galactoside-binding proteins of relative molecular masses 14 500 and 29 000, termed RL-14.5 and RL-29, which represent potential ligands for lactoseries oligosaccharides. RL-14.5 is expressed by the majority of DRG neurons whereas RL-29 is restricted to the subset of small DRG neurons that express surface *N*-acetyllactosamine structures. RL-14.5 and RL-29 are expressed soon after the differentiation of DRG neurons and appear to be released from cultured DRG neurons. Rat brain cDNA clones encoding RL-14.5 have been isolated. The nucleotide and predicted amino acid sequence of RL-14.5 has confirmed that this lectin is highly homologous to soluble β-galactoside-binding proteins in other vertebrate species. Northern blot analysis and *in situ* hybridization indicate that

RL-14.5 mRNA is selectively expressed in sensory and motor neurons in the rat nervous system. The selective expression of lactoseries oligosaccharides and complementary β-galactoside-binding lectins may contribute to the differentiation and/or development of these two classes of neurons.

1989 Carbohydrate recognition in cellular function. Wiley, Chichester (Ciba Foundation Symposium 145) p 189–218

The possibility that cellular interactions within the developing vertebrate nervous system are mediated by cell surface carbohydrates has received continued attention over the past 20 years. The first suggestion for an involvement of carbohydrate recognition in neural development derived from the characterization of complex oligosaccharides, in particular gangliosides, on neural cell surfaces and from the detection of cell surface glycosyltransferases that were proposed to function in a non-enzymic manner to bind surface oligosaccharides (Roseman 1970, Roth et al 1971). Progress in elucidating the functional role of cell surface carbohydrates has, however, been slow, largely for three reasons. First, there are technical difficulties in purifying and characterizing complex oligosaccharides from small numbers of neural cells and even greater problems in generating these structures synthetically. Second, the idea that glycosyltransferases are present on the cell surface and contribute to cell recognition has not been generally accepted. Third, the characterization of cell surface molecules such as N-CAM, N-cadherin and integrins (Edelman 1986, Takeichi 1988, Ruoslahti & Pierschbacher 1987) that mediate neural cell adhesion via protein–protein interactions has, to some extent, diverted attention from the potential role of carbohydrate recognition in neuronal development.

In contrast to the situation in the nervous system, there has been accumulating evidence that complex oligosaccharides on a variety of mammalian cells do function in cell adhesion and recognition. Hepatic asialoglycoprotein receptors have been shown to participate in the clearance of circulating serum glycoproteins (Ashwell & Harford 1982) and remain the best characterized surface proteins with a defined physiological role in carbohydrate recognition (see Drickamer, this volume and Lee, this volume). In addition, the stage- and species-specific binding of sperm to the zona pellucida coat that surrounds mammalian oocytes has been shown to result from the interaction of sperm with an O-linked oligosaccharide that is attached to the zona pellucida glycoprotein ZP3 (see Wassarman, this volume). There is also evidence that the poly-*N*-acetyllactosamine oligosaccharides present on the blastomeres of preimplantation mouse embryos are involved in adhesive interactions that occur during compaction (Fenderson et al 1984, Rastan et al 1985). Other studies, in particular on the specificity of homing of recirculating lymphocytes to peripheral lymphoid targets (Rosen & Yednock 1986, Rosen et al 1985, Brandley et al 1987) have begun to establish the contribution of carbohydrate recognition to cell interactions in the adult animal.

The structural characterization of neural oligosaccharides and glycoproteins has revealed that many complex oligosaccharides, lectins and glycosyltransferases that appear to mediate recognition and adhesion of non-neural cells are also expressed in the developing nervous system (see Hynes et al 1989). These molecules may therefore serve similar adhesive and recognition functions on developing neural cells. The most detailed information on the potential contribution of carbohydrate recognition in neural systems has derived from the analysis of the development of the retina (Barbera et al 1973, Roth et al 1971, Marchase 1977, Blackburn et al 1986, Balsamo et al 1986), the neuromuscular junction (Scott et al 1988) and of primary sensory neurons (Dodd & Jessell 1986). In this article, we focus on the structure and expression of cell surface oligosaccharides and complementary carbohydrate-binding proteins by primary sensory neurons and discuss the possible roles of carbohydrate recognition in the differentiation and development of this set of neurons.

Development of dorsal root ganglion neurons

Primary sensory neurons in the dorsal root ganglion (DRG) transmit peripheral sensory information from cutaneous and muscle receptors to second-order neurons in the spinal cord. Analysis of the receptive properties and morphology of the peripheral endings of sensory neurons has indicated that there are more than a dozen functional classes of DRG neurons (Perl 1983). It is clear that these neurons differentiate from precursor cells that migrate from the neural crest (Le Douarin 1982), although little is known about the cellular or molecular mechanisms that influence the differentiation of subsets of DRG neurons. Soon after sensory neuroblast differentiation, neurons extend axons both peripherally and centrally (Altman & Bayer 1984). The central terminal arbor of each functional class of DRG neurons is located in a distinct domain in the spinal cord (Brown 1981). The anatomical segregation of the central projections of different classes of DRG neurons occurs during the period that afferent fibres first enter the embryonic spinal cord. Because different classes of afferents enter the spinal cord over the same period of development, it is unlikely that the specificity in sensory axonal projections results simply from temporal constraints on axonal outgrowth. The existence of specific targeting mechanisms for sensory axons is further supported by studies on the development of connections between sensory and motor neurons in the amphibian spinal cord and by observations that sensory afferents that enter the mammalian spinal cord ectopically still terminate in appropriate target regions (see Dodd & Jessell 1986).

Oligosaccharide expression by subsets of DRG neurons

One approach to the analysis of mechanisms underlying the development and extension of sensory axons has been to identify cell surface molecules that are expressed on functionally and anatomically distinct subsets of DRG neurons. These studies have demonstrated that cell surface oligosaccharide structures define subsets of DRG neurons during development. Monoclonal antibodies (mAbs) directed against defined carbohydrate epitopes have identified three classes of oligosaccharides that are expressed on subsets of DRG neurons (Table 1) (Dodd et al 1984, Dodd & Jessell 1985, Jessell & Dodd 1985, Chou et al 1989). These oligosaccharides are defined by their backbone and have been classed as globoseries, ganglioseries and lactoseries structures (Hakomori 1981, Feizi 1985). The expression of these oligosaccharides on subsets of DRG neurons correlates with the central projection sites of physiologically defined subsets of DRG neurons (Dodd & Jessell 1985).

Monoclonal antibodies anti-SSEA-3 and anti-SSEA-4 recognize globoseries oligosaccharides that are expressed on DRG neurons of both intermediate and large diameter (Dodd et al 1984) (Fig. 1a,c). The central terminals of SSEA-3$^+$ and SSEA-4$^+$ DRG neurons are located in lamina III and the medial part of lamina IV (Fig. 1a), with a sparse projection to lamina I. This subset of DRG neurons does not contain neuropeptides or other cytochemical markers that define subpopulations of small diameter DRG neurons (Dodd et al 1984). The location of afferent terminals that express globoseries oligosaccharides suggests that they define myelinated primary afferents involved in the transmission of low-threshold cutaneous information to deeper laminae and possibly high-threshold mechanoreceptive or thermoreceptive information to lamina I (Dodd & Jessell 1985).

A distinct subset, constituting approximately 50% of adult DRG neurons, expresses the type-2 lactoseries structure detected by mAbs A5 and 1B2 (Dodd & Jessell 1985) (Fig. 1b,d). This antigen appears on the surface of DRG neurons during embryonic development (Fig. 2) and is present on about 10% of neurons in E18 (Embryonic Day 18) DRG. During the first postnatal week, the proportion of neurons expressing this oligosaccharide structure increases to its adult value. In the adult, this *N*-acetyllactosamine structure is restricted to small and intermediate diameter DRG neurons (Fig. 1d) that have central terminals in laminae I and II of the dorsal horn. Neurons that express type-2 lactoseries structures also contain neuropeptides and other selective markers of nociceptive sensory neurons (Dodd & Jessell 1985).

The expression of a distinct set of α-galactose-extended ganglioseries or lactoseries structures is associated with a more restricted subset of small diameter DRG neurons (Dodd & Jessell 1985, Chou et al 1989) (Fig. 1b). The α-galactose extended structure identified by mAb LD2 is expressed by a subpopulation of

TABLE 1 Surface oligosaccharides expressed by subsets of developing dorsal root ganglia and motor neurons in the rat

Oligosaccharide structure	*mAb*	*Neuronal subclass*	*Laminar termination in spinal cord*
Lactoseries			
Gal(β1–4)GlcNAc–R	A5, 1B2	Small and intermediate diameter DRG neurons	I, II
Gal(α1–3)Gal(β1–4)GlcNAc–R	2C5	Small diameter DRG neurons	II
SO_4-3GlcUA(β1–3)Gal(β1–4)GlcNAc–R	HNK-1 (L2)	Subset of embryonic DRG neurons	
Gal(β1–4)GlcNAc(β1–3)Gal(β1–4)GlcNAc(β-1)	FH6	Embryonic motor neurons	
Fuc(α1–3) Fuc(α1–3)			
Globoseries			
R-3GalNAc(β1–3)Gal(α1–4)R	SSEA-3	Intermediate diameter DRG neurons	I, III, IV
NeuAc(α2–3)Gal(β1–3)GlcNAc(β1–R)	SSEA-4	Intermediate diameter DRG neurons	III, IV
Ganglioseries			
Gal(α1–3)Gal(β1–3)GalNAc(β1–4)Gal(β1–4)Glc(β1–R)	LA4, LD2, TC6	Small diameter DRG neurons	LD2, II outer
Fuc(α1–2)			LA4, II inner

For further details of the structure and expression of oligosaccharides by developing DRG neurons see Dodd & Jessell (1985), Jessell & Dodd (1985) and Chou et al (1989).

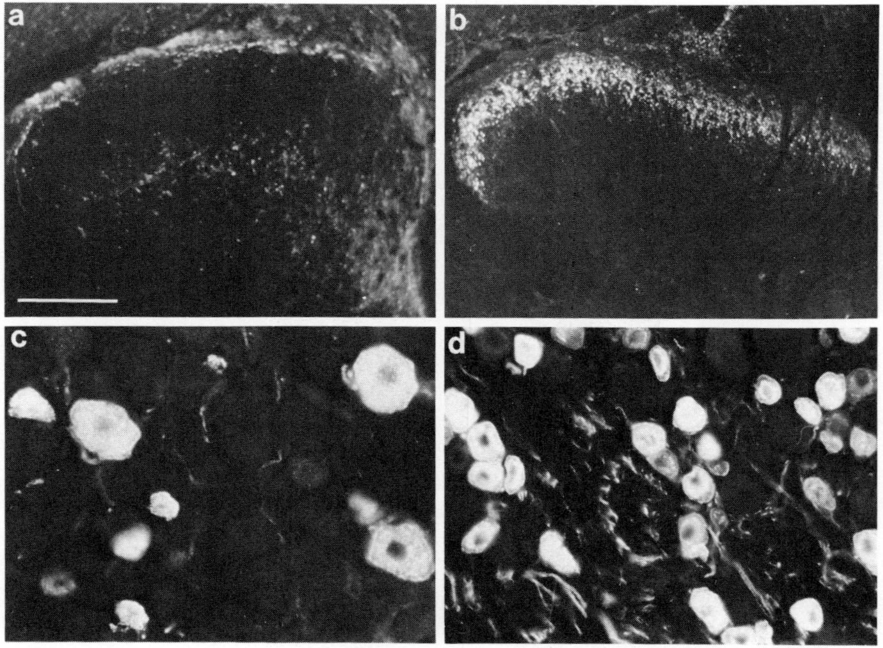

FIG. 1. Immunocytochemical localization of complete oligosaccharides in subsets of adult rat DRG neurons and in the central terminals of primary afferents in the dorsal horn of the spinal cord. (a) Localization of afferent terminals expressing the SSEA-3 globoseries oligosaccharide structure. Labelled axons and terminals are found in laminae I, III and IV. (b) Localization of afferent terminals expressing the LD2 oligosaccharide structure. Afferent terminals are largely confined to lamina II. (c) Localization of the SSEA-3 antigen in a subset of primary afferent neurons and in the axons of these neurons within the DRG. (d) Localization of the A5 lactoseries oligosaccharide antigen within subsets of small and intermediate diameter primary sensory neurons and the axons of these neurons within the DRG. Scale bar: a, 80 μm; b, 80 μm; c, 100 μm; d, 200 μm.

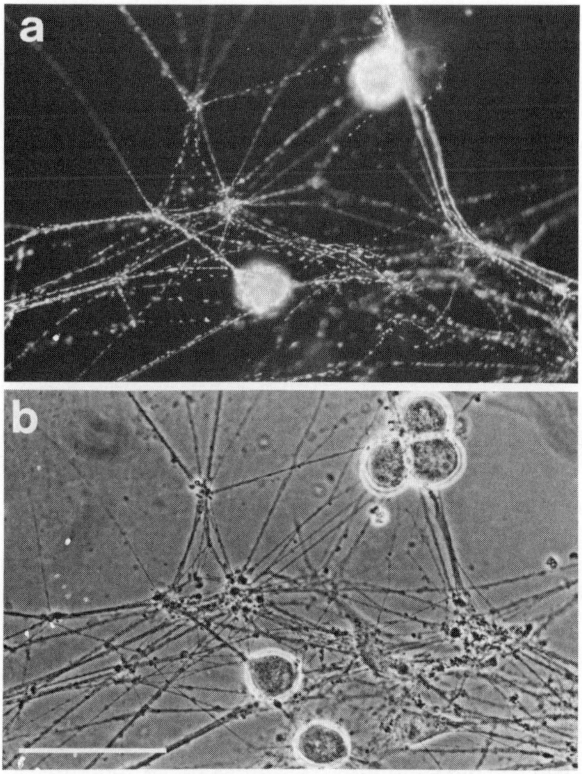

FIG. 2. Localization of the A5-reactive lactoseries oligosaccharide epitope on neonatal rat DRG neurons maintained in dissociated cell culture. (a) Fluorescence micrograph showing two labelled neurons and a network of labelled axons. (b) Phase contrast micrograph of the same field showing other unlabelled sensory neuron cell bodies. Scale bar, 80 μm.

small DRG neurons that project to the dorsal region of lamina II (Fig. 1d). The LD2-reactive sensory neurons include all those that express somatostatin but not those that contain substance P (Dodd & Jessell 1985). A closely related α-galactose-extended structure, recognized by mAb LA4, is found on a separate population of small and intermediate diameter DRG neurons (constituting 40–50% of total number DRG neurons) that project predominantly to the ventral part of lamina II (Dodd & Jessell 1985). Other complex lactoseries oligosaccharide structures delineate additional subsets of small diameter DRG neurons (Jessell & Dodd 1985). The distribution of afferent terminals expressing these oligosaccharides indicates that they define C fibres and also some A-δ fibres.

These observations provide evidence that there is a high degree of specificity in the expression of complex oligosaccharides on functional subsets of DRG neurons. Because similar or identical molecules, in particular those based on the *N*-acetyllactosamine backbone sequence, have been implicated in cell adhesion and recognition of other embryonic cells, it is possible that the oligosaccharides on DRG neurons serve the same general functions.

Oligosaccharide functions in development appear to be mediated by interactions with complementary receptors that function as carbohydrate-binding proteins. A large number of proteins that are capable of binding to surface oligosaccharides on vertebrate cells have been characterized (Barondes 1984, Bayna et al 1986, see Drickamer, this volume). Many of them recognize β-galactoside linkages contained within the lactoseries structures described above (Sarkar et al 1975, Leffler & Barondes 1986). Moreover, biochemical studies have shown that carbohydrate-binding proteins with β-galactoside binding specificity are present in neural tissues (Kobiler & Barondes 1977, Eisenbarth et al 1978, Joubert et al 1987). These lectins appear to exist in soluble form and represent one class of proteins with the potential to interact with lactoseries oligosaccharides found on the surface of DRG neurons.

Expression of β-galactoside-binding lectins by DRG neurons

In order to examine whether lectins are expressed in sensory or spinal neurons, antibodies raised against β-galactoside-binding lectins isolated from rat lung (Cerra et al 1984, 1985) have been used to localize two related or identical proteins in DRG and spinal cord (Regan et al 1986). These lectins, termed RL-14.5 and RL-29, have subunit M_r values of 14 500 and 29 000 (Cerra et al 1985). Western blot analysis with antibodies specific for these lectins indicates that DRG and spinal cord from embryonic, neonatal, and adult rats contain both RL-14.5 and RL-29 as the only detectable immunoreactive species (Regan et al 1986) (Fig. 3). Both lectins can be detected in DRG neurons soon after the differentiation of sensory neuroblasts and are present in central sensory axons that terminate in the dorsal horn of the spinal cord (Figs 4 and 5). RL-14.5 is detected in DRG by Embryonic Day 14 (E14) and continues to be expressed at later stages of development, with the highest levels of immunoreactivity detected from E20 to Postnatal Day 5 (P5) (Fig. 3). DRG neurons that express highest levels of RL-14.5 are those with small or intermediate diameters, although larger diameter neurons do appear to express low levels of the lectin (Fig. 6). In contrast, RL-29 is not detectable by immunocytochemistry in DRG neurons until E16 (Fig. 3). From this developmental stage onwards, about 50% of DRG neurons, predominantly those of small diameter, exhibit intense RL-29 immunoreactivity, with an additional 10% expressing lower levels of RL-29 (Regan et al 1986) (Fig. 6).

There is substantial overlap between subsets of DRG neuronal cell bodies that contain RL-14.5 and/or RL-29 and those that express sensory neuron-specific

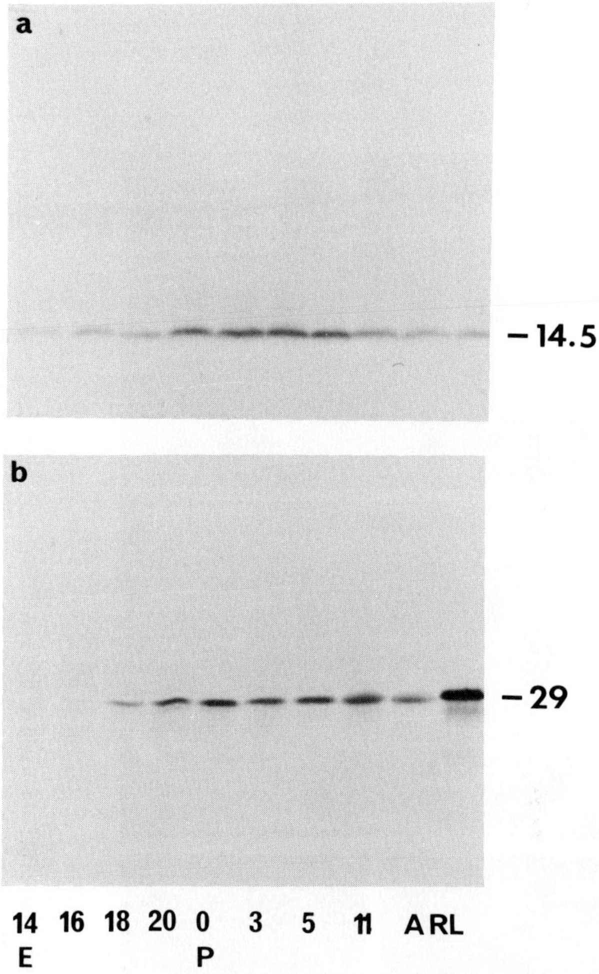

FIG. 3. Identification and developmental regulation of (a) RL-14.5 and (b) RL-29 in rat DRG. Expression of RL-14.5 and RL-29 in soluble extracts of rat DRG during embryonic and postnatal development. Numbers under each lane refer to the embryonic (E) and postnatal (P) age of rats from which DRG extracts were prepared. A, extract prepared from adult rat DRG. Markers indicate approximate molecular masses (shown as $M_r \times 10^{-3}$) of immunoreactive bands calculated from protein standards of known molecular mass.

lactoseries oligosaccharides (Regan et al 1986) (Fig. 6). The coincident distribution of RL-14.5, RL-29, and lactoseries antigens in the superficial dorsal horn provides further evidence for the preferential association of these lectins with sensory neurons that express lactoseries carbohydrate determinants.

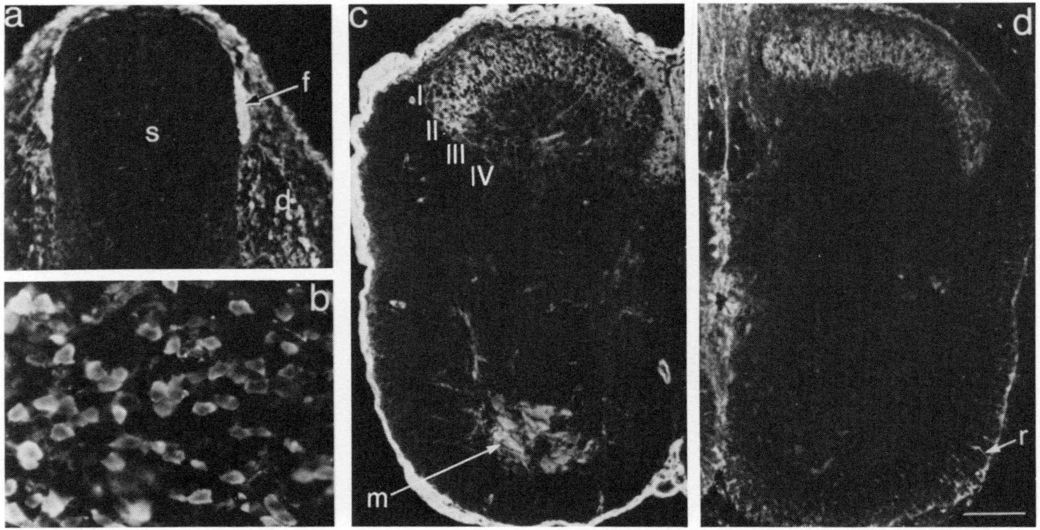

FIG. 4. Immunocytochemical localization of RL-14.5 and RL-29 in rat DRG and spinal cord during development. (a) Localization of RL-14.5 in the DRG and dorsal funiculus of E14 rat spinal cord. Immunoreactive sensory fibres are not detectable in the dorsal horn of the spinal cord at this stage. Labelling of connective tissue elements surrounding the spinal cord is also apparent. d, DRG. f, dorsal funiculus; s, spinal cord. (b) Localization of RL-29 in E20 rat DRG. At this stage 40–50% of neurons contain the lectin. (c) Localization of RL-14.5 in the neonatal rat spinal cord. An intense band of immunoreactive sensory fibres is present in laminae I and III, with less intense labelling in laminae III and IV. The collaterals of primary afferents in the dorsal columns and the cell bodies and processes of motoneurons (m) in the ventral horn are also heavily labelled. Roman numerals indicate dorsal horn laminae. (d) Localization of RL-29 in the neonatal spinal cord. Immunoreactive sensory afferents are restricted to laminae I and II. Light staining of radial glial fibres (r) in the ventral white matter can be observed. Scale bar: a, 150 μm; b, 28 μm; c and d, 100 μm.

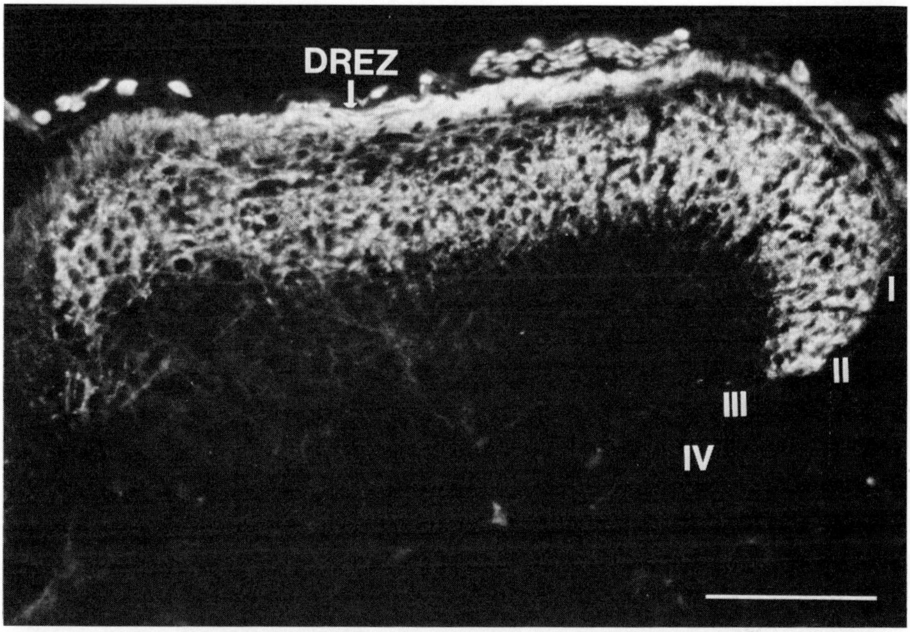

FIG. 5. Immunocytochemical localization of RL-29 in the superficial dorsal horn of neonatal rat spinal cord. Intense labelling can be observed in the dorsal root entry zone (DREZ) and in the axons and terminals of afferent fibres in laminae I and II. A few fibres can also be observed in laminae II and IV although the intensity of labelling is much less. The intracellular profiles of spinal cord neurons in laminae I and II can also be seen. Roman numerals indicate approximate locations of laminae of Rexed. Scale bar, 100 μm.

RL-14.5 is present in the central processes of DRG neurons from E13–14 and the projection of sensory afferents from the dorsal root into the dorsal horn can be monitored by expression of RL-14.5 (Fig. 4). At P0, afferent fibres containing RL-14.5 are distributed at highest density in the superficial dorsal horn, with a lower intensity of labelling associated with afferents in laminae III and IV and in the dorsal columns. The distribution of RL-29 in the DRG and spinal cord is more restricted than that of RL-14.5 and is absent from deeper laminae of the dorsal horn (Figs 4 and 5). The intensity of RL-14.5 and RL-29 immunoreactivity in the dorsal horn of adult rats is much lower than that at P0 and appears restricted to laminae I and II. RL-14.5 and RL-29, therefore, appear to be expressed at maximal levels in sensory neurons over the period that afferent fibres extend axons into the spinal cord. RL-14.5 is also expressed in motoneurons in the spinal cord and brainstem from E13, with the intensity of labelling increasing during later embryonic and early postnatal periods (Figs 4

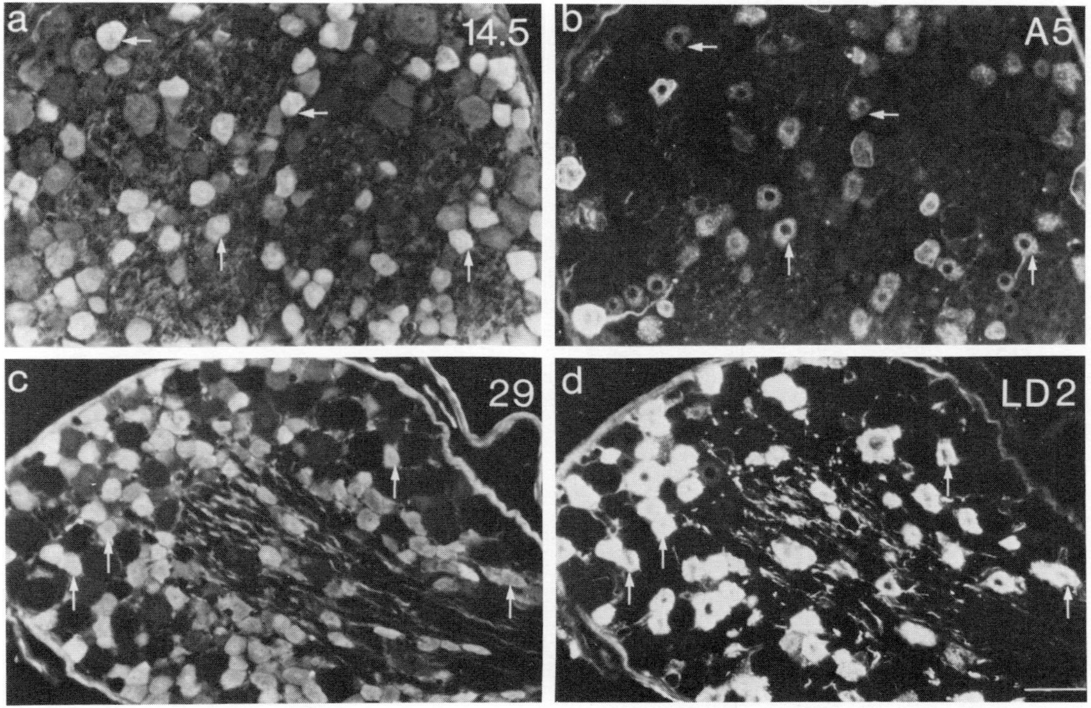

FIG. 6. Overlap in expression of RL-14.5 and RL-29 and sensory neuron-specific oligosaccharides in adult rat DRG neurons. (a) and (b) show localization of RL-14.5 (a) and the lactoseries epitope recognized by mAb A5 (b) on the same section of DRG using dual-colour immunofluorescence labelling. Extensive, but not complete, overlap is apparent between neurons that express RL-14.5 and A5-reactive antigens. (c) and (d) show localization of RL-29 (c) and the LD2 epitope (d) on the same DRG section. Most DRG neurons labelled by mAb LD2 also contain RL-29. Arrows identify the same cells in each pair of micrographs. Scale bar, 60 μm in a and b; 50 μm in c and d.

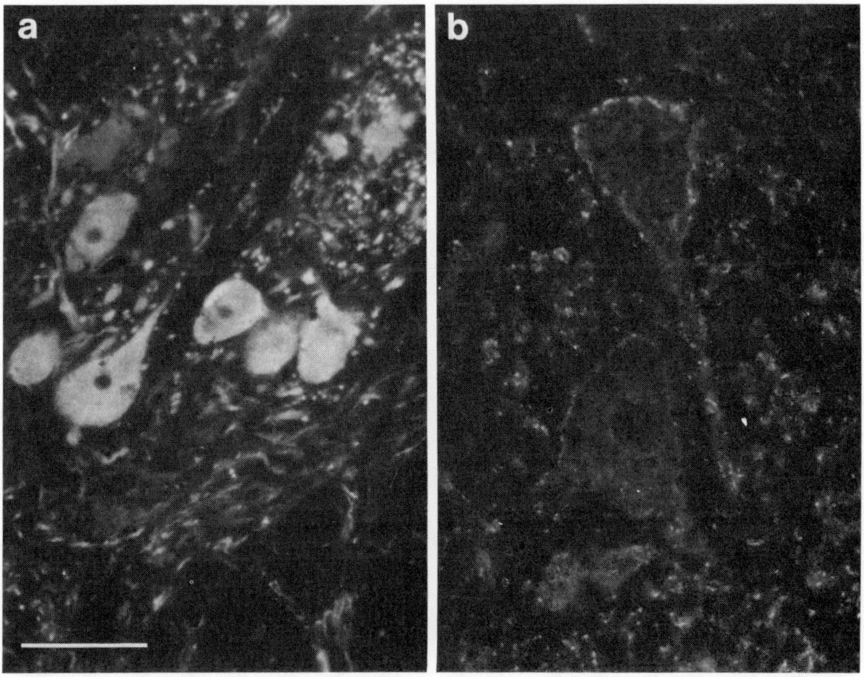

FIG. 7. Immunocytochemical localization of RL-14.5 in motoneuron cell bodies and axons in the ventral horn of adult rat spinal cord. (a) Localization of RL-14.5 in motoneurons in sections of rat spinal cord fixed with 4% paraformaldehyde. RL-14.5 immunoreactivity is detected within the cytoplasm of motoneuron cell bodies and axons. (b) Localization of RL-14.5 in the ventral horn of rat spinal cord in fresh frozen sections. RL-14.5 immunoreactivity is located predominately in association with the plasma membrane of the two motoneurons shown in the section and also is associated with the membranes of motoneuron axons. These observations suggest that RL-14.5 is bound to complementary β-galactoside-containing oligosaccharides on the surface or in the environment of motor axons. The majority of RL-14.5 within the neuronal cytoplasm appears to be present in diffusible form since it is lost from fresh frozen sections. Scale bar: a, 50 μm; b, 30 μm.

and 7), whereas RL-29 is not present in motoneurons (Regan et al 1986). These findings thus indicate that there is a striking specificity in the expression of these two soluble β-galactoside-binding lectins in subsets of developing mammalian neurons.

Cloning and expression of RL-14.5

cDNA clones that encode RL-14.5 have been isolated from rat brain cDNA libraries to provide further information on the structure and function of RL-14.5

Rat brain	M A C G L V A - S N L N L K P G E C L K V R G E L A P D A K S F V L N L
Mouse fibroblast	M A C G L V D Q Q A E S Q T G A M S Q S S G Q A W T S D D K S F V L N L
Mouse 3T3	. . . . E V A S D A K S F V L N L
Human placenta	M A C G L V A - S N L N L K P G E C L A V R G E V A P D A K S F V L N L
Chick skin	M S C Q G P V C T N L G L K P G Q R L T V K G I I A P N A K S F V M N L
HEP-1	. . . . E V K N M D M K P G S T L K I T G S I A D G T D G F V I N L
EEL	. . . . . G V V D E A M S F K A G Q N L T V K G V P S I D S T N G A I N V
CBP-35	L T V P Y D L P L P G G V M P A M L I T I M G T V K P N A N R I V L D F
Rat brain	G K D S N N L C L H F N P R F N A H G D A N T I V C N S K D D G T W G T
Mouse fibroblast	G K D S N N L C L H F N P R F N A H G D A N T I V C N T K E D G T W G T
Mouse 3T3	G K D S N N L C L H F N P R F N A H G D A N R I V C N T K E D G T W G T
Human placenta	G K D S N N L C L H F N P R F N A H G D A N T I V C N S K D G G A W G T
Chick skin	G K D S T H L G L H F N P R F D A H G D V N L I V C N S K K M E E W G T
HEP-1	G Q G T D K L N L H F N P R F S G S T . . . . I V C N S L D G S N W G Q
EEL	G N S A E D L A L H I N P R F D A H G D Q Q A V V V N S F Q G G N W G -
CBP-35	- R R G N D V A F H F N P R F N - E N N R R V I V C N T K Q D N N W G K

```
Rat brain            [illegible]
Mouse fibroblast     E H R E P A F P F Q P G S I T E C A C T F D Q A D L T M Q A A R R T . .
Mouse 3T3            E H R E P A F P F Q P G S I T E V C I T F D Q A D L T I K L P D G H E F
Human placenta       E Q R E V A F P F Q P G S V A E V C I T F D Q A N L T V K L P D G Y E F
Chick skin           E Q R E T V F P F Q K G A P I E I T F S I N P S D L T V H L P - G H Q F
HEP-1                E Q R E D H L C F S P G S E V K F T V T F G S D K F K V K L P D G H E L
EEL                  E Q R E G G F P F K Q G E D F K I Q I T F N S E E F R I I L P D G S E I
CBP-35               E E R Q S A F P F E S G K P F K I Q V L V E A D H F K V A V N D A H L L

Rat brain            K F P N R L N M E A I N Y M A A D G D F K I K C V A F E . .
Mouse fibroblast     . . . . . . . . . . . . . . . . . . . . . . . . . . . .
Mouse 3T3            . . . . . . . . . . . . . . . . . . . . . . . . . . . .
Human placenta       K F P N R L N L E A I N Y M A A D G D F K I K C V A F D . .
Chick skin           S F P N R L G L S V F D Y F D T H G D F T L A S V S W E . .
HEP-1                T F P N R L G H S H L S Y E S I R G G F N M S S F K L K E .
EEL                  H F P N N R Y M H F E G E A R I Y S - I E I K . . . . . . .
CBP-35               Q Y N H R M K N L R E I S Q L G I S G D I T L T S A N H A M
```

FIG. 8. Amino acid sequence of RL-14.5 deduced from cDNA clones isolated from a rat brain cDNA library. The sequence of RL-14.5 is compared with other β-galactoside-binding proteins isolated from other tissues and species. Conserved amino acids are boxed. Details of cloning and sequences of the various β-galactoside-binding proteins are given in the following references; rat brain (Hynes et al 1989, see also Clerch et al 1988); mouse fibroblasts (Raz et al 1987); human placenta (Hirabayashi & Kasai 1988); chick skin (Ohyama et al 1986, Ohyama & Kasai 1988); HEP-1 human hepatoma line (see Gitt & Barondes 1986, Couraud et al 1989); eel lectin (Paroutaud et al 1987) and CBP-35 isolated from mouse 3T3 cells (Jia & Wang 1988). Amino acid single letter coding is used.

within primary sensory neurons (Hynes et al 1989). The sequence of RL-14.5 brain cDNA predicts the synthesis of a protein of 135 amino acids which is identical to that predicted from a rat lung cDNA clone (Clerch et al 1988). This protein is homologous to 14 kDa β-galactoside-binding proteins that have been isolated from other tissues and species (Gitt & Barondes 1986, Ohyama et al 1986, Raz et al 1987) (Fig. 8). These soluble β-galactoside-binding lectins constitute a large family of structurally related proteins (Southan et al 1987, Paroutaud et al 1987, see Drickamer, this volume). The sequence of RL-14.5, contains a series of conserved amino acids that is shared with other members of this lectin family (Paroutaud et al 1987) (Fig. 8). Most of the soluble β-galactoside-binding proteins so far characterized have molecular masses of approximately 14 kDa. However, it has become apparent that this family includes at least one larger protein, termed CBP-35, which has been identified as a 35 kDa β-galactoside-binding protein isolated from mouse fibroblasts (Jia & Wang 1988). Sequence comparisons (H. Leffler & S. H. Barondes, unpublished, Jia & Wang 1988) suggest that the mouse protein CBP-35 is closely related to RL-29, the lectin expressed in subsets of rat DRG neurons.

The availability of cDNA clones encoding RL-14.5 has permitted an analysis of the expression pattern of the lectin in the developing rat nervous system (Hynes et al 1988, 1989). RNA blot analysis indicates that high levels of RL-14.5 mRNA are present in embryonic and postnatal DRG and spinal cord, with much lower levels in other regions of the central nervous system (Hynes et al 1989). It has also been possible to localize RL-14.5 mRNA at the cellular level by *in situ* hybridization. Both DRG neurons and motoneurons express RL-14.5 mRNA soon after their differentiation and this selective expression pattern persists in the adult nervous system (Fig. 9). All embryonic and adult DRG neurons appear to express RL-14.5 mRNA, but the level of expression varies about fivefold between individual neurons. In the adult spinal cord, RL-14.5 is detected at high levels in motoneurons and there is no detectable expression in other cells. Within the adult brain, hybridization is largely restricted to motoneuron nuclei in the brain stem (Hynes et al 1989). These observations therefore confirm the selective expression pattern of RL-14.5 detected by immunocytochemistry and Western blot analysis.

Possible functions of sensory neuron-specific lactoseries oligosaccharides and β-galactoside-binding proteins

The studies outlined above indicate that developing and mature primary sensory neurons selectively express surface lactoseries oligosaccharides and soluble β-galactoside-binding proteins. Previous studies of this family of β-galactoside-binding proteins in non-neural cells have provided some evidence that these lectins can be released from cells (Beyer & Barondes 1982, Barondes 1984).

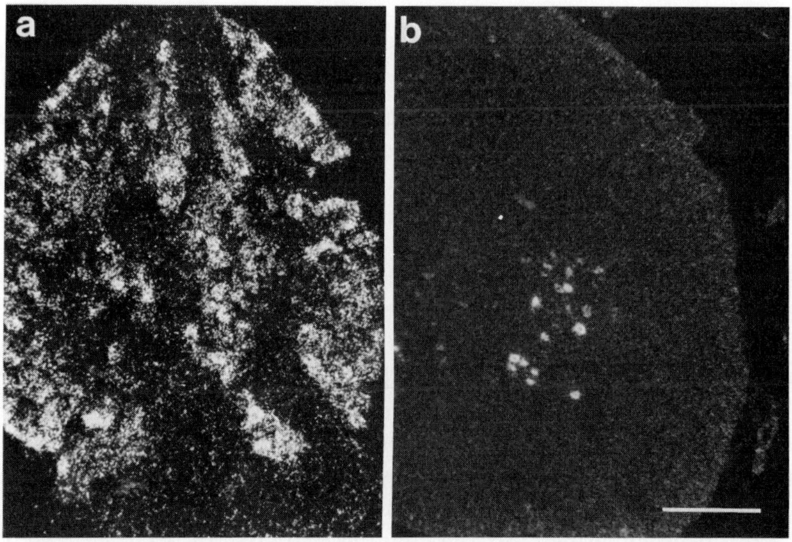

FIG. 9. Localization of RL-14.5 mRNA in DRG neurons and ventral horn motoneurons of adult rat, detected by *in situ* hybridization. ^{35}S-labelled antisense RNA probes were used to localize RL-14.5 mRNA. (a) Intense hybridization is detected over sensory neurons in the DRG. (b) Selective expression of RL-14.5 mRNA in spinal motoneurons. No other neuronal or non-neuronal cells in the spinal cord exhibit detectable hybridization signals. Scale bar: a, 300 μm; b, 400 μm.

Preliminary experiments have also demonstrated that both RL-14.5 and RL-29 are secreted from rat DRG neurons maintained in dissociated cell culture (M. A. Hynes et al, unpublished observations). While the extracellular location of RL-14.5 is consistent with a role for these lectins in cell adhesion, the mechanism by which the lectins are released from cells remains unclear. Sequence analysis of cDNAs encoding RL-14.5 and CBP-35 have revealed that there is no signal sequence at the N-terminus of the proteins (Clerch et al 1988, Hynes et al 1988, 1989, Jia & Wang 1988). The mechanism of secretion of these lectins may therefore be unorthodox. The absence of a signal sequence has, however, also been noted for several other proteins with presumed extracellular functions.

The predominant neural expression of RL-14.5 and RL-29 is in sensory and motor neurons, both of which extend axons into the periphery. One possible function of these lectins may therefore be to cross-link carbohydrate structures on the surface of axons with those on cell surface or matrix molecules in the peripheral environment through which these axons migrate. Lactoseries oligosaccharides that are potential ligands for RL-14.5 and RL-29 have been demonstrated on a variety of peripheral cell surface and extracellular matrix components, including the axonal adhesion molecule NILE (Margolis et al 1986), the glycoprotein laminin (Fujiwara et al 1988) and the glycosaminoglycan chains on chondroitin sulphate proteoglycans (Krusius et al 1986). In addition, the HNK-1 carbohydrate epitope that has been identified on a variety of cell and matrix adhesion molecules (see Schachner, this volume) is a sulphated glucuronic acid derivative of lactoseries oligosaccharides (Chou et al 1986), suggesting that this structure may be recognized by soluble β-galactoside-binding lectins. These observations suggest that these lectins may interact with other classes of neural adhesion molecules that also mediate cell binding by direct protein–protein interactions (see Schachner, this volume). In preliminary studies (M. A. Hynes et al, unpublished) affinity-purified RL-14.5 immobilized on nitrocellulose substrates appears to promote the outgrowth of axons from neonatal DRG neurons in a manner similar to that of laminin.

Although RL-14.5 and RL-29 have similar, but not overlapping, expression patterns within the nervous system, they may mediate different functions. Detailed studies of the binding of these two lectins indicate that saccharide substitutions of the poly-*N*-acetyllactosamine backbone structure can have markedly different effects on the binding affinities of RL-14.5 and RL-29 (Leffler & Barondes 1986). Moreover, *in vitro* studies on the binding of RL-14.5 and RL-29 to oligosaccharides on neural cells have provided evidence that these lectins have different binding specificities. RL-14.5 binds to the surface of DRG neurons but not to spinal cord neurons, whereas RL-29 binds to a subset of DRG and spinal cord neurons (J. Dodd & T. M. Jessell, unpublished). These two lectins may therefore be capable of mediating distinct cellular interactions even if released from the same set of sensory neurons. The demonstration that these lectins will bind to appropriate saccharides on embryonic neurons should open the way for functional assays to test their contribution to the extension and fasciculation of axons of sensory and motoneurons.

It is also possible that RL-14.5 and RL-29 have intracellular functions in sensory and motor neurons. Both lectins can be detected in the cytoplasm and in the nucleus of DRG neurons and other cells (Barondes 1984, Regan et al 1986). Since RL-29 resembles CBP-35 it may share a nuclear function (Moutsatsos et al 1987) with it. While there is no direct evidence that such lectins function within the nucleus, the observations that many nuclear proteins (Hart et al 1987), including some transcription factors (Jackson & Tjian 1988), are glycosylated raises the possibility that this class of lectins may be involved in regulating the function of some nuclear proteins. The structure of CBP-35 has provided support for this idea. The amino acid sequence of CBP-35, deduced from cDNA clones, has revealed that the protein consists of two domains with the C-terminal portion resembling the other soluble β-galactoside-binding proteins whereas the N-terminus exhibits homology with proteins of the heterogeneous ribonucleoprotein complex (Jia & Wang 1988). CBP-35 does appear to be a component of ribonucleoprotein complexes in mouse 3T3 cells

(Laing & Wang 1988). The possibility of nuclear functions of lectins in neurons, however, remains to be determined.

The studies described in this chapter illustrate only one possible set of interactions between complex oligosaccharides and carbohydrate-binding proteins on neural cells. Other studies have also demonstrated the existence of ganglioside receptors on neural retina cells (Blackburn et al 1986), the presence of ganglioside and lactoseries saccharide gradients in many regions of the developing central nervous system (Blum & Barnstable 1987, Yamamoto et al 1985), and the existence of synapse-specific oligosaccharides and glycosyltransferases at the neuromuscular junction (Scott et al 1988). In addition to the soluble β-galactoside-binding proteins described here, other soluble and membrane-bound lectins have been described, many of which have different saccharide-binding specificities (Zanetta et al 1987). The large and distinct family of cation-dependent carbohydrate-binding proteins exemplified by the hepatic asialoglycoprotein receptor (see Drickamer, this volume) may also have structural homologues within the nervous system.

Analysis of the role of carbohydrate recognition in neural development is clearly complicated by problems of tissue complexity and the contribution of other adhesion and recognition molecules. However, the recent progress in elucidating the role of cell surface oligosaccharides in simpler vertebrate tissues may set the stage for understanding functional roles of similar or identical molecules on developing neurons.

Acknowledgements

This work is supported by grants from NIH, the McKnight Foundation and the Howard Hughes Medical Institute. TMJ is an HHMI Investigator. We thank R. Lenertz for help in preparing the manuscript.

References

Altman J, Bayer SA 1984 The development of the rat spinal cord. Springer Verlag, New York, p 166

Ashwell G, Harford J 1982 Carbohydrate-specific receptors of the liver. Annu Rev Biochem 51:531–554

Balsamo J, Pratt RS, Lillien J 1986 Chick neural retina N-acetylgalactosaminyltransferase/acceptor complex: catalysis involves transfer of N-acetylgalactosamine phosphate to endogenous acceptors. Biochemistry 25:5402–5407

Barbera AJ, Marchase RB, Roth S 1973 Adhesive recognition and retinotectal specificity. Proc Natl Acad Sci USA 70:2482–2486

Barondes SH 1984 Soluble lectins: a new class of extracellular proteins. Science (Wash DC) 223:1259–1264

Bayna EM, Runyan RB, Scully NF, Reichner J, Lopez LC, Shur BD 1986 Cell surface galactosyltransferase as a recognition molecule during development. Mol Cell Biochem 72:141–151

Beyer EC, Barondes SH 1982 Secretion of endogenous lectin by chicken intestinal goblet cells. J Cell Biol 92:28–33

Blackburn CC, Swank-Hill P, Schnarr RL 1986 Gangliosides support neural retina cell adhesion. J Cell Biol 261:2873–2881

Blum AS, Barnstable CJ 1987 O-Acetylation of a cell-surface carbohydrate creates discrete molecular patterns during neural development. Proc Natl Acad Sci USA 84:8716–8720

Brandley BK, Ross RS, Schnarr RL 1987 Multiple carbohydrate receptors on lymphocytes revealed by adhesion to immobilized polysaccharides. J Cell Biol 105:991–997

Brown AG 1981 Organization in the spinal cord. Springer Verlag, Berlin & New York

Cerra RF, Haywood-Reid PL, Barondes SH 1984 Endogenous mammalian lectin localized extracellularly in lung elastic fibers. J Cell Biol 98:1580–1589

Cerra RF, Gitt MA, Barondes SH 1985 Three soluble rat β-galactoside-binding lectins. J Biol Chem 260:10474–10477

Chou DK, Ilyas AA, Evans JE, Costello C, Quarles RH, Jungalwala FB 1986 Structure of sulfated glucuronyl glycolipids in the nervous system reacting with HNK-1 antibody and some IgM paraproteins in neuropathy. J Biol Chem 261:11717–11725

Chou DK, Dodd J, Jessell TM, Costell CE, Jungalwala FB 1989 Identification of α-galactose (α-Fucose)-asialo GMI glycolipid expressed by subsets of rat dorsal root ganglion neurons. J Biol Chem 264:3409–3415

Clerch LB, Whitney P, Hass M et al 1988 Sequence of a full-length cDNA for rat lung β-galactoside-binding protein: primary and secondary structure of the lectin. Biochemistry 27:692–699

Couraud P, Casentini-Borocz D, Bringman T-S, Griffith J, McGregor M, Nedwin GE 1989 Molecular cloning, characterization and expression of a human 14 kDa lectin. J Biol Chem 264:1310–1317

Dodd J, Jessell TM 1985 Lactoseries carbohydrates specify subsets of dorsal root ganglion neurons projecting to superficial dorsal horn of rat spinal cord. J Neurosci 5:3278–3294

Dodd J, Jessell TM 1986 Cell surface glycoconjugates and carbohydrate-binding proteins: possible recognition signals in sensory neuron development. J Exp Biol 129:225–238

Dodd J, Solter D, Jessell TM 1984 Monoclonal antibodies against carbohydrate differentiation antigens identify subsets of primary sensory neurons. Nature (Lond) 311:469–472

Drickamer K 1989 Multiple subfamilies of carbohydrate recognition domains in animal lectins. In: Carbohydrate recognition in cellular function. Wiley, Chichester (Ciba Found Symp 145) p 45–61

Edelman GM 1986 Cell adhesion molecules in the regulation of animal form and tissue pattern. Annu Rev Cell Biol 2:81–116

Eisenbarth GS, Ruffolo RR, Walsh FS, Nirenberg M 1978 Lactose sensitive lectin of chick retina and spinal cord. Biochem Biophys Res Commun 83:1246–1252

Feizi T 1985 Demonstration by monoclonal antibodies that carbohydrate structures of glycoproteins and glycolipids are onco-developmental antigens. Nature (Lond) 314:53–57

Fenderson BA, Zehavi U, Hakomori SI 1984 A multivalent lacto-N-fucopentose III-lysyllysine conjugate decompacts preimplantation mouse embryos, while the free oligosaccharide is ineffective. J Exp Med 160:1591–1596

Fujiwara K, Shinkai H, Deutzmann R, Paulsson M, Timpl R 1988 Structure and distribution of N-linked oligosaccharide chains on various domains of mouse tumour laminin. Biochem J 252:453–461

Gitt MA, Barondes SH 1986 Evidence that a human soluble β-galactoside-binding lectin is encoded by a family of genes. Proc Natl Acad Sci USA 83:7603–7607

Hakomori S 1981 Glycosphingolipids in cellular interaction, differentiation and oncogenesis. Annu Rev Biochem 50:733–764
Hart GD, Snow CM, Senior A, Haltiwanger RS, Gerace L, Hart GW 1987 Nuclear pore complex glycoproteins contain cytoplasmically disposed O-linked N-acetylglucosamine. J Cell Biol 104:1157–1164
Hirabayashi J, Kasai KI 1988 Complete amino acid sequence of a β-galactoside binding lectin from human placenta. J Biochem 104:1–4
Hynes MA, Buck LB, Casano FI, Huang KK, Barondes SH, Jessell TM 1988 Cloning and expression of a soluble 14 kDa β-galactoside binding lectin in rat nervous system. Soc Neurosci Abstr
Hynes MA, Dodd J, Jessell TM 1989 Carbohydrate recognition, cell interactions and vertebrate neural development. In: Margolis RK, Margolis RU (eds) Neurobiology of glycoconjugates. Raven Press, New York p 337–365
Jackson SP, Tijan R 1988 O-glycosylation of eukaryotic transcription factors: implications for mechanisms of transcriptional regulation. Cell 55:125–133
Jessell TM, Dodd J 1985 Structure and expression of differentiation antigens on functional subclasses of primary sensory neurons. Philos Trans R Soc Lond B Biol Sci 308:271–281
Jia S, Wang JL 1988 Carbohydrate binding protein 35. Complementary DNA sequence reveals homology with proteins of the heterogeneous nuclear RNP. J Biol Chem 263:6009–6011
Joubert R, Caron M, Bladier D 1987 Brain lectin-mediated agglutinability of dissociated cells from embryonic and post-natal mouse brain. Dev Brain Res 36:146–150
Kobiler D, Barondes SH 1977 Lectin activity from embryonic chick brain, heart and liver: changes with development. Dev Biol 60:326–330
Krusius T, Finne J, Margolis RK, Margolis RU 1986 Identification of an O-glycosidic mannose-linked sialylated tetrasaccharide and keratan sulfate oligosaccharides in the chondroitin sulfate proteoglycan of brain. J Biol Chem 261:8237–8242
Laing JG, Wang JL 1988 Identification of carbohydrate binding protein 35 in heterogeneous nuclear ribonucleoprotein complex. Biochemistry 27:5329–5334
Le Douarin N 1982 The Neural Crest. Cambridge University Press, Cambridge, p 259
Lee YC 1989 Binding modes of mammalian hepatic Gal/GalNAc receptors. In: Carbohydrate recognition in cellular function. Wiley, Chichester (Ciba Found Symp 145) p 80–95
Leffler H, Barondes SH 1986 Specificity of binding of three soluble rat lung lectins to substituted and unsubstituted mammalian β-galactosides. J Biol Chem 261:10119–10126
Marchase RB 1977 Biochemical investigations of retinotectal adhesive specificity. J Cell Biol 75:237–257
Margolis RK, Greene LA, Margolis RU 1986 Poly(N-acetyllactosaminyl) oligosaccharides in glycoproteins of PC12 pheochromocytoma cells and sympathetic neurons. Biochemistry 25:3463–3468
Moutsatsos IK, Wade M, Schindler M, Wang JL 1987 Endogenous lectins from cultured cells: nuclear localization of carbohydrate-binding protein 35 in proliferating 3T3 fibroblasts. Proc Natl Acad Sci USA 84:6452–6456
Ohyama Y, Kasai KI 1988 Isolation and characterization of the chick 14K β-galactoside-binding lectin gene. J Biochem 104:173–177
Ohyama Y, Hirabayashi J, Oda Y, Ohno S, Kawasaki H, Suzuki K, Kasai K 1986 Nucleotide sequence of chick 14K β-galactoside-binding lectin mRNA. Biochem Biophys Res Commun 134:51–56
Paroutaud P, Levi G, Teichberg VI, Strosberg AD 1987 Extensive amino acid sequence homologies between animal lectins. Proc Natl Acad Sci USA 84:6345–6348

Perl ER 1983 Characterization of nociceptors and their activation of neurons in the superficial dorsal horn: first steps for the sensation of pain. Adv Pain Res Ther 6:23–51

Rastan S, Thorpe SJ, Scudder DP, Brown S, Gooi HC, Feizi T 1985 Cell interactions in preimplantation embryos: evidence for involvement of saccharides of the poly-N-acetyllactosamine series. J Embryol Exp Morphol 87:115–128

Raz A, Carmi P, Pazerini G 1987 Expression of two different endogenous galactoside-binding lectins sharing sequence homology. Cancer Res 48:645–649

Regan L, Dodd J, Barondes SH, Jessell TM 1986 Selective expression of endogenous lactose-binding lectins and lactoseries glycoconjugates in subsets of rat sensory neurons. Proc Natl Acad Sci USA 83:2248–2252

Roseman S 1970 The synthesis of complex carbohydrates by multiglycosyltransferase systems and their potential function in intercellular adhesion. In: Chemistry and physics of lipids. North Holland Publishing Company, Amsterdam

Rosen SC, Yednock TA 1986 Lymphocyte attachment to high endothelial venules during recirculation: a possible role for carbohydrates as recognition determinants. Mol Cell Biochem 72:153–164

Rosen SD, Singer MS, Yednock TA, Stoolman LM 1985 Involvement of sialic acid on endothelial cells in organ-specific lymphocyte recirculation. Science (Wash DC) 228:1005–1007

Roth S, McGuire EJ, Roseman J 1971 Evidence for cell-surface glycosyltransferases: their potential role in cellular recognition. J Cell Biol 51:526–547

Ruoslahti E, Pierschbacher M 1987 New perspectives in cell adhesion: RGD and integrins. Science (Wash DC) 238:491–497

Sarkar M, Liao J, Kabat EA, Tanake T, Ashwell G 1975 The binding site of rabbit hepatic lectin. J Biol Chem 254:3170–3174

Schachner M 1989 Families of neural adhesion molecules. In: Carbohydrate recognition in cellular function. Wiley, Chichester (Ciba Found Symp 145) p 156–172

Scott LJC, Bacon F, Sanes JR 1988 A synapse-specific carbohydrate of the neuromuscular junction association with both acetylcholinesterase and a glycolipid. J Neurosci 8:932–944

Southan C, Aitken A, Childs RA, Abbott WM, Feizi T 1987 Amino acid sequence of β-galactoside-binding bovine heart lectin. Member of a novel class of vertebrate proteins. FEBS (Fed Eur Biochem Soc) Lett 214:301–304

Takeichi M 1988 The cadherins: cell–cell adhesion molecules controlling animal morphogenesis. Development 102:639–655

Wassarman PM 1989 Role of carbohydrates in receptor-mediated fertilization in animals. In: Carbohydrate recognition in cellular function. Wiley, Chichester (Ciba Found Symp 145) p 135–155

Yamamoto M, Boyer AM, Schwarting GA 1985 Fucose-containing glycolipids are stage and region-specific antigens in developing embryonic brain of rodents. Proc Natl Acad Sci USA 82:3045–3049

Zanetta JP, Meyer A, Kuchler S, Vincendon G 1987 Isolation and immunochemical study of a soluble cerebellar lectin delineating its structure and function. J Neurochem 49:1250–1257

DISCUSSION

Wassarman: The cDNAs encoding RL-14.5 and CBP-35 show no predicted signal sequence. Is that true for the other β-galactoside-binding lectins?

Jessell: It is certainly true for the spinal cord and DRG lectins that I described.

Wassarman: Are all those lectins secreted like RL-14.5?

Jessell: There is some confusion about the localization of these proteins. They are clearly in the cytoplasm, but Douglas Cooper has results suggesting that cells are capable of secreting them. There is no apparent canonical signal sequence, although there are stretches of hydrophobic residues. This is not unprecedented; interleukin 1 and fibroblast growth factors are thought to be secreted, yet do not have signal sequences. There may be mechanisms for secretion other than this classical pathway, but there is no evidence for that at the moment.

Wassarman: Do the antibodies against these lectins label cell surface glycoproteins?

Barondes: In skeletal muscle cell lines, which have been studied extensively, differentiated cells have lectins in the extracellular matrix. Some may also be present on the cell surface, presumably bound to complementary glycoconjugates.

Feizi: One of the three highly conserved regions in these lectins contains WGAE or a similar sequence. It appears to be highly immunogenic and is recognized by two hybridoma antibodies (Abbott et al 1989). Myelin basic protein also has this tetrapeptide sequence and immunoreactivity. Do any of your histochemical findings correspond to myelin basic protein?

Jessell: So far, immunocytochemistry and *in situ* hybridization under high stringency conditions have given us essentially the same pattern. Similarly, Western blots show only the 14.5 and 29 kDa proteins. So we do not think we are dealing with a cross-reactivity. But one reason that we went to the *in situ* experiments was to confirm that we were looking at the right thing.

Fukuda: One useful control for experiments on promotion of neurite outgrowth might be to use the other lectin when it is not supposed to be active.

Jessell: On spinal cord cells, that would be a good experiment because, at least from our cell surface binding studies, the 14.5 kDa lectin doesn't bind to spinal cord, but the 29 kDa one does. We are studying the cell-type specificity of these assays.

Fukuda: The 14.5 kDa lectin binds to poly-*N*-acetyllactosamine. So you may need poly-*N*-acetyllactosamine to inhibit interaction.

Jessell: We are looking for multivalent saccharides to get a more potent inhibition of this effect. Most of the epitopes we have identified on sensory neurons seem to be associated with glycolipid structures, although we also have some evidence for poly-*N*-acetyllactosamine oligosaccharides.

Ruoslahti: You said you used the lectin at the same molar concentration as laminin. Laminin is very active in promoting neurite outgrowth. Was the laminin titrated to the lowest possible concentration?

Jessell: We are studying dose–response curves, both for lectin and for laminin.

Hughes: Does immobilized lectin induce neurite outgrowth as well as bundling?

Jessell: We are not certain what is happening at the moment. The experiment is simply to immobilize the lectin and then place the ganglion or dissociated neurons on it. The lectin clearly promotes adhesion but, in addition, it promotes neurite outgrowth. Fasciculation could happen in a number of ways. It could be that adhesion between axonal integrins and the laminin substrate is much tighter than adhesion between carbohydrate and lectin. Therefore, the interactions between adjacent axons, perhaps mediated by molecules such as L1, will promote fasciculation. Another interpretation would be that only a subpopulation of the neurons is capable of extending on the lectin. Thus you get a few axons that grow out and other axons that extend by fasciculating on the pioneer axons. One way of addressing this question is to move to the dissociated cell culture system where we can look at individual neurons and ask whether only those neurons that have complementary saccharide structures are capable of extending.

Hughes: What happens if you plate out the neurons on mixed substrates? For example, if you let the cells stretch out on laminin and then added the lectin, would those neurons bundle?

Jessell: If the laminin was titrated to a point where it doesn't provide a highly adhesive substrate, one could study whether in addition to promoting outgrowth you have effects on direct fibre–fibril interaction. We haven't done that, but it would be a good experiment to do.

Schachner: You coupled your lectin to nitrocellulose. Can you be sure that none is in the supernatant?

Jessell: We are testing for dissociation of the protein from the nitrocellulose substrate under the conditions that we have been using for the binding. It is certainly possible that small amounts are released and are cross-linking surface oligosaccharides, which could contribute to the fasciculation.

Schachner: What happens when you put soluble lectin into that type of assay?

Jessell: If we use a collagen substrate and then add soluble lectin we see very little neurite outgrowth.

Lee: Is the mechanism by which laminin supports neurite outgrowth well known? Can one even suggest that laminin is acting as a lectin?

Jessell: It has certainly been suggested that laminin acts as a lectin, and it may have that function, but in the presence of this chairman I wouldn't say that was the primary mechanism of action! It is now clear that the primary mechanism by which neurites grow axons on laminin is by an interaction with a heterodimer of proteins which belongs to the integrin family. It is interesting that laminin has an α-galactose linkage. It is conceivable that the soluble β-galactoside-binding proteins could interact with laminin at sites other than those responsible for neurite outgrowth. The carbohydrate substitution on laminin is probably far away from the RGD or YIGSR sites that are thought to be involved in adhesion and axon outgrowth.

Ruoslahti: Human laminin does not have that α-galactosyl structure, as far as we can tell. Yet it is quite active in promoting neurite outgrowth.

Feizi: The α-galactose is present in bovine tissues and the bovine 14 kDa lectin reacts strongly with the *N*-acetyllactosamine sequence substituted with α-Gal (Abbott et al 1988).

Barondes: Tom Jessell has pointed out that RL-14.5 and RL-29 have distinct carbohydrate-binding sites: both bind β-galactosides, but they have different specificities when reacted with a series of mammalian glycoconjugates in a competitive binding assay (Leffler & Barondes 1986). For example, adult erythrocyte poly-*N*-acetyllactosamine binds twenty times as well to RL-29 as to RL-14.5. A third rat β-galactoside-binding lectin, RL-18, has different specificity and three human β-galactoside-binding lectins also have distinct specificities (Sparrow et al 1987).

We have evidence for additional rat β-galactoside-binding lectins. For example, in rat intestine, Hakon Leffler and I have resolved 10 distinct β-galactoside-binding lectins by affinity chromatography followed by ion-exchange chromatography. We are characterizing these lectins and determining their carbohydrate-binding specificities. Obviously, this large number of proteins provides the possibility of a variety of specific protein–carbohydrate interactions that could have many biological functions.

Gahmberg: Have you looked at other types of lectins with other specificities?

Barondes: No.

Gahmberg: Has anybody else? Kurt Drickamer has been looking only at proteoglycans that have lectin ability.

Ruoslahti: There are the mannose-specific lectins.

Fukuda: Has anybody isolated lectins from brain tissue?

Barondes: Tom Jessell has done that, as has my research group. In rat brain we find small amounts of RL-14.5, RL-18 and RL-29.

Wassarman: Have you used your probes to look at the copy number of these genes? Also, where are they located?

Barondes: We don't know if there are multiple copies of the same gene. We have analysed a number of genomic clones that encode a human 14 kDa lectin and it appears that there is a single copy. We also have a distinct but related human genomic clone, but we do not know what it encodes.

Wassarman: Is there any evidence from lectin purification for molecules of higher molecular mass than 35 kDa?

Barondes: We have one of 67 kDa. It is a bifunctional protein with an elastin-binding site as well as a carbohydrate-binding site (Hinek et al 1988).

Feizi: By immunoblotting, using the two monoclonal anti-lectin antibodies I mentioned earlier, we have detected a number of immunoreactive polypeptide bands ranging from 14 kDa to greater than 200 kDa in several human and bovine cell lines (Carding et al 1985a,b). We have to be cautious in interpreting these results because we are detecting a tetrapeptide epitope which may occur on totally unrelated proteins. However, because the epitope is so regularly found in the soluble galactose-binding proteins, we think some of the polypeptides of mass greater than 14 kDa may be lectins.

Jessell: We have made antibodies against synthetic peptides encoding the entire sequence of the 14.5 kDa lectin. Many of those antisera react with other proteins on Western blots in different patterns. I think it is very likely that there are other lectins.

Ruoslahti: Is there any correlation between reactivity with other proteins and inhibition of the lectin activity? One might expect an antibody against the active site to be cross-reactive. Dr Feizi, do your antibodies inhibit the lectin activity?

Feizi: Ours do not influence the carbohydrate-binding activity at all. The epitope must be at a distant site.

Jessell: Most of the antibodies against synthetic peptides, even though they react with a native protein, don't inhibit activity.

Ruoslahti: The sequence of that 67 kDa protein is known. Is there any homology to your protein, Dr Jessell?

Jessell: It's not homologous.

Ruoslahti: Does that sequence come from the right protein?

Barondes: We don't know.

Jessell: There are several different laminin-binding proteins of similar molecular masses. Two of them have been sequenced and they seem to have different structures.

Wang: From mouse 3T3 fibroblasts, we have isolated three lectins, designated CBP-35, CBP-16, and CBP-13.5 according to the molecular mass of the polypeptide chains (Roff & Wang 1983). All three are galactose-specific and thiol-dependent; they belong to the S-type family introduced by Kurt Drickamer (this volume). CBP-16 and CBP-13.5 correspond to the lactose-specific lectins identified by Sam Barondes in chicken tissues and by Ten Feizi and her colleagues, and by Stuart Kornfeld and co-workers, in bovine tissues (for reviews, see Barondes 1981, 1988). CBP-35 is the murine counterpart of HL-29 and RL-29 that Sam Barondes described.

Using a highly specific antibody directed against CBP-35, immunoblotting and immunofluorescence studies have shown the lectin to be in the nucleus of 3T3 fibroblasts (Moutsatsos et al 1986). The level of expression and its nuclear localization are dependent on the proliferation state of the cell. In quiescent 3T3 cells, there is a small amount of CBP-35, mostly in the cytoplasm. Mitogenic stimulation of these cells results in increased expression of CBP-35 and an apparent translocation to the nucleus (Moutsatsos et al 1987). Treatment of permeabilized cells with ribonuclease A removed the immunofluorescence due to CBP-35 in the nucleus, whereas parallel treatment with deoxyribonuclease I did not have that effect. These results suggest that CBP-35 is associated with some ribonucleoprotein element of the 3T3 cell nuclei.

This notion is supported by experiments that have identified CBP-35 in a heterogeneous nuclear ribonucleoprotein complex (hnRNP) (Laing & Wang 1988): (a) Nucleoplasm derived from 3T3 cells was fractionated on a caesium sulphate gradient and immunoblotting analysis localized CBP-35 in fractions with densities in the range 1.30–1.35 g/ml, which corresponds to the density of hnRNP on caesium sulphate gradients; (b) Sucrose gradient centrifugation yielded hnRNP as a 40S particle, and immunoblotting with anti-CBP-35 showed that the lectin is found in the same fractions; (c) When nucleoplasm was fractionated on a column of Sepharose derivatized with *N*-ϵ-aminocaproyl-D-galactosamine, the bound and eluted fraction yielded CBP-35, as well as a set of polypeptides whose molecular masses matched those reported for hnRNP. One of these polypeptides was identified as a *bona fide* hnRNP protein on the basis of immunoblotting with a monoclonal antibody specifically directed against hnRNP protein A1. Moreover, the complex isolated on the saccharide affinity column also contained RNA. These results suggest that CBP-35 in hnRNP can bind to the galactose affinity column, which, in turn, co-isolates the other hnRNP polypeptides as a unitary structure.

The final line of evidence is derived from the primary structure of the polypeptide, deduced from the sequence of a cDNA clone for CBP-35 (Jia & Wang 1988). A hydrophilicity plot of the amino acid sequence clearly delineates two domains for the polypeptide. The N-terminal portion is homologous to certain regions of proteins of the hnRNP and the C-terminal portion is homologous to the β-galactose-specific lectins that Tom and Sam described.

Because hnRNP is thought to function in the processing, packaging and transport of messenger RNA, the identification of CBP-35 as a component of this complex suggests a new direction for studying the activities of intracellular lectins.

Ruoslahti: Have you tested for this protein in culture medium?

Wang: Yes. We collect culture medium, much as Tom did with dorsal root ganglia, and subject it to immunoblotting. Occasionally we see a positive immunoblot with anti-CBP-35. Alternatively, we radiolabel the cells with [^{35}S]methionine, take the conditioned medium, and pass it over a galactose-Sepharose column. We can isolate radioactive CBP-35. Therefore there is CBP-35 in the medium, but it is a very small fraction.

Barondes: Douglas Cooper and I have performed extensive experiments with cultured mouse muscle cells to examine the secretion of a cytoplasmic lectin. We find that myoblasts, which contain large amounts of a β-galactoside-binding lectin with subunit relative molecular mass of 14 000 (the mouse form of RL-14.5), retain the lectin in their cytoplasm. However, as the cells differentiate into myotubes, the lectin is secreted into the extracellular matrix. The mechanism for secretion appears to be unconventional and we are now trying to determine it.

Wassarman: Does CBP-35 bind to RNA?

Wang: If you pass a nuclear extract or purified CBP-35 over a single-stranded (ss) DNA-cellulose column, you do get binding. In general, ssDNA rather than ssRNA is a preferable substrate for chromatography of ss nucleic acid-binding proteins because the ssDNA column is more stable (Piñol-Roma et al 1988).

Wassarman: Does the lectin bind ribose?

Wang: No.

Hart: There was a paper about a lectin in frog skin where frogs were treated with adrenaline.

Barondes: That work was done in my laboratory (Bols et al 1986). We showed that a soluble dimeric β-galactoside-binding lectin, which has a subunit relative molecular mass of about 16 000, and is abundant in the cytoplasm of granular glands of *Xenopus laevis*, is secreted by a holocrine mechanism. Secretion is stimulated by the injection of adrenaline which presumably simulates fright. Adrenaline causes contraction of the muscles surrounding the granular gland cells, leading to breaks in their plasma membranes and release of certain cytoplasmic contents, including the lectin. The cells then reseal and regenerate their contents. So this is another example of an unconventional secretion of a cytoplasmic lectin which might prove to be related to the secretion of the cytoplasmic mouse muscle lectin that I mentioned earlier.

Schachner: Can the binding of CBP-35 to RNA compete with that of highly charged extracellular matrix components, such as heparin?

Wang: Our preliminary experiments show that CBP-35 bound to ssDNA-cellulose cannot be completely eluted by heparin.

Ruoslahti: And what about the galactosyl component?

Wang: We haven't tried that.

Lee: Is the degree of hydrophobicity of your galactose-binding zone comparable to that of similar β-galactoside-binding proteins?

Wang: The sequence in the C-terminal portion of CBP-35 is very close to what Sam and Tom have found for their 14–16 kDa lectins. So I would expect the hydrophobicity plots for the corresponding regions to be quite similar.

Fukuda: Is CBP-35 also present in chromosome puffs? Gerry Hart showed that the puff doesn't bind to GlcNAcs, so you may have CBP-35 in there.

Wang: I would like it to be in puffs, but we have not done the experiments.

Stanley: Does the 29 kDa lectin have hnRNP homology in the N-terminal half?

Barondes: Yes, it does.

Feizi: Do any other hnRNP proteins bind galactose? Is there sequence homology or is this just a mere co-aggregation?

Wang: Two hnRNP proteins have been sequenced. One is hnRNP protein A1 (Riva et al 1986). There is no sequence homology, either with the lectin,

C-terminal portion of CBP-35, or with the N-terminal portion. The second is hnRNP protein C1 (Swanson et al 1987). There is sequence homology between C1 and the N-terminal portion of CBP-35. To the best of my knowledge, none of the other hnRNP proteins for which the structures are known have the carbohydrate-binding domain.

Feizi: Are there immunological cross-reactions with the lectin?

Wang: We did affinity chromatography of nucleoplasm and isolated CBP-35 as well as the other polypeptides associated with hnRNP on the basis of galactose-specific binding. We identified CBP-35 in the bound fraction by Western blotting. We also identified a *bona fide* hnRNP protein, designated A1, on the basis of reactivity with a specific monoclonal antibody. So among the polypeptides that we co-isolated with CBP-35 is a *bona fide* hnRNP protein. There was no cross-reactivity between anti-CBP-35 and the other hnRNP proteins.

Feizi: Using our two anti-lectin monoclonal antibodies we (Carding et al 1985b) detected in human lymphocytes immunoreactive proteins of molecular masses 130, 80, 65, 36 and 13 kDa. The amounts and the relative proportions of these proteins changed in association with transformation or after stimulation with plant lectins. These antibodies gave strong immunofluorescence of the nuclei of the stimulated cells.

References

Abbott WM, Hounsell EF, Feizi T 1988 Further studies of oligosaccharide recognition by the soluble 13 kDa lectin of bovine heart muscle. Biochem J 252:283–287

Abbott WM, Mellor A, Edwards Y, Feizi T 1989 Soluble bovine galactose-binding lectin cDNA cloning reveals the complete amino acid sequence and an antigenic relationship with the major encephalitogenic domain of myelin basic protein. Biochem J, in press

Barondes SH 1981 Lectins: their multiple endogenous cellular functions. Annu Rev Biochem 50:207–231

Barondes SH 1988 Bifunctional properties of lectins: lectins redefined. Trends Biochem Sci 13:480–482

Bols NC, Roberson MM, Haywood-Reid PL, Cerra RF, Barondes SH 1986 Secretion of a cytoplasmic lectin from *Xenopus laevis* skin. J Cell Biol 102:492–499

Carding SR, Childs RA, Thorpe R, Spitz M, Feizi T 1985a Multiple proteins related to the soluble galactose-binding animal lectin revealed by a monoclonal anti-lectin antibody. Biochem J 228:147–153

Carding SR, Thorpe SJ, Thorpe R, Feizi T 1985b Transformation and growth related changes in levels of nuclear and cytoplasmic proteins antigenically related to mammalian β-galactoside-binding lectin. Biochem Biophys Res Commun 127:680–686

Drickamer K 1989 Multiple subfamilies of carbohydrate recognition domains in animal lectins. In: Carbohydrate recognition in cellular function. Wiley, Chichester (Ciba Found Symp 145) p 45–61

Hinek A, Wrenn DS, Mecham RP, Barondes SH 1988 The elastin receptor: a galactoside-binding protein. Science (Wash DC) 239:1539–1541

Jia S, Wang JL 1988 Carbohydrate binding protein 35. Complementary DNA sequence reveals homology with proteins of the heterogeneous nuclear RNP. J Biol Chem 263:6009–6011

Laing JG, Wang JL 1988 Identification of carbohydrate binding protein 35 in heterogeneous nuclear ribonucleoprotein complex. Biochemistry 27:5329–5334

Leffler H, Barondes SH 1986 Specificity of binding of three soluble rat lung lectins to substituted and unsubstituted mammalian β-galactosides. J Biol Chem 261:10119–10126

Moutsatsos IK, Davis JM, Wang JL 1986 Endogenous lectins from cultured cells: subcellular localization of carbohydrate-binding protein 35 in 3T3 fibroblasts. J Cell Biol 102:477–483

Moutsatsos IK, Wade M, Schindler M, Wang JL 1987 Endogenous lectins from cultured cells: nuclear localization of carbohydrate-binding protein 35 in proliferating 3T3 fibroblasts. Proc Natl Acad Sci USA 84:6452–6456

Piñol-Roma S, Choi YD, Matunis MJ, Dreyfuss G 1988 Immunopurification of heterogeneous nuclear ribonucleoprotein particles reveals an assortment of RNA-binding proteins. Genes & Develop 2:215–227

Riva S, Morandi C, Tsoulfas P et al 1986 Mammalian single-stranded DNA binding protein UPI is derived from hnRNP core protein A1. EMBO (Eur Mol Biol Organ) J 5:2267–2273

Roff CF, Wang JL 1983 Endogenous lectins from cultured cells. Isolation and characterization of carbohydrate-binding proteins from 3T3 fibroblasts. J Biol Chem 258:10657–10663

Sparrow CP, Leffler H, Barondes SH 1987 Multiple soluble β-galactoside-binding lectins from human lung. J Biol Chem 262:7383–7390

Swanson MS, Nakagawa TY, LeVan K, Dreyfuss G 1987 Primary structure of human nuclear ribonucleoprotein particle C proteins: conservation of sequence and domain structures in heterogeneous nuclear RNA, mRNA, and pre-rRNA-binding proteins. Cell Mol Biol 7:1731–1739

General discussion II

Expression of developmentally regulated carbohydrates

Stanley: One group of carbohydrates looking for a function are those known as developmentally regulated carbohydrates. Nobody has described a lectin that might be specific for structures of that type. We have isolated mutant Chinese hamster ovary (CHO) cells that express large amounts of these developmentally regulated structures and they might be useful in a search for lectins with a more stringent specificity than simply terminal galactose residues. One might find a new class of lectins. We are not doing such experiments, but the properties of these cells are summarized in Table 1.

TABLE 1 (Stanley) Chinese hamster ovary cell mutants expressing developmentally regulated carbohydrates

CHO mutant	*New carbohydrate structures*[a]	*Consequences for lectin interactions*[b]	
LEC10	Gal(β1–4)GlcNAc(β1–2)Man(α1–6) \\ **GlcNAc(β1–4)**Man— / Gal(β1–4)GlcNAc(β1–2)Man(α1–3)	RIC (R_{20})	E-PHA (S_{15})
LEC11	**Fuc(α1–3)** | Gal(β1–4)GlcNAc(β1–2)Man— **Fuc(α1–3)** | NeuAc(α2–3)Gal(β1–4)GlcNAc(β1–2)Man—	WGA (R_8)	RIC (S_{25})
LEC12	**Fuc(α1–3)** | Gal(β1–4)GlcNAc(β1–2)Man—	WGA (R_{50})	RIC (S_4)

[a]The structures shown represent the non-reducing termini of antennae of N-linked, lactosamine-containing carbohydrates. Bold type, new residues.
[b]WGA, wheat germ agglutinin; E-PHA, erythroagglutinin from *P. vulgaris*; RIC, ricin

The new residues added to N-linked carbohydrates and expressed at the cell surface of these dominant CHO mutants are due to the apparently *de novo* expression of the glycosyltransferases GlcNAc-TIII (LEC10; Campbell & Stanley 1984), (α1–3)Fuc-T1 (LEC11; Howard et al 1987) and (α1–3)Fuc-T2 (LEC12; Howard et al 1987). The presence of the new residues (bold type on Table 1) has profound effects on the interaction of each mutant with plant lectins. Thus LEC10 cells become 20-fold more resistant to ricin and 15-fold more sensitive to the erythroagglutinin from *P. vulgaris*, compared with parental cells (Campbell & Stanley 1984). The changes in LEC11 and LEC12 cells cause them to become more resistant to wheat germ agglutinin and hypersensitive to ricin (Stanley 1983).

It can be seen that the new sugar residue in each mutant has a profound effect on the carbohydrate recognition properties of the cells. The addition of each new residue generates a developmentally regulated carbohydrate that is typical of those expressed in a restricted pattern during development and differentiation and is often found in cancerous cells. One might be able to use these cells to find lectins in animal tissues that have a preference for these types of structures. In addition to using glycoproteins from these cells as affinity reagents, one could grow the cells on beads and fix them with glutaraldehyde, thereby generating a matrix which might allow one to fish out interesting molecules.

Sharon: LEC11 exhibits both fucose changes. Even when there is sialic acid, do you still get an increased sensitivity to ricin?

Stanley: Yes, the interesting thing about LEC11 is that when glycoproteins have (α2–3)sialic acid and (α1–3)fucose, the cell becomes much more sensitive to ricin.

Sharon: Increase in sialic acid levels does not necessarily lead to an increase in resistance to ricin. If the sialic acid is attached (α2–6) to galactose, the ricin may still bind (Baenziger & Fiete 1979). It is not clear why LEC11 is more resistant to wheat germ agglutinin.

Stanley: It has also lost some sialic acid.

Hughes: Why is LEC11 sensitive to ricin?

Stanley: That's a good question. Building a model of that structure with the sialylated, fucosylated lactosamine might be interesting.

Hindsgaul: Does LEC11 have glycolipids with X activity?

Stanley: We have not looked at minor glycolipids, because the major one (92%) in these cells is GM_3. Fucosylated glycolipids could be among the minor glycolipids.

Sharon: Has LEC12 completely lost its sialic acid?

Stanley: No, not completely. The cells are 50-fold more resistant to wheat germ agglutinin than the parental cells. However, I have not measured directly the sialic acid content of the LEC12 glycoproteins.

Ruoslahti: What is the missing enzyme?

Stanley: It is acquired, not missing. These are all dominant mutants that have acquired an enzyme activity not expressed by the parent cell. LEC12 cells have acquired the ability to add (α1–3)fucose. In a single step, presumably by controlling one gene, the array of cell surface carbohydrates can be changed. It's an interesting model system from that point of view.

Watkins: Has LEC11 only one (α1–3)-fucosyltransferase?

Stanley: I believe it is one enzyme, because it is completely inhibited by *N*-ethylmaleimide; all (α1–3)-Fuc-T activity is abolished. The LEC12 enzyme, however, is resistant to *N*-ethylmaleimide. There could be two enzymes in LEC11 that are both equally sensitive to *N*-ethymaleimide, but it is not a mixture of the LEC12 enzyme and another enzyme.

Watkins: In normal granulocytes we find an enzyme that appears to behave like the one in LEC11 cells; it adds fucose to the O-3 position of *N*-acetylglucosamine in *N*-acetyllactosamine end-groups with β-galactosyl residues that are either unsubstituted or sialylated at the O-3 position. In leukaemic granulocytes the (α1–3)-fucosyltransferase behaves like your LEC12 enzyme in that it transfers fucose only to the non-sialylated structures (Johnson & Watkins 1987).

Stanley: Minoru Fukuda has found structures that correlate with LEC-like transferases in chronic myelogenous leukaemic cells.

Barondes: What is the mechanism for the mutations?

Stanley: We have absolutely no idea. I wish to study the molecular genetics of this enzyme system. The forward mutation rates are very low; for LEC11 and LEC12 we appear to need mutagens, but the LEC10 mutant can be obtained spontaneously. The reverse mutation frequencies are very high—about 1000 to 10 000 times greater than the forward frequency. Something happens that allows these dominant mutants to revert readily.

Barondes: But is it clear that just a single gene is involved?

Stanley: Until they are cloned, we cannot be sure. Each mutant was isolated in a single step, and they were not exposed to a long period of selection. Therefore they should result from single genetic events. We try not to do anything that would produce multiple changes.

Finne: We have similar cell lines in a mouse melanoma cell system, and we also get the variant in a single step. There is an increase of 60 to 70-fold in the specific (α1–3)-fucosyl transferase (Finne et al 1982). Therefore we have an increased amount of fucose in N-linked glycans and less sialic acid. We also obtain revertants in a single step. Cells with this developmental antigen are less metastatic than the original cells, whereas the revertants are again more metastatic (Finne et al 1989). We also think there must be a developmental switch which is turned on and off. It would be nice to isolate those genes to find the basis of the regulation.

Watkins: That is also what happens in leukaemic cells. An (α1–3)-

fucosyltransferase that does not act on the sialylated structures is present at an early stage of development and then a second enzyme must be switched on as the cells differentiate. But we have never been sure that there are not two different enzymes in normal granulocytes, although we haven't been able to separate them.

Feizi: Are there any differences in the growth or metastatic characteristics of the CHO mutant cells?

Stanley: We haven't studied metastasis directly. Siminovitch's research group in Toronto are now looking at experimental metastasis with the CHO glycosylation mutants. We did look at tumorigenicity. Following a subcutaneous injection, there was no change in the tumorigenicity of these mutants compared with the parental CHO cells.

Fukuda: In leukaemic cells, if fucose is added, sialic acid is not added. That's why they are sensitive to ricin. In leukaemic cells, apparently, this sialyltransferase activity is very high, compared to that in normal granulocytes. Sialylation happens very fast so that fucose is added after sialic acid. In this case, fucose is first added to internal *N*-acetylglucosamine and then to external *N*-acetylglucosamine. This is because even (α2–3)-sialylation hinders the binding of fucosyltransferase.

Stanley: Would the internal fucose react with the monoclonal antibody CSLEX-1?

Fukuda: Yes, if it also contains external fucose.

Ruoslahti: So do the LEC12 cells still have the normal amount of sialyltransferase?

Stanley: Yes, as far as we know. But I have not checked every transferase; we use fairly non-specific sialyltransferase substrates. There are monoclonal antibodies, for example SSEA-1, which recognize the fucosylated structures that are not sialylated. The CSLEX-1 antibody recognizes the sialylated, fucosylated structure. LEC11 cells bind both monoclonal antibodies whereas LEC12 cells only bind the SSEA-1 antibody.

With a glycolipid substrate, nLc_6, where there are two lactosamine groups in a neutral glycolipid, LEC11 prefers to add the fucose to the internal *N*-acetylglucosamine and LEC12 prefers to add it to the external one. Faced with the same substrate, they act differently. If sialic acid is added to the terminal GlcNAc, LEC12 will not fucosylate it, whereas LEC11 will use it as a substrate. If vesicular stomatitis virus (VSV) is grown in the cells, the biantennary carbohydrates of the G glycoprotein have the sialylated, fucosylated structure. So in that case it is not on an extended polylactosamine chain.

Watkins: Does LEC11 like the sialyl substrate and unsialylated substrates equally well?

Stanley: Yes, *in vitro*.

Ruoslahti: Those cell lines might be good for cloning those enzymes.

Stanley: We have been trying to do that for some years. If you take wild-type CHO cells and transfect in human DNA, transfectants can be selected with wheat germ agglutinin and screened for the presence of (α1–3)fucose residues with monoclonal antibodies. You get positives in every experiment. The problem appears to be activation of the endogenous CHO genes encoding (α1–3)-Fuc-T1 and (α1–3)-Fuc-T2 with the transfection procedure.

Ruoslahti: I wouldn't do it that way. I would make a subtraction cDNA library.

Stanley: Chris Campbell, who defined the enzymes present on these mutants, has spent the past few years trying to do that, with little success, despite help from experienced molecular biologists. It is probably difficult because of the amount of mRNA encoding glycosyltransferases.

Ruoslahti: So even though there is a 50-fold increase, is it still a minor mRNA?

Stanley: There could be ten mRNA molecules per cell.

Kobata: How about the sialyltransferase activity in LEC10? E-PHA should not act on the sialylated complex-type sugar chains.

Stanley: I haven't measured that. When we did NMR of the carbohydrates associated with VSV grown in LEC10 cells, there was certainly sialic acid present, probably on the lower arm. It's probably reduced, but I don't know for sure.

References

Baenziger JU, Fiete D 1979 Structural determinants of *Ricinus communis* agglutinin and toxin specificity for oligosaccharides. J Biol Chem 254:9795–9799

Campbell C, Stanley P 1984 A dominant mutation to ricin-resistance in CHO cells induces UDP-GlcNAc: glycopeptide β-4-N-acetylglucosaminyltransferase III activity. J Biol Chem 261:13370–13378

Finne J, Burger MM, Prieels J-P 1982 Enzymatic basis for a lectin-resistant phenotype: increase in a fucosyltransferase in mouse melanoma cells. J Cell Biol 92:277–282

Finne J, Castori S, Feizi T, Burger MM 1989 Lectin-resistant variants and revertants of mouse melanoma cells: differential expression of a fucosylated cell surface antigen and altered metastasizing capacity. Int J Cancer 43:300–305

Howard DR, Fukuda M, Fukuda MN, Stanley P 1987 The GDP-fucose: N-acetylglucosaminide 3-α-L-fucosyltransferases of LEC11 and LEC12 Chinese Hamster ovary mutants exhibit novel specificities for glycolipid substrates. J Biol Chem 262:16830–16837

Johnson PH, Watkins WM 1987 Sialyl compounds as acceptor substrates for fucosyltransferases in normal and leukaemic granulocytes. Biochem Soc Trans 15:396

Stanley P 1983 Lectin-resistant CHO cells. Selection of new lectin-resistant phenotypes. Somatic Cell Genet 9:593–608

Function and pathology of the sugar chains of human immunoglobulin G

Akira Kobata, Tsuguo Mizuochi, Tamao Endo and Kiyoshi Furukawa

Department of Biochemistry, Institute of Medical Science, University of Tokyo, 4-6-1 Shirokanedai, Minato-ku, Tokyo 108, Japan

Abstract. Human immunoglobulin G (IgG) is unique among serum glycoproteins because it contains more than 30 different biantennary complex-type asparagine-linked oligosaccharides. This extremely high microheterogeneity is probably produced because human individuals have a series of B cell clones equipped with different sets of glycosyltransferases. Despite this complex composition, IgG samples purified from whole human sera have the same mole ratios of oligosaccharides, indicating that the ratio of B cell clones synthesizing IgGs with different sugar chains is constant in healthy individuals. We found that the glycosylation patterns of whole serum IgGs obtained from patients with rheumatoid arthritis (RA) are quite different from those of whole serum from healthy individuals. Structural studies of the oligosaccharides revealed that the sugar chains of the IgGs obtained from patients with RA are depleted of the β-galactose residues. The sugar chains of transferrin from patients with RA are fully galactosylated. Therefore the galactose deletion from IgG is probably brought about by a decrease in galactosyltransferase activity in B cells rather than by degradation by galactosidase during circulation. Enzymic study revealed that human B cells contain various β-galactosyltransferases which form the Gal(β1–4)GlcNAc groups in the sugar chains of different glycoproteins. Among these enzymes, abnormality in patients with RA was found only in the one that transfers β-galactose residues specifically to degalactosylated IgG. This enzyme showed lower affinity toward UDP-Gal in B cells of patients with RA than that in healthy individuals.

1989 Carbohydrate recognition in cellular function. Wiley, Chichester (Ciba Foundation Symposium 145) p 224–240

Human immunoglobulin G (IgG) is a glycoprotein composed of two types of polypeptide chains, heavy (H) and light (L), with a stoichiometry of H_2L_2. Each of the H chains contains an asparagine-linked sugar chain at Asn-297 in the C_H2 domain of the Fc region (Sutton & Phillips 1982). Although L chains usually lack sugars, approximately 29% of the IgG molecules in human sera contain sugars in the L chains as well as in the variable region of the H chains (Abel et al 1968, Spiegelberg et al 1970). Structural study of the oligosaccharides released by hydrazinolysis (Takasaki et al 1982) from IgG samples purified from human sera revealed that the sugar chains of IgG differ in many ways from those of other human serum glycoproteins. The oligosaccharide mixtures released from IgG samples obtained from healthy individuals (normal IgGs) all separated into a neutral (N) and two acidic fractions (A1 and A2) on paper electrophoresis. Sialidase digestion revealed that the A1 and A2 fractions are mono- and disialylated oligosaccharides, respectively. A characteristic feature revealed by this study is that approximately 75–76% of all the oligosaccharides released from an IgG sample are found in fraction N. This is unusual: the sugar chains of all other serum glycoproteins so far investigated are highly sialylated and only a few percent of the oligosaccharides are neutral. Complex-type sugar chains with fucosylated outer chains are resistant to sialylation (Yamashita et al 1980), but no fucosyl residue is found in the outer chain moieties of the sugar chains of IgG (see below).

Study of the neutral oligosaccharide mixtures obtained by sialidase digestion revealed that the basic structure of the oligosaccharides is rather simple (Fig. 1). However, the variety afforded by the presence or absence of the two galactoses, of the bisecting *N*-acetylglucosamine and of the fucose residue produces sixteen different sugar chains. This extremely high structural multiplicity is another distinctive feature of the sugar chains of human IgG, compared with those of other serum glycoproteins. Although the basic structures of the sugar chains of bovine and rabbit IgGs are the same as those of human IgG (Taniguchi et al 1985), the sugar chains of mouse IgG completely lack the bisecting *N*-acetylglucosamine (Mizuochi et al 1987).

Separation of the sixteen oligosaccharides of human IgG

The structures of the sixteen asialo-oligosaccharides of human IgG are shown in Table 1. This oligosaccharide mixture was separated into fourteen fractions by serial chromatography on three immobilized lectin columns (Harada et al 1987) (Fig. 2). A column containing *Aleuria aurantia* lectin (AAL)-Sepharose retains those oligosaccharides with the Fuc(α1–6)GlcNAc disaccharide unit (Yamashita et al 1985). Therefore, oligosaccharides 1–8 in Table 1 are retained in the column, while oligosaccharides 9–16 pass through the column and are recovered in the AAL(–) fraction. The bound oligosaccharides are eluted from the column by buffer containing 1 mM L-fucose. Both fractions are separated

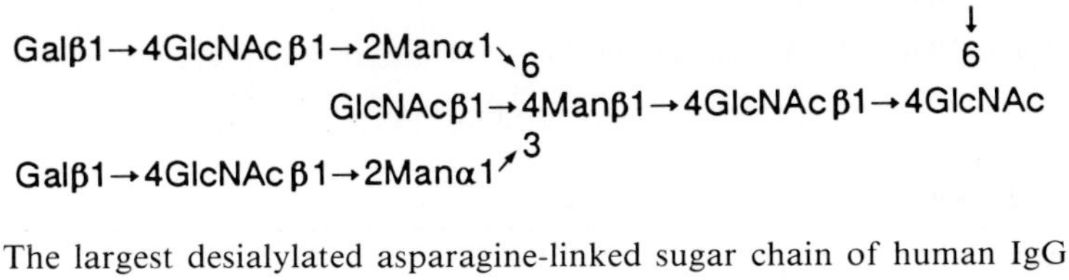

FIG. 1. The largest desialylated asparagine-linked sugar chain of human IgG.

TABLE 1 Structures of desialylated asparagine-linked sugar chains found in human IgG

		R	
Fractions[a]	*Structures*	*[Fuc-(α1-6)] GlcNAc*	*GlcNAc*
I	GlcNAc(β1-2)Man(α1-6) \\ Man(β1-4)GlcNAc(β1-4)R / GlcNAc(β1-2)Man(α1-3)	1	9
I	GlcNAc(β1-2)Man(α1-6) \\ GlcNAc(β1-4)Man(β1-4)GlcNAc(β1-4)R / GlcNAc(β1-2)Man(α1-3)	2	10
IV	GlcNAc(β1-2)Man(α1-6) \\ Man(β1-4)GlcNAc(β1-4)R / Gal(β1-4)GlcNAc(β1-2)Man(α1-3)	3	11
II	GlcNAc(β1-2)Man(α1-6) \\ GlcNAc(β1-4)Man(β1-4)GlcNAc(β1-4)R / Gal(β1-4)GlcNAc(β1-2)Man(α1-3)	4	12
III	Gal(β1-4)GlcNAc(β1-2)Man(α1-6) \\ Man(β1-4)GlcNAc(β1-4)R / GlcNAc(β1-2)Man(α1-3)	5	13
V	Gal(β1-4)GlcNAc(β1-2)Man(α1-6) \\ GlcNAc(β1-4)Man(β1-4)GlcNAc(β1-4)R / GlcNAc(β1-2)Man(α1-3)	6	14
VI	Gal(β1-4)GlcNAc(β1-2)Man(α1-6) \\ Man(β1-4)GlcNAc(β1-4)R / Gal(β1-4)GlcNAc(β1-2)Man(α1-3)	7	15
VII	Gal(β1-4)GlcNAc(β1-2)Man(α1-6) \\ GlcNAc(β1-4)Man(β1-4)GlcNAc(β1-4)R / Gal(β1-4)GlcNAc(β1-2)Man(α1-3)	8	16

[a]The roman numerals correspond to those in Fig. 2

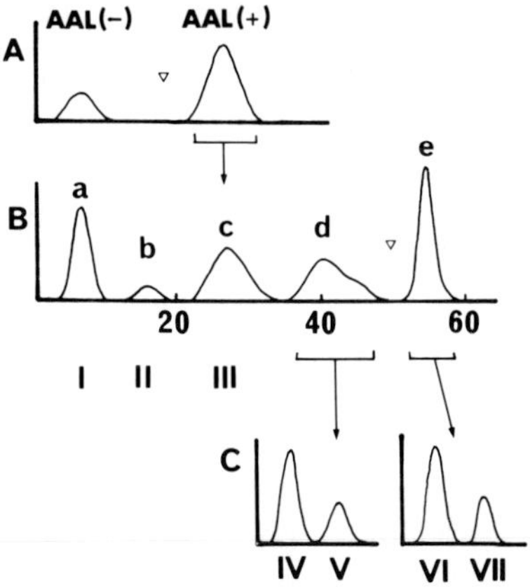

FIG. 2. Fractionation scheme for asialo-oligosaccharides of human IgG by serial lectin column chromatography (Harada et al 1987). Asialo-oligosaccharides were subjected to *Aleuria aurantia* lectin-Sepharose column chromatography (A). The fraction bound to the column, AAL(+), was eluted with buffer containing 1 mM L-fucose and separated into five fractions by chromatography on an RCA-1-agarose column (B). Upon chromatography on an E_4-PHA-agarose column, the peaks d and e were further separated into two fractions (C). By serial chromatography on an RCA-1-agarose column and an E_4-PHA-agarose column, the fraction AAL(−) gave exactly the same fractionation patterns as (B) and (C). The y axis of each figure represents radioactivity, and the numbers on the x axis of (B) represent the retention times (min).

into five peaks by affinity chromatography on a *Ricinus communis* agglutinin 1 (RCA-1) column. Because RCA-1 interacts with β-galactosyl residues, the column is useful for separating oligosaccharides with different numbers of β-galactose residues. Peaks d and e are further fractionated into two peaks by passing through an E_4-PHA-agarose column. The seven fractions labelled from I to VII in Fig. 2 contain the oligosaccharides shown in Table 1. If necessary, oligosaccharides 1 and 2 or oligosaccharides 9 and 10, which are recovered in fraction I of AAL(+) or AAL(−) respectively, can be mutually separated on an immobilized concanavalin A column. Oligosaccharides 2 and 10 pass through the column, while oligosaccharides 1 and 9 are bound to the column and eluted with buffer containing 0.1 M α-methyl mannopyranoside (Ogata et al 1975). Therefore, all sixteen neutral oligosaccharides from human IgG can be completely separated by this serial affinity chromatography. Because the procedure is simple, and the sugar pattern of an IgG can be revealed by using only 1–2 nmol of sample, this method has been widely used in our laboratory.

Comparative study of the sugar chains of IgG myeloma proteins

Despite the extremely high microheterogeneity, the molar ratios of the sixteen neutral oligosaccharides in normal IgGs are quite constant. The molar ratio of oligosaccharides in the three fractions, N, A1 and A2, are almost the same (Table 2). When the three neutral oligosaccharide mixtures obtained by sialidase digestion of these oligosaccharide fractions were subjected to Bio-Gel P-4 column chromatography, they all gave the fractionation pattern shown in Fig. 3A. More detailed analysis of the oligosaccharides by serial affinity chromatography revealed that the molar ratios of bisected oligosaccharides and of fucosylated oligosaccharides were 18% and 84–86% respectively (Table 2).

In contrast, the molar ratios of radioactive oligosaccharides released from thirteen IgG myeloma proteins are not constant (Mizuochi et al 1982): the molar ratios of fractions N, A1 and A2 are quite different in each sample (Table 2). Three of the myeloma proteins lack the A2 fraction. Although these three

TABLE 2 Characteristics of carbohydrate moieties of normal and myeloma IgG molecules

		Percent molar ratio			*Percent of total oligosaccharides with*		
IgG molecule	*Subclass*	*N*	*A1*	*A2*	*M↘GN—M—, M↗* (bisecting GN)	*Fuc*	*Number of Asn-linked sugar chains per mole of IgG*
Normal IgG	—	75.7	18.0	6.3	18.2	86.4	2
Normal IgG	—	75.9	18.3	5.8	18.4	84.5	2
Normal IgG	—	75.2	18.7	6.1	18.1	86.2	2
Yot	IgG1	75.2	23.0	1.8	7.8	75.2	2
Kyo	IgG1	77.6	21.4	1.0	14.3	82.6	2
Han	IgG1	76.6	21.1	2.3	23.3	83.4	2
Ogo	IgG1	83.7	15.7	0.5	28.8	85.8	2
Jir	IgG3	88.8	11.2	0	9.3	81.2	2
Tam	IgG1	88.1	11.9	0	18.2	88.8	2
Ike	IgG1	54.3	42.2	3.5	10.2	85.8	2
Gab	IgG3	89.3	10.7	0	8.9	77.5	3
Saw	IgG1	57.6	33.7	8.7	29.8	86.6	5
RφV	IgG1	81.6	17.7	0.7	18.1	73.3	2
Dom	IgG2	65.8	26.5	7.7	58.4	92.2	4
Heb	IgG4	77.1	18.9	4.0	10.3	82.3	2
Til	IgG2	89.9	9.4	0.7	5.7	92.3	2

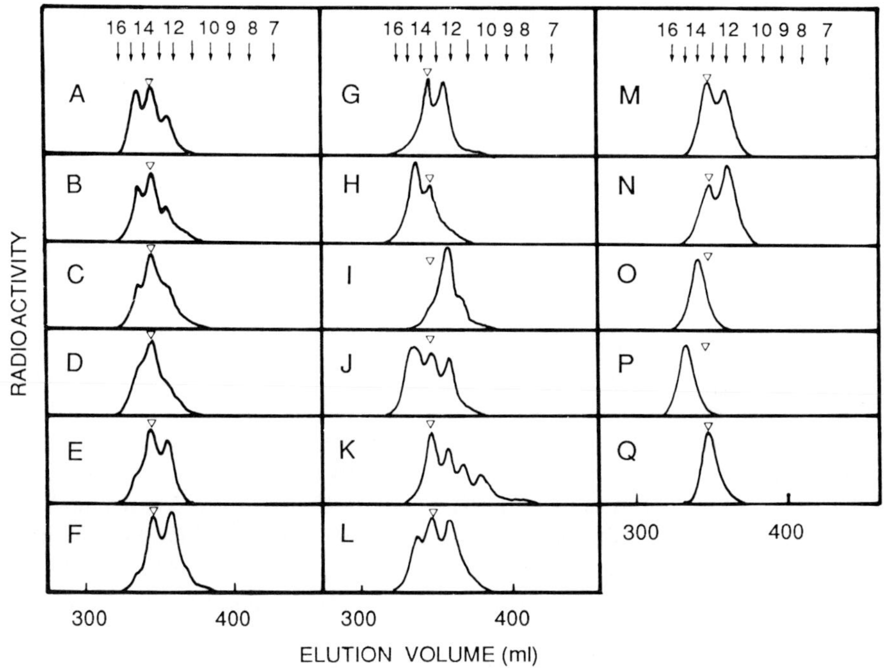

FIG. 3. Bio-Gel P-4 column chromatography of radioactive asialo-oligosaccharide fractions obtained from normal IgG (A), myeloma IgGs (B–N) and Bence-Jones proteins Sm λ (O), Wh λ (P) and Nei λ (Q). The black arrows and numbers indicate the elution positions and number of residues of glucose oligomers added to each fraction as internal standards. The white triangles indicate the elution position of authentic Gal(β1–4)GlcNAc(β1–2)Man(α1–6)[GlcNAc(β1–2)Man(α1–3)]Man(β1–4)GlcNAc(β1–4)[Fuc(α1–6)]GlcNAc$_{OT}$. The data were taken from Mizuochi et al (1982), Endo et al (1989) and Ohkura et al (1985).

proteins contain the A1 fraction, the molar ratio is lower than for the other proteins, indicating that the myeloma cells producing these three proteins might have less sialyltransferase activity than the other cells. The structural variation among the sugar chains of IgG myeloma proteins was more clearly shown when the neutral portions obtained by sialidase digestion were analysed by Bio-Gel P-4 column chromatography. The elution patterns of the oligosaccharide fractions from thirteen myeloma proteins are shown in Fig. 3B–N. The great variation is obvious from these data. Analysis of the structures of oligosaccharides in each fraction revealed that all myeloma proteins contain biantennary complex-type sugar chains with the GlcNAc(β1–2)Man(α1–6)[GlcNAc(β1–2)Man(α1–3)]Man-(β1–4)GlcNAc(β1–4)GlcNAc group as a common structure but, unlike the proteins from normal IgG, differ in the contents of the Gal(β1–4) residues, the bisecting *N*-acetylglucosamine residue and the Fuc(α1–6) residue (Table 2). It is also interesting that some of the IgG myeloma proteins contain more than two asparagine-linked sugar chains. Further detailed structural analysis is needed to elucidate the mechanism producing these odd numbers of asparagine-linked sugar chains. However, most of the extra sugar chains are probably linked to the asparagine residues in the hypervariable region of the L and/or the H chains of myeloma proteins.

Some of the Bence-Jones proteins were reported to contain sugars. Comparative study of the structures of the asparagine-linked sugar chains of three glycosylated Bence-Jones proteins (Sm λ, Wh λ and Nei λ) revealed new evidence, which had not been found in the study of the sugar chains of whole myeloma IgGs (Chandrasekaran et al 1981, Ohkura et al 1985). These Bence-Jones proteins belong to subgroup II and have an Asn-X-Ser/Thr sequence in their hypervariable region. Sm λ has a deletion of 81 amino acid residues in its variable region and contains an asparagine-linked sugar chain at Asn-25 and an O-linked sugar chain at Ser 21 (Kiefer et al 1980). Both Wh λ and Nei λ contain an asparagine-linked sugar chain at Asn-93 but no O-linked sugar chain. A prominent feature is that these two Bence-Jones proteins are 84% homologous in their variable regions, indicating that they have originated from the same germ line (Garver & Hilschmann 1973, Garver et al 1981).

Paper electrophoresis of the radioactive oligosaccharide fractions obtained from the three Bence-Jones proteins by hydrazinolysis revealed that they are mostly sialylated, whereas more than half of the sugar chains of whole myeloma IgGs, are non-sialylated (Table 2). Another important finding, commonly observed for the sugar chains of the three Bence-Jones proteins, is that they give a single peak upon Bio-Gel P-4 column chromatography after desialylation (Fig. 3O, P and Q), in contrast to the several peaks obtained for whole myeloma IgGs (Fig. 3B–N). The structures of the neutral oligosaccharides from Sm λ, Wh λ and Nei λ were found to be 7, 8 and 15 respectively in Table 1 (Ohkura et al 1985).

Two important conclusions can be drawn from these data on myeloma IgGs and glycosylated Bence-Jones proteins. Because the sugar chains in the L chains are homogeneous, the microheterogeneities found in the sugar chains of myeloma IgGs must be restricted to the sugar chains linked to the Fc regions. This discrepancy in the sugar chain structures might have been produced by the structural difference between the L and H chains of the IgG molecule. Because the L chain is small in molecular size and the glycosylation site is located at the variable region, this site may be more accessible for glycosyltransferases than the hinge region of the H chain, which has a more rigid molecular structure. That the three Bence-Jones proteins have different sugar chains is of particular importance. Because the peptide moieties of Wh λ and Nei λ are very similar and their glycosylation sites are the same, the structural difference is probably not caused by differences in the steric effects of their polypeptide moieties. Instead, these differences might be ascribed to the complement of glycosyltransferases in each myeloma cell, which may indicate that B cells are a mixture of cells equipped with different sets of such enzymes, as suggested by the variation in the oligosaccharide patterns of IgG myeloma proteins. If this is the case, IgG molecules may be classified by their sugar chain structures even though microheterogeneity does exist in the whole IgG molecule. It must be stressed that the oligosaccharide patterns of myeloma IgGs cannot be correlated with their currently known subclasses (Table 2). It is possible that such a new subclass of IgG has a different function.

Altered glycosylation of IgG in diseases

As already described, the oligosaccharide patterns of whole serum IgGs isolated from healthy individuals are quite constant. While investigating the sugar chains of more than thirty serum samples, we met with a sample that gave a markedly different oligosaccharide pattern from the others (Fig. 4B compared to 4A). We investigated the physiological background of the donor and found that he was a patient with rheumatoid arthritis. In collaboration with Dr Dwek's research group at Oxford University, we started analysing the sugar patterns of IgGs purified from sera of patients with rheumatoid arthritis. Panels C–G and I–N in Fig. 4 are the fractionation patterns of radioactive oligosaccharides released from eleven rheumatoid arthritis IgG samples. All samples showed a similar tendency to be enriched in smaller oligosaccharides compared with those of normal IgGs.

Detailed structural study of the oligosaccharides by serial affinity chromatography on immobilized lectin columns revealed that the content of galactose residues in the sugar chains of serum IgGs from patients with rheumatoid arthritis is much less than in those from healthy individuals. In contrast, no difference in the incidence of the bisecting *N*-acetylglucosamine residue and of the fucose residue is observed among the sugar chains of both IgG groups (Parekh et al 1985).

As shown above, the desialylated radioactive oligosaccharides released from IgG samples by hydrazinolysis can be easily separated into non-galactosylated, mono-galactosylated, and di-galactosylated oligosaccharides by affinity chromatography on an RCA-1 agarose column. Thus the value of the percent molar ratio of mono-galactosyl plus di-galactosyl oligosaccharides could be calculated for each serum IgG sample. The values for samples from patients with rheumatoid arthritis are significantly lower than those from healthy individuals. This abnormality in sugar chains was not found in the serum IgGs from patients with systemic lupus erythematosis, asthma, upper respiratory infection and Shögren disease (Endo & Kobata 1986, Nagano et al 1986). Serum

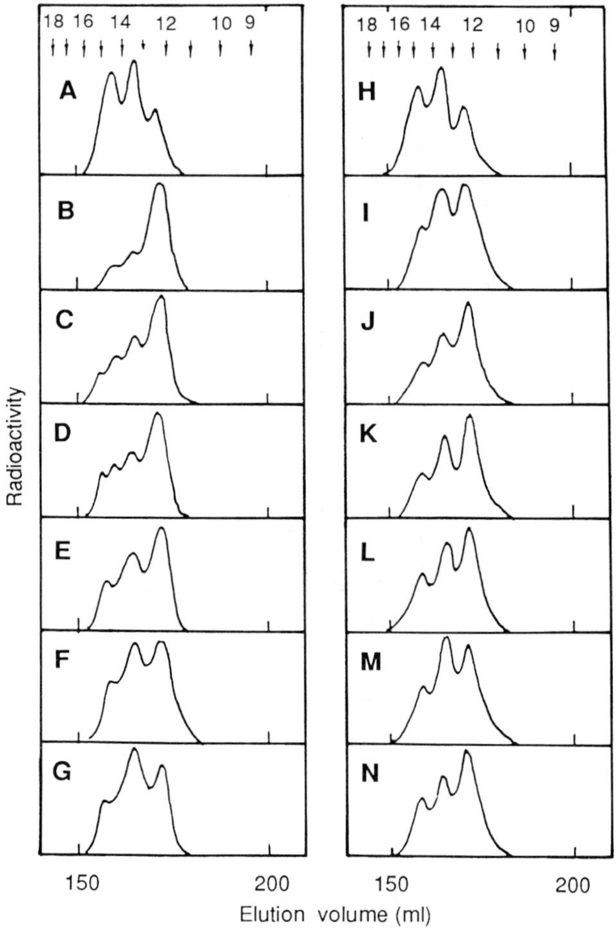

FIG. 4. Bio-Gel P-4 column chromatography of radioactive asialo-oligosaccharide fractions obtained from normal IgGs (A and H) and from rheumatoid arthritis IgGs (B–G and I–N). Black arrows are the same as in Fig. 3.

IgGs from patients with osteoarthritis showed moderately, but significantly, lower values than normal IgGs (Parekh et al 1985). Interestingly, very low content of galactose residues was found in the serum IgGs from patients with tuberculosis.

Enzymic basis for the altered glycosylation of serum IgG from patients with rheumatoid arthritis

The sugar chains of transferrin samples purified from the sera of patients with rheumatoid arthritis are fully galactosylated, as are those from healthy individuals (T. Endo & A. Kobata, in preparation). Therefore, galactose deficiency in rheumatoid arthritis may be limited to the sugar chains of IgG molecules. This also indicated that the galactose deficiency is caused by a change in the galactosyltransferase of B cells rather than by degradation by galactosidase during circulation.

Therefore, the activities of galactosyltransferases in detergent-activated B cell homogenates were assayed by using asialo-agalacto-IgG (AsAg-IgG), asialo-agalacto-transferrin (AsAg-Tf) and asialo-ovine submaxillary mucin (As-OSM) as acceptors (Furukawa et al 1987). B cells were obtained from blood of patients with rheumatoid arthritis and of healthy individuals. These cells, and cell lines obtained by immortalization of B cells from normal individuals and rheumatoid arthritis patients with Epstein–Barr virus (EBV), and a line of B cell lymphoma (BALL-1) were used to prepare homogenates.

No significant difference in the galactosyltransferase activity was observed between normal and rheumatoid arthritis B cell homogenates when As-OSM was used as an acceptor (Table 3). When assayed with AsAg-Tf as an acceptor, the galactosyltransferase activities in B cells from rheumatoid arthritis are slightly lower than those in normal B cells. A significant difference was observed by using AsAg-IgG as an acceptor: B cells collected from blood of patients with rheumatoid arthritis were found to have 60% of the galactosyltransferase activity of those of normal individuals. Experiments with EBV-immortalized B cells gave similar results to those with intact B cells, indicating that EBV transformation does not affect the activity of galactosyltransferases of both B cell groups. The enzyme activities in BALL-1 cells, which do not produce any IgG, were approximately 25% of those in normal B cells, irrespective of the three acceptors.

Kinetic studies uncovered important evidence about the cause of the low galactosyltransferase activity toward AsAg-IgG. The K_m values of the galactosyltransferases for AsAg-IgG in the four different cell homogenates were

TABLE 3 Galactosyltransferase activities in human B cell homogenates

Acceptor (250 µg)	*Specific activities (pmol Gal transferred/mg protein per hour)*				
	Normal		*Rheumatoid arthritis*		
	—	*EBV*[a]	—	*EBV*[a]	*BALL-1*[b]
AsAg-IgG	329 ± 64 (9)[c]	330 ± 72 (10)	186 ± 21 (5)	222	95
AsAg-Tf	1014 ± 74 (3)	1243 ± 390 (2)	786	1194	345
As-OSM	1784 ± 389 (3)	1085 ± 122 (2)	1526	n.d.[d]	700

[a]B cells immortalized with Epstein–Barr virus.
[b]B cell lymphoma.
[c]Values are means ± standard deviations (number of persons examined).
[d]Not determined.

almost the same; 1.4–1.5 nM for the sugar chains. In contrast, the K_m for UDP-Gal, determined by using AsAg-IgG, was 36 μM in rheumatoid arthritis B cell homogenates, which is more than twice the K_m measured in the homogenates of normal B cells (16 μM), EBV-immortalized normal B cells (15 μM) and BALL-1 cells (17 μM). The K_m measured in EBV-immortalized rheumatoid arthritis B cells was even higher (43 μM). Therefore, the low galactosyltransferase activities in BALL-1 cell homogenate did not arise from a lowered affinity for UDP-Gal or the acceptors, but rather the low concentrations of enzymes in the homogenate.

These results indicated that antibody-secreting B cells contain a specific galactosyltransferase that transfers galactose from UDP-Gal to AsAg-IgG, and the affinity of this enzyme for UDP-Gal in the B cells of patients with rheumatoid arthritis is decreased. This change leads to the production of IgG depleted of galactose residues. The data also suggested that there is some genetic background in the outbreak of this complicated disease.

Acknowledgements

The authors express their gratitude to Yumiko Kimizuka for her expert secretarial assistance. Research described in this paper has been supported by Grant-in-Aid for Scientific Research from the Ministry of Education, Science and Culture of Japan and a grant from the Yamanouchi Foundation for Research on Metabolic Disorders.

References

Abel CA, Spiegelberg HL, Grey H 1968 The carbohydrate content of fragments and polypeptide chains of human γG-myeloma proteins of different heavy-chain subclasses. Biochemistry 7:1271–1278

Chandrasekaran EV, Mendicino A, Garver FA, Mendicino J 1981 Structures of sialylated O-glycosidically and N-glycosidically linked oligosaccharides in a monoclonal immunoglobulin light chain. J Biol Chem 256:1549–1555

Endo T, Kobata A 1986 Structure and function of the carbohydrate chains of immunoglobulin G. (in Japanese) Modern Media 32:423–433

Endo T, Kochibe N, Kobata A 1989 Structural study of the carbohydrate moieties of two human immunoglobulin subclasses (IgG2 and IgG4). Glycoconjugate J 6:57–67

Furukawa K, Matsuta K, Takeuchi F et al 1987 Alteration of a galactosyltransferase in the B cells of rheumatoid arthritis patients. Proc IXth Int Symp Glycoconjugates Lille France 6–11 July E56, A Lerouge, Tourcoing, France

Garver FA, Hilschmann N 1972 The primary structure of a monoclonal human λ-type immunoglobulin L-chain of subgroup II (Bence-Jones protein Nei). Eur J Biochem 26:10–32

Garver FA, Chang LS, Kiefer CR et al 1981 Localization of the carbohydrate units in a human immunoglobulin light chain, protein Sm λ. Eur J Biochem 115:643–652

Harada H, Kamei M, Tokumoto Y et al 1987 Systematic fractionation of oligosaccharides of human IgG by serial affinity chromatography on immobilized lectin columns. Anal Biochem 164:374–381

Kiefer CR, Patton HM Jr, McGuire BS Jr, Garver FA 1980 The V region sequence of λ Bence-Jones protein Wh: evidence for separate germ-line sets within λ-subgroups. J Immunol 124:301–306
Mizuochi T, Taniguchi T, Shimizu A, Kobata A 1982 Structural and numerical variations of the carbohydrate moiety of immunoglobulin G. J Immunol 129:2016–2020
Mizuochi T, Hamako J, Titani K 1987 Structures of the sugar chains of mouse immunoglobulin G. Arch Biochem Biophys 257:387–394
Nagano Y, Mizuochi T, Taniguchi T, Matsuta K, Miyamoto T, Kobata A 1986 Structural difference of the sugar chains of IgGs purified from sera of patients with Shögren syndrome with or without complication of rheumatoid arthritis. (in Japanese) Proceeding of the XXXth Annual Meeting of Jpn Rheum Assoc (Yokohama)
Ogata S, Muramatsu T, Kobata A 1975 Fractionation of glycopeptides by affinity chromatography on concanavalin A-Sepharose. J Biochem (Tokyo) 78:687–696
Ohkura T, Isobe T, Yamashita K, Kobata A 1985 Structures of the carbohydrate moieties of two monoclonal human λ-type immunoglobulin light chains. Biochemistry 24:503–508
Parekh RB, Dwek RA, Sutton BJ et al 1985 Association of rheumatoid arthritis and primary osteoarthritis with changes in the glycosylation pattern of total IgG. Nature (Lond) 316:452–457
Spiegelberg HL, Abel CA, Fishkin BC, Grey H 1970 Localization of the carbohydrate within the variable region of light and heavy chains of human γG myeloma proteins. Biochemistry 9:4217–4223
Sutton BJ, Phillips DC 1982 The three-dimensional structure of the carbohydrate within the Fc fragment of immunoglobulin G. Biochem Soc Trans 11:130–132
Takasaki S, Mizuochi T, Kobata A 1982 Hydrazinolysis of asparagine-linked sugar chains to produce free oligosaccharides. Methods Enzymol 83:263–268
Taniguchi T, Mizuochi T, Beale M, Dwek RA, Rademacher TW, Kobata A 1985 Structure of the sugar chains of rabbit immunoglobulin G: occurrence of asparagine-linked sugar chains in Fab fragment. Biochemistry 24:5551–5557
Yamashita K, Tachibana Y, Nakayama T, Kitamura M, Kobata A 1980 Structural studies of the sugar chains of human parotid α-amylase. J Biol Chem 255:5635–5642
Yamashita K, Kochibe N, Ohkura T, Ueda I, Kobata A 1985 Fractionation of L-fucose-containing oligosaccharides on immobilized *Aleuria aurantia* lectin. J Biol Chem 260:4688–4693

DISCUSSION

Finne: In rheumatoid arthritis, is the galactosyltransferase changed in all the B cells? Or could there be enzymes with different properties in different subsets of B cells, the observed change being due to the expansion of one subset of the cells?

Kobata: That is hard to answer. In the study using the Epstein–Barr virus (EBV) transfectant, the possibility of picking up some particular clones cannot be eliminated. However, the data obtained with whole B cell populations from a patient with rheumatoid arthritis gave the same low enzyme activity. Therefore, it is more probable that all B cells have an altered enzyme.

Ruoslahti: Have you done any genetic studies?

Kobata: We have started family studies. The sample is small, but there seems to be some genetic background to this phenomenon. Rheumatoid arthritis has long been considered an autoimmune disease, but these findings may soon change our ideas.

Feizi: Have you looked at any other glycosyltransferase activities, such as *N*-acetylglucosaminyltransferase, in the same cells?

Kobata: No, we have not.

Feizi: Do B cells from patients with rheumatoid arthritis produce immunoglobulins?

Kobata: Yes.

Feizi: Do you think the anti-IgG, which these cells may well be producing, is in some way blocking the glycosylation sites on your acceptor IgG?

Kobata: I cannot answer that question. The undergalactosylated sugar chains are also found in serum IgG of healthy individuals, but in much smaller amounts. So there might also be a mechanism to inhibit galactosylation to some extent in normal IgG.

Feizi: How about doing your galactosyltransferase assay using IgG as acceptor and adding rheumatoid factor (anti-IgG) to the reaction mixture? You might find that because the rheumatoid factor binds to the IgG acceptor, galactose transfer is inhibited.

Ruoslahti: That might explain the *in vitro* results, but it is hard to explain the *in vivo* results.

Feizi: I have wondered whether the altered oligosaccharide profile in rheumatoid arthritis is due to an increased clearance of IgG molecules with terminal galactose residues. We know that aggregated IgGs are present in sera of patients with rheumatoid arthritis; there will therefore be clustered oligosaccharides terminating with galactose. Are they not likely to be cleared by the liver?

Rademacher: What we are looking at in whole serum is a change which seems to be affecting a large number of B cells in these patients. Professor Kobata (and Ivan Roitt's group also) has obtained data which suggest that galactosyltransferase activities in these cells are changed. That does not tell you what IgG rheumatoid factor is, nor does it tell you what is in immune complex. Those are separate issues. We have been looking at what is left after immune complex formation.

Feizi: Of course, the glycosyltransferase issue needs to be resolved. But I would like to suggest that the enzyme assays might be repeated in the presence of rheumatoid factors.

Lindahl: Dr Feizi, if your suggestion about impaired accessibility of the acceptor site is correct, wouldn't that show up as a change in the K_m for the sugar acceptor rather than for the UDP-Gal?

Feizi: I would like to refer that to the experts.

Kobata: So you think the rheumatoid factor will bind to the sugar moiety and cover the acceptor site for the transferase, or may interact with the peptide and just systemically inhibit? I think the data will be hard to interpret.

Feizi: It would be easy to test. Dr Lee, what is your view on the possible accelerated clearance of aggregated IgG by the liver? Baynes (1983) retracted the results in Day et al (1980) and Thornburg et al (1980) because IgG antibody complexes with the originally reported properties could not be reproduced and therefore the clearance experiments could not be repeated.

Lee: I think the issue has been resolved by the author's admission that the results cannot be reproduced.

Feizi: So is aggregated IgG not likely to be cleared by the hepatic receptor?

Lee: I don't work on IgG or rheumatoid factors so I cannot say. But I now think the exposure of a galactose as an immune defence mechanism is incorrect.

Fukuda: The selective galactosylation of transferrin rather than immunoglobulin is interesting. The glycosyltransferase produced in rheumatoid arthritis must detect some change in the immunoglobulin, and therefore does not add galactose. Dr Kobata did you use oligosaccharides as substrates to look for K_m or V_{max} differences between the healthy and disease states?

Kobata: We haven't done that. The sugar chains of transferrin are also present on IgG. Therefore the differences cannot be detected by using oligosaccharides or glycopeptides.

Sharon: You are using mature agalacto-IgG as the acceptor. This is probably not the substrate for the galactosyltransferase in the biosynthetic pathway; *in vivo* the individual polypeptide chains are glycosylated and processed first. Using the individual chains is therefore an important control experiment. What would happen?

Kobata: We are looking at the final maturation stage of the sugar chains. We have analysed a human liver enzyme which works on the agalactosylated transferrin and also agalactosylated IgG. However, in this case transfer of galactose to transferrin was more efficient than to IgG.

Sharon: That supports your notion of multiple galactosyltransferases, the specificity of which may be affected by the polypeptide chain.

Kobata: The mechanism is interesting and we would like to study that in more detail, but the phenomenon is quite clear.

Sharon: It may be similar to the reaction catalysed by *N*-acetylglucosaminyl phosphotransferase which is also specified by the protein part of the acceptor; transfer occurs only to the mannose residues of lysosomal enzymes.

Lee: Dr Kobata, you prepare substrate from healthy patients. If you could prepare asialo-agalacto-IgG from patients with rheumatoid arthritis, would you expect a greater difference?

Kobata: I don't think so. The degalactosylation of IgG by the new β-galactosidase was complete.

Lee: You said that even for a healthy patient, removal of galactose is difficult. Thus you might see an even greater degalactosylation for patients with rheumatoid arthritis. The converse is that if it is difficult to remove the galactose, it is probably also difficult to add galactose.

Kobata: We think precursor sugar chains are added at a very early stage. Perhaps maturation occurs after the two heavy chains are coupled. There may be a large steric effect from the polypeptide moiety at this galactosylation stage. The study of galactosylation using degalactosylated IgG is therefore considered to mimic the last stage of sugar maturation in IgG.

Schachner: What is the function of the sugars on immunoglobulin? How is the abnormality of their composition implicated in the disease?

Kobata: Rheumatoid arthritis is a complicated disease, and modification of the sugar moiety of IgG is not the only cause. However, we have some evidence that degalactosylation of IgG modifies its function.

Ruoslahti: How is the function modified?

Kobata: The functional modifications correlate to the Fc fragment.

Ruoslahti: Wouldn't it also have a different half-life?

Kobata: We haven't analysed that yet.

Schachner: What functions do the sugars on the variable region have? That region is strategically important.

Kobata: We have not studied the sugars on the variable region in detail. The sugar chains linked to the variable region could inhibit the binding of the IgG to antigen.

Gahmberg: Using tunicamycin in a B cell making immunoglobulin shows that the sugars are needed for complement fixation.

It might be interesting to look at the sugar composition of the EBV-transformed cells. Is there any difference between the sugars of membrane and soluble glycoproteins?

Kobata: We have not looked at the membrane glycoproteins of EBV-transformed cells.

Rademacher: IgG sugar-dependent functions can be separated into fluid phase and tissue-binding events. The ability of IgG to bind protein A, to bind antigen and to fix complement is not affected by the sugars (Leathbarrow et al 1985). It should be noted that slight changes in the C1q binding constant occur when the oligosaccharide is absent from IgG. This leads to differences in binding being most noticeable at low C1q concentrations. Since it is usually at low effective C1q concentrations that whole complement fixation assays are performed, this may lead to the differences noted by other workers when assaying their carbohydrate-depleted IgG (Koide et al 1977, Nose & Wigzell 1983). This effect could be magnified by the nature of the complement cascade process. This indicates the necessity of performing a complete binding curve and using individual components if a meaningful comparison of binding ability is to be made. But everybody agrees that the ability of IgG, whether in monovalent form or aggregated form, to bind to Fc receptors on monocytes and macrophages is dependent on the sugars. Dr Kobata's results suggest that those sugar-dependent interactions of IgG with cellular receptors are likely to be changed in rheumatoid arthritis. There is no evidence for abnormalities in complement fixation in rheumatoid arthritis.

Kobata: Only degalactosylation can induce the same functional changes as nonglycosylated IgG produced by tunicamycin treatment.

Sharon: Has agalacto-IgG completely lost its Fc receptor functions?

Kobata: Not completely. The tunicamycin experiment also did not kill the activity totally; it just modified the Fc functions.

Rademacher: In lectin fractionation of IgG from patients with rheumatoid arthritis, the fraction containing the agalactose structures appeared to have little cytophilic ability (Malaise et al 1987).

Lee: Dr Kobata, can you extend your ideas to the very low degree of sialylation of IgG sugar chains?

Kobata: Sutton & Phillips (1982) found by X-ray analysis of IgG that the galactose residues are buried within the polypeptide pockets. So the galactose residues may not be readily accessible to sialyltransferases. However, some sialylation does occur. How the galactose residues escape from the protecting effect is interesting; the IgG molecule might be more complicated than we think.

Lee: But could those sialylated residues be from Fc regions or variable regions?

Kobata: They could be. That is a good point.

Rademacher: The sialylated sugars come from the upper half, the Fab.

Carver: In the myeloma IgGs which only have a sugar at the 297 asparagine, you still get some monosialylated sugars, but always with the sialic acid residue attached to the mannose (α1–3) arm.

What is known about the order of addition in the transferase reaction? Does the UDP-Gal bind to the enzyme first, before the substrate? Rather than have a different enzyme, you may have a 'different' substrate—the same sugar, but with a different three-dimensional structure, by virtue of its interaction with the protein. If the order of reaction is such that the UDP-Gal binds last, the complex that forms between the enzyme and the substrate may have a different K_m for UDP-Gal by virtue of the different three-dimensional structures.

Kobata: That's an important point which I'll keep in mind in our future studies.

Fukuda: In the *in vitro* study, Dr Kobata uses the same substrate and sees a difference in K_m. Therefore I don't think the difference arises just from how the acceptor molecules are recognized.

Carver: I wasn't addressing that issue as much as whether there's a unique transferase for IgG.

Kobata: We have to think about that. There are many possibilities for how the discrimination of the substrate occurs. The galactosyltransferase itself might be the same and another protein entity may participate in the reaction.

Watkins: What is known about the state of maturation of the lymphocytes in rheumatoid arthritis? My colleague, Dr Patricia Skacel has been looking at fucosyl- and sialyltransferase levels in normal granulocytes and leukaemic cells (Skacel & Watkins 1988, and unpublished results). She concluded that although there are differences in leukaemia they are probably not connected directly with the leukaemic process, but rather with the stage of maturation arrest of the cells. Do all B lymphocytes go through this phase of having a different β-galactosyltransferase and, if so, are they pushed out from the marrow early in rheumatoid arthritis?

Kobata: I don't think anything is known about that. That is an interesting target for cellular immunologists.

References

Baynes JW 1983 Retraction. J Biol Chem 258:8519

Day JF, Thornburg RW, Thorpe SR, Baynes JW 1980 Carbohydrate-mediated clearance of antibody–antigen complexes from the circulation. J Biol Chem 255:2360–2365

Koide N, Nose M, Muramatsu T 1977 Recognition of IgG by Fc receptor and complement: effects of glycosidase digestion. Biochem Biophys Res Commun 75:838–844

Leathbarrow RJ, Rademacher TW, Dwek RA et al 1985 Effector functions of a monoclonal aglycosylated mouse IgG2a: binding and activation of complement component C1 and interaction with human monocyte Fc receptor. Mol Immunol 22:407–415

Malaise MG, Franchimont P, Gomez F, Bouillene C, Mahieu PR 1987 The spontaneous ability of normal human IgG to inhibit the Fc receptors of normal human monocytes is related to their binding capacity to lectins. Clin Immunol Immunopathol 45:1–16

Nose M, Wigzell H 1983 Biological significance of carbohydrate chains on monoclonal antibodies. Proc Natl Acad Sci USA 80:6632–6636

Skacel PO, Watkins WM 1988 Significance of altered α-2-L-fucosyltransferase levels in serum of leukaemic patients. Cancer Res 48:3998–4001

Sutton BJ, Phillips DC 1982 The 3-dimensional structure of the carbohydrate within the Fc-fragment of immunoglobulin-G. Biochem Soc Trans 11:130–132

Thornburg RW, Day JF, Baynes JW, Thorpe SR 1980 Carbohydrate-mediated clearance of immune complexes from the circulation. J Biol Chem 255:6820–6825

The role of oligosaccharides in modifying protein function

Thomas W. Rademacher and Raymond A. Dwek

Oxford Glycobiology Unit, Department of Biochemistry, University of Oxford, South Parks Road, Oxford OX1 3QU, UK

Abstract. It has been proposed that protein-bound oligosaccharides interact with the protein to which they are attached to up- or down-regulate the bioactivity of the 'composite' glycoprotein. Oligosaccharide analyses of the glycoproteins Thy-1, tissue plasminogen activator and immunoglobulin G are presented. Correlations between particular glycoforms and enzymic activities are demonstrated for tissue plasminogen activator. The change in the prevalence of particular immunoglobulin G glycoforms is shown to correlate with disease activity in rheumatoid activity.

1989 Carbohydrate recognition in cellular function. Wiley, Chichester (Ciba Foundation Symposium 145) p 241–256

The function of protein-bound oligosaccharides and the 'apparent' necessity for oligosaccharide microheterogeneity remain the major unanswered questions in the field of glycobiology (Rademacher et al 1988b). Convincing data that carbohydrates act as ligands and are involved in recognition events with carbohydrate-binding receptors are lacking. We have been examining an alternative function for protein-bound oligosaccharides—that the carbohydrate primarily interacts with the protein to which it is attached and up- or down-regulates its activity, i.e. there is an interactive relationship between the peptide and its attached oligosaccharide. Glycoproteins therefore bind their own carbohydrates and any interactions with other molecules, via their carbohydrates, will affect their function. Carbohydrate-binding proteins therefore may be considered vehicles to either *trap* or modify particular functional modes of glycoproteins.

To approach this question experimentally, we have been attempting to relate oligosaccharide heterogeneity to its expression at the glycoprotein level. This approach necessitates: (1) the structural analysis of all oligosaccharides at each unique glycosylation 'sequon'; (2) the determination of the molar prevalence of each structure at each sequon; and (3) knowledge of the pairing of oligosaccharides between sites. This type of analysis permits reconstruction of the 'composite' glycoprotein into its set of glycoforms according to their prevalence. Physical fractionation of these glycoforms then allows for experimental analysis of the effects of differential glycosylation on protein function via analysis of the individual glycoforms. This approach is necessary because functional information cannot be obtained at the level of the oligosaccharide itself.

In addition, information about the activity of individual glycoforms can be gained from following the changes in the prevalence of each glycoform in disease states. This follows from the findings that while oligosaccharide sequences remain constant, the prevalence of individual glycoforms is continually changing in both normal and disease states. This has changed fundamentally our perception of the nature and importance of glycosylation. Three examples are detailed here to illustrate these concepts.

Thy-1 as a glycoprotein

Thy-1 is the smallest member of the immunoglobulin superfamily, and in rats and mice is a major cell surface glycoprotein of thymocytes and brain. The Thy-1 polypeptide has 111 amino acids, is attached to the membrane by a glycosyl-phosphatidylinositol anchor, and is N-glycosylated at three sites (Asn-23, -74, and -98, in rat). Fig. 1 demonstrates the relative sizes of the polypeptide and oligosaccharide parts. It is clear that the glycan moieties make a significant contribution to the surface area of the molecule. Indeed, when the Thy-1 molecule is bound by its anchor to the membrane, the polypeptide appears to 'float' within an oligosaccharide shell (Fig. 2). It is also interesting to note that the solubilization of Thy-1 by release from its lipid tail (e.g. by phosphatidylinositol-phospholipase C) would effectively expose a nascent fourth glycan moiety (Homans et al 1988).

A set of 'composite' glycoforms of rat Thy-1 based largely on a determination of the class of oligosaccharide at each site is shown in Fig. 3. Any actual glycoform would be derived from one of these composite forms by taking into account the microheterogeneity of the structures within each class. A comparison of the set of glycoforms constituting rat brain Thy-1 with that constituting rat thymocyte Thy-1 shows that these glycoproteins do not have any glycoforms in common: the two tissues have quite different glycosylation capacities (Parekh et al 1987a).

Fractionation of thymocyte Thy-1 on immobilized lentil lectin is determined predominantly by the structures present at site 74. Analysis of the oligosaccharides at the 'passenger' sites (sites 23 and 98) after lectin fractionation of the whole Thy-1 glycoprotein, showed that these structures were not independent of those present at site 74. These results suggest that some site interaction may take place during oligosaccharide biosynthesis and that the glycoforms are more restricted than would be expected if each site were acting

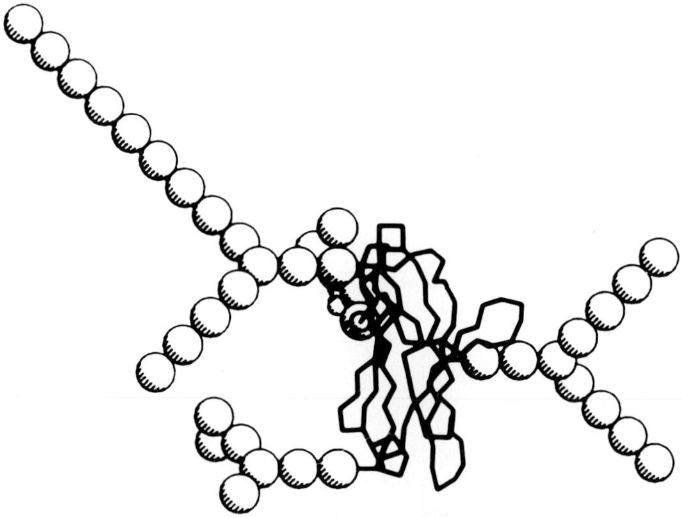

FIG. 1. Computer-generated model of the Thy-1 glycoprotein (minus C-terminal glycan). The carbohydrate residues are represented by spheres and the peptide by a solid black line (from Rademacher et al 1988b).

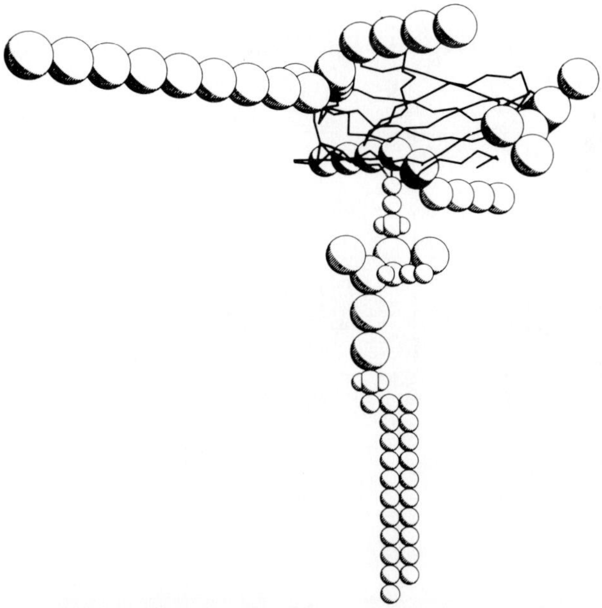

FIG. 2. Computer-generated model of the membrane form of the Thy-1 glycoprotein. The membrane anchor is represented by the vertical assembly of spheres. The soluble form of Thy-1 generated by phosphatidylinositol-phospholipase C action would be a molecule with four attached oligosaccharide chains (from Rademacher et al 1988b and Homans et al 1988).

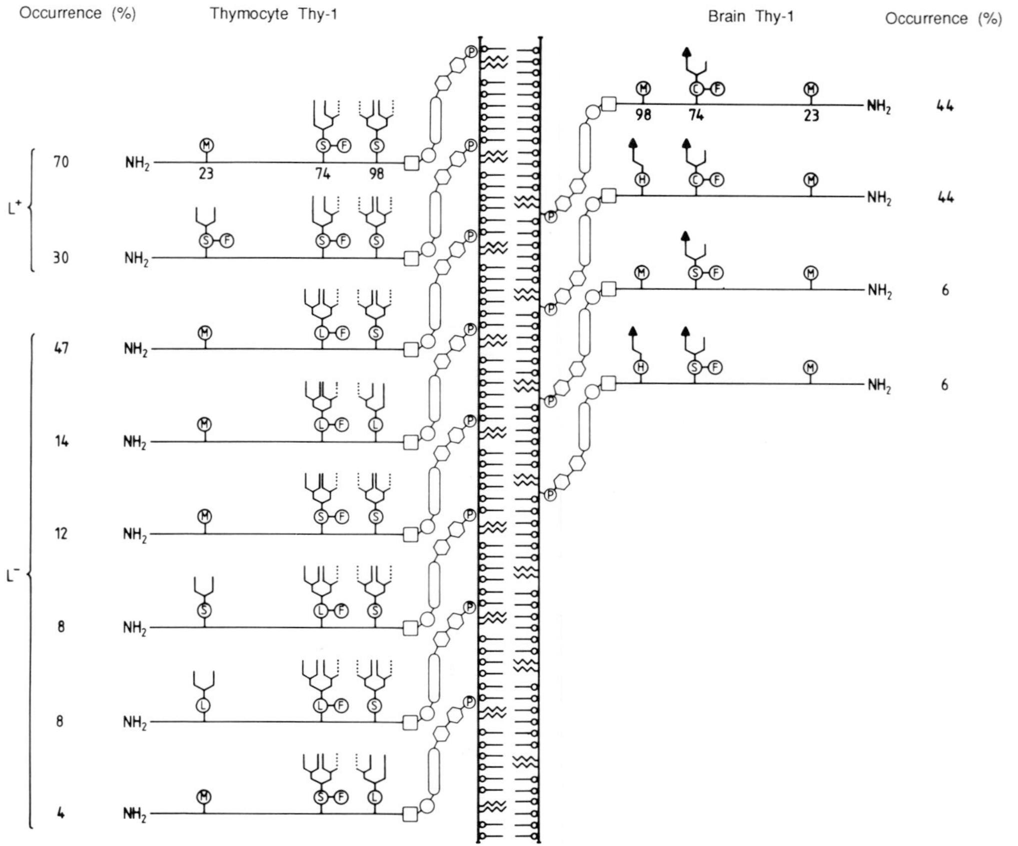

FIG. 3. 'Composite' glycoforms found in brain and thymus Thy-1. Symbols represent the type of structures present at each site (Parekh et al 1987a). L^+, L^-, lectin-positive and -negative respectively.

independently of each other. Alternatively, it may be that the lentil lectin-positive and -negative glycoproteins are made by different cell types. In either case, distinct glycoforms with particular sets of oligosaccharides are being synthesized.

Tissue plasminogen activator as a glycoprotein

Extrinsic tissue plasminogen activator (tPA) is a serine protease which converts plasminogen into plasmin and can thereby induce clot lysis (fibrinolysis). Its value as a thrombolytic agent (compared with plasminogen activators such as urokinase and streptokinase) lies in its clot-specific binding to fibrin. tPA is a glycoprotein of 527 amino acid residues (molecular mass ~70 kDa), with four potential N-glycosylation sites. Cells produce it, however, in at least two variant forms, I and II, which differ in the number of attached oligosaccharides. Type I tPA has three N-linked oligosaccharides (Fig. 4) whereas type II tPA carries only two. tPA is therefore a glycoprotein which illustrates the property of variable glycosylation site occupancy.

Both types of tPA have similar kinetic constants in amidolytic assays using small chromogenic substrates (Wittwer et al 1989). However, the two types differ in the rate at which they form an active complex with fibrin that is able to cleave plasminogen. This activation is characterized by a kinetic transition for type II tPA but not type I. When the oligosaccharides are removed, tPA cleaves plasminogen in the absence of fibrin, although the affinity for fibrin is unchanged. These data together imply that a physiological consequence of tPA N-glycosylation is to down-regulate the tPA (i.e. to prevent tPA from converting plasminogen to plasmin in the absence of a fibrin clot). Furthermore, the different extents of glycosylation of types I and II tPA, with the consequent difference in their kinetic transition in the presence of fibrin, may confer a greater control and sensitivity to tPA-mediated thrombolysis (i.e. slow- and fast-acting forms). It may also be relevant that plasminogen itself also exists in two forms, 1 and 2, with either one or more oligosaccharide units attached.

The complete set of glycoforms for both type I and type II tPA have been determined (Fig. 5) (Parekh et al 1989a). No site interaction during oligosaccharide biosynthesis is evident, in contrast to Thy-1 where site interactions take place. The tPA glycoforms can be separated by their ability to interact with lysine-Sepharose which mimics the ability of these molecules to bind to lysine residues on fibrin. Gradient elution of tPA from lysine-Sepharose with arginine demonstrates that tPA is not a single homogeneous molecule but a population of molecules with different affinities for lysine (Fig. 6). Because the amino acid sequence is the same for all molecules, this heterogeneity of binding reflects the influence of the attached carbohydrates on this interaction. Kinetic analysis of the various fractions indicates a continuous variation in the kinetic properties of the molecules (Fig. 6). Molecules which bind tightest to Sepharose-lysine have

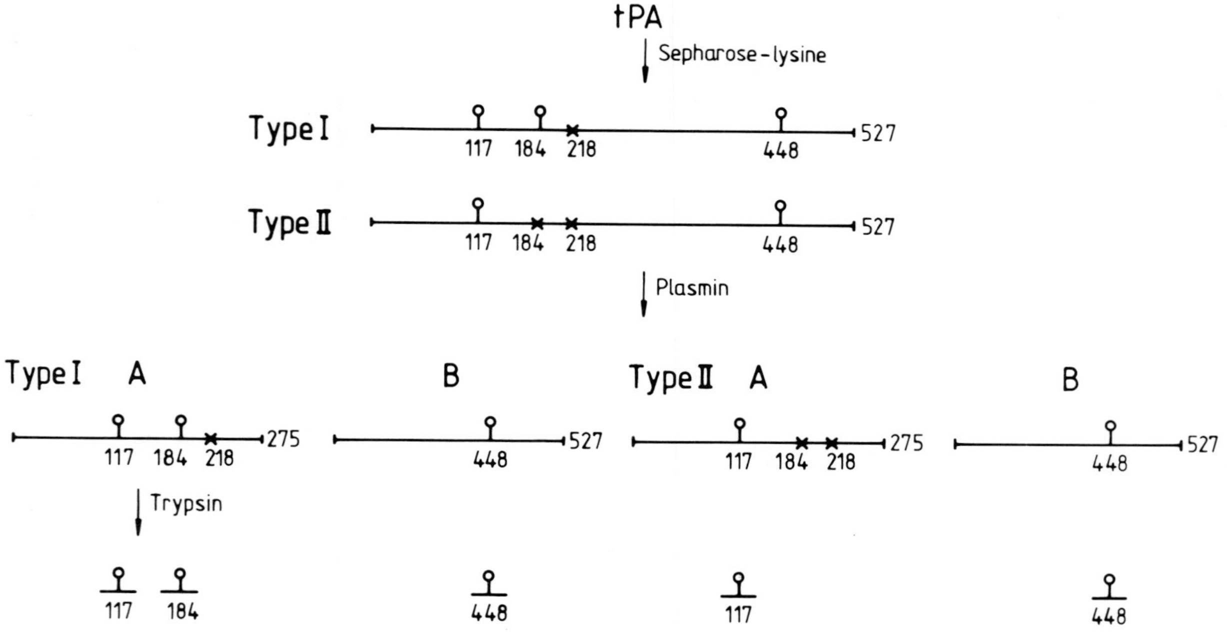

FIG. 4. Schematic diagram of the tissue plasminogen activator (tPA) molecule showing the location of the oligosaccharides on type I and type II variants.

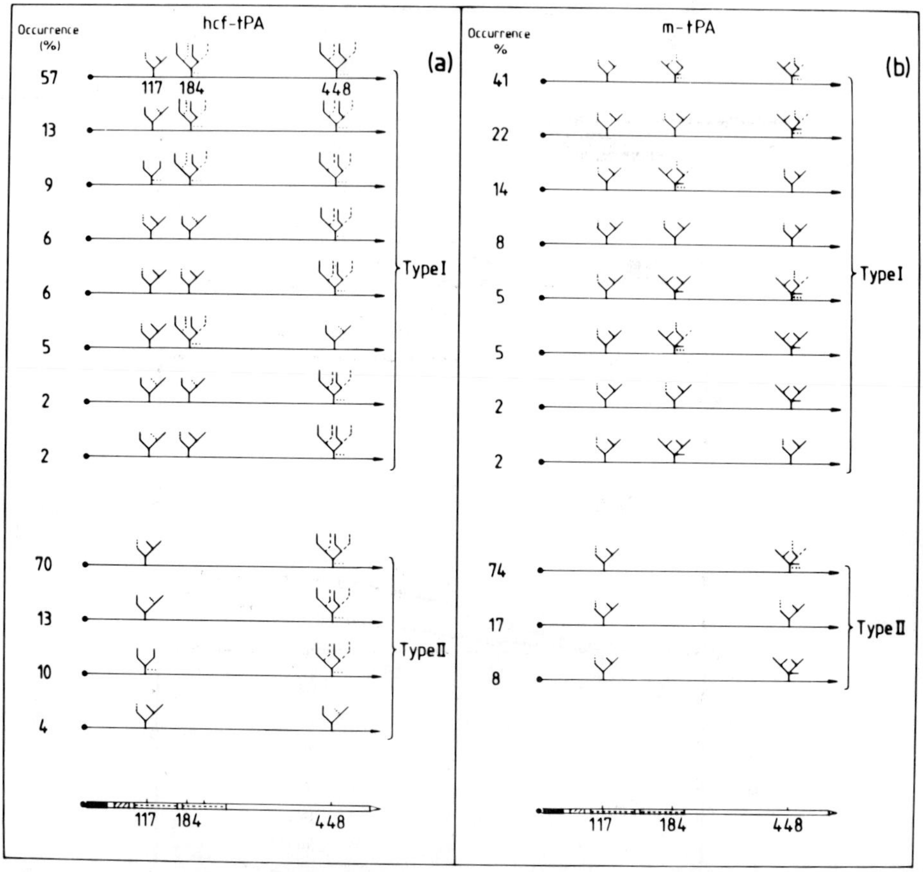

FIG. 5. Set of glycoforms and their prevalence for human colon fibroblast-derived tPA (hcf-tPA) and Bowes melanoma-derived tPA (m-tPA) (From Parekh et al 1989a).

the most rapid activation kinetics, while those weakly binding to lysine are activated poorly by fibrin. This variation in activities and its balance may be important in controlling the seemingly contradictory actions of tPA—dissolving fibrin clots but allowing new fibrin to be deposited. Figures 7 and 8 illustrate the physiological properties of these molecules, their roles in haemostasis and the proposed molecular basis for the down-regulating effect of glycosylation at site 184. Deviations from this critical balance would predispose individuals to thrombosis or haemorrhagic disorders. Alternatively, a change in the prevalence of specific tPA glycoforms may be an important physiological response to excessive fibrin deposition (by analogy to the changes in glycosylation which affect liver-derived glycoproteins during the acute phase response) and an

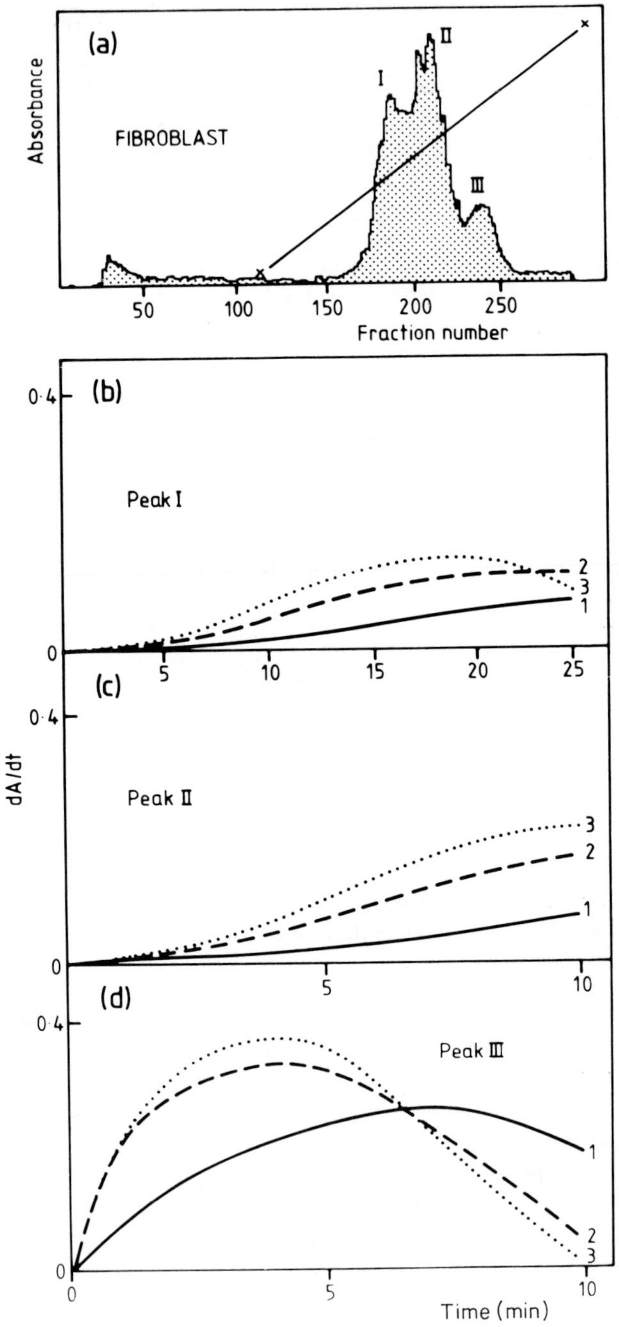
(a)
Absorbance
FIBROBLAST
I
II
III
50 100 150 200 250
Fraction number
(b)
0·4
Peak I
0
5 10 15 20 25
1 2 3
(c)
0·4
dA/dt
Peak II
0
5 10
1 2 3
(d)
0·4
Peak III
0
0 5 10
1 2 3
Time (min)

inadequate response may dispose individuals to the sequelae of thrombosis (myocardial infarction, disseminated intravascular coagulopathy, stroke, etc.).

Immunoglobulin G as a glycoprotein

An analysis of the immunoglobulin G (IgG) molecule provides clear evidence for direct interactions between N-linked oligosaccharides, with the result that only certain IgG glycoforms appear to be allowed. In IgG there is a conserved N-glycosylation site at Asn-297 in each of the C_H2 domains, and the two oligosaccharide units are in direct contact with each other, forming a bridge across the two domains. At least 30 complex-type biantennary oligosaccharides are associated with serum IgG, but X-ray crystallography and oligosaccharide sequence studies have shown that there is a restriction in pairing of the two oligosaccharides across the domains (Rademacher et al 1985) (Fig. 9). This leads to there being different monosaccharide sequences for the (α1–3) arms of the two oligosaccharides in some Fc moieties. One of these arms must always be devoid of galactose, thereby exposing the (β1–2)GlcNAc residue, which then interacts with the Man(β1–4)GlcNAc core segment of the opposing oligosaccharide chain. The (α1–3) antenna of this latter chain can extend outward between the domains with no apparent steric restrictions on the primary sequence up to, and including, the terminal sialic acid residue. Despite having identical amino acid sequences, therefore, the two heavy chains of some IgG molecules carry N-linked oligosaccharides of different primary sequence. N-linked oligosaccharides are also attached to the Fab region of IgG with a frequency and location dependent on the occurrence of glycosylation sites in the hypervariable region. The occurrence and nature of Fab N-glycosylation will lead to further diversification of an IgG. Although the role of the Fab oligosaccharide is not known, there is increasing evidence that IgG forms containing Fab oligosaccharides may be preferentially involved in IgG self-association, aggregation, and cryoprecipitation (Parekh et al 1988b).

Serum IgG from patients with rheumatoid arthritis (RA) contains the same set of biantennary oligosaccharides found in healthy individuals, but in very different proportions (Fig. 10). The occurrence of structures with outer-arm galactose is dramatically decreased, and the occurrence of those structures terminating in outer-arm *N*-acetylglucosamine correspondingly increased (Fig. 9) (Parekh et al 1985, Rademacher et al 1988a).

FIG. 6. (a) Elution of hcf-tPA from Sepharose-lysine with an arginine gradient. The tPA activity (absorbance) was monitored by the direct amidolytic assay. (b–d) Various fractions of tPA from (a) were activated with fibrin fragments and the kinetics of plasmin production monitored (dA/dt). Equivalent amounts of tPA (by active-site titration) were used in (b–d) and the activity was monitored at three concentrations (Wittwer et al 1989).

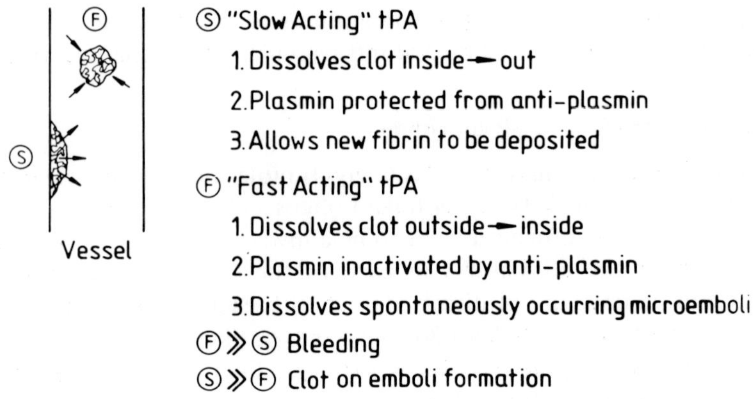

FIG. 7. Possible physiological roles and actions of the slow- and fast-acting tPA glycoforms. *In vivo* there is a continuous distribution of tPA glycoforms with activities ranging between the two idealized extremes of 'slow' and 'fast'.

The serum level of IgG(0) (IgG molecules containing only *N*-acetylglucosamine-terminating sugars) is controlled by at least two signals. Elevated levels of IgG(0) occur only in diseases characterized by an acute phase response and chronic T cell activation. (Dube et al 1989, Filley et al 1989) At present this restricts IgG glycosylation changes to patients with adult and juvenile arthritis, Crohn's disease (not ulcerative colitis) and erythema nodosum leprosum (Rademacher et al 1988a, Parekh et al 1987b, 1989b). However, transient rises in IgG(0) occur as a response to a number of infectious agents, which may indicate that the increase in this particular IgG glycoform is part of our repertoire of effector functions. Interestingly, during pregnancy, women with and without RA show a negligible concentration of IgG(0) at partum and post-partum return to either the abnormal or normal level of IgG(0), respectively (G. Rook and T. Rademacher, unpublished data). The presence of even small quantities of IgG(0) at birth may be deleterious.

Recent data have indicated that, to achieve this control over a particular IgG glycoform population, the B cell may have compartmentalized the biosynthesis of the IgG molecules (T. Rademacher, unpublished). This would ensure that only a fraction of the IgG is affected by changes in the regulated galactosyltransferase activity (Axford et al 1988, Roitt et al 1988) while other glycoproteins can be glycosylated normally (i.e. IgA, IgM, etc.).

The carbohydrate heterogeneity of IgG may therefore reflect to some extent cellular compartmentalization of IgG biosynthesis and may cast some doubt on the theory that oligosaccharide heterogeneity is due to the incomplete action of glycosyltransferases.

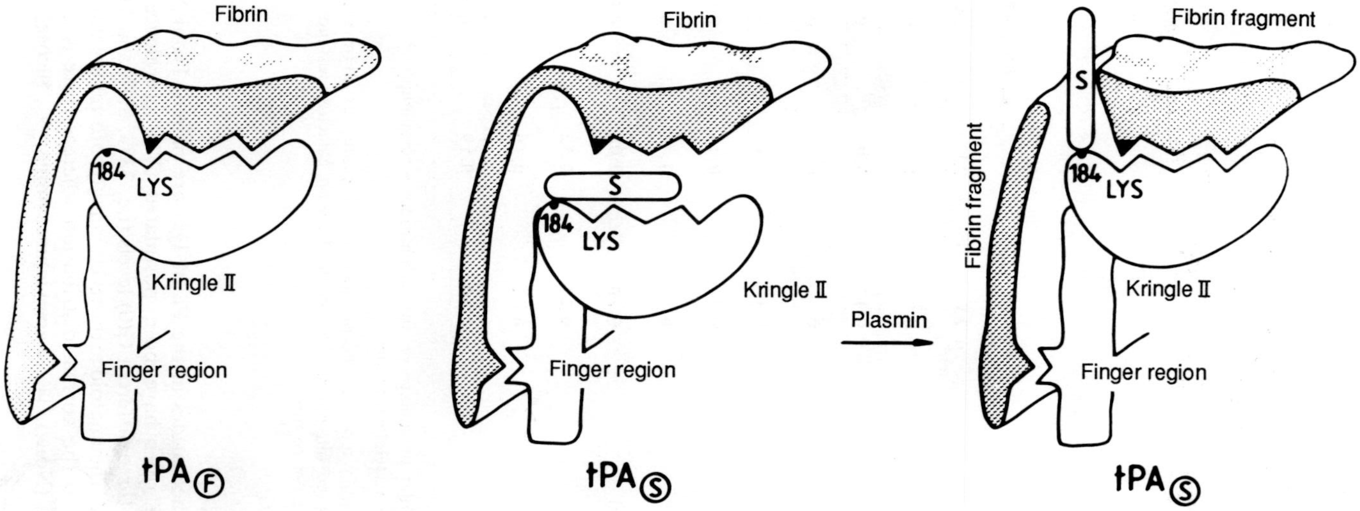

FIG. 8. Possible molecular mechanism for the down regulation of tPA activity by the presence of an oligosaccharide at site 184 in the 'slow' population. For fibrin-dependent activation kinetics to occur, binding to fibrin via the finger region and kringle II is required. The oligosaccharide at 184 (S) physically covers the kringle II fibrin-binding site and activation only occurs when small plasmin-derived fibrin fragments (generated by fast tPA) are able to displace the oligosaccharde. *In vivo* this synergy between tPA glycoforms on the same fibrin clot would result in a controlled, slow release of plasmin on the clot surface which itself would be under the control of antiplasmin.

Pairing of Human Fc Oligosaccharides

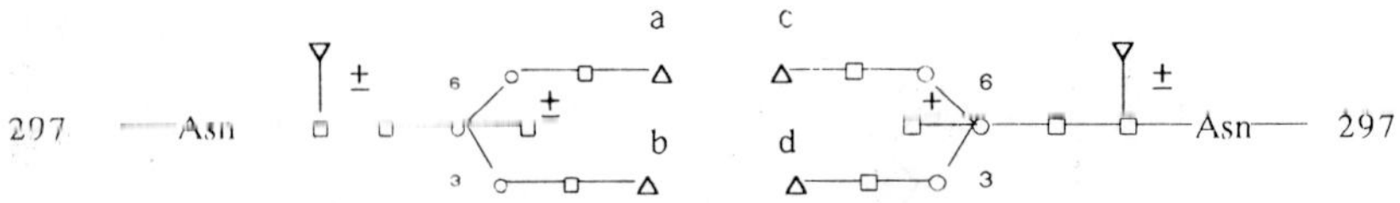

	Normal ◄	———	► Rheumatoid Arthritis
G(0) †	23	33	51
abcd	0 ‡	0	0
abc(=acd)	0.17	0.08	0
abc(=bcd)	0	0	0
ab(=cd)	0.17	0.11	0
ac	0.10	0.11	0.08
bd	0	0	0
ad(= bc)	0.10	0.08	0.05
a(=c)	0.27	0.32	0.38
b(=d)	0.09	0.10	0.10
None§	0.10	0.19	0.40

† Molar percentage of oligosaccharides terminating with N–acetylglucosamine

‡ Values represent the prevalence of a particular glycoform containing paired oligosaccharides in the Fc with galactose residues **a → d** present or absent.

§ IgG(0)

FIG. 9. The oligosaccharides in the Fc of IgG are paired. From X-ray diffraction data certain pairing rules have been postulated. The prevalence of the various glycoforms found at three different G(0) levels (i.e. 23%, 33% and 51%) is shown. In IgG from patients with rheumatoid arthritis certain glycoforms are diminished (e.g. abc, ab) when compared to the normal distribution whereas there is a large increase in the IgG(0) glycoform (None). □, *N*-acetylglucosamine; ▽, fucose; ○, mannose; △, galactose.

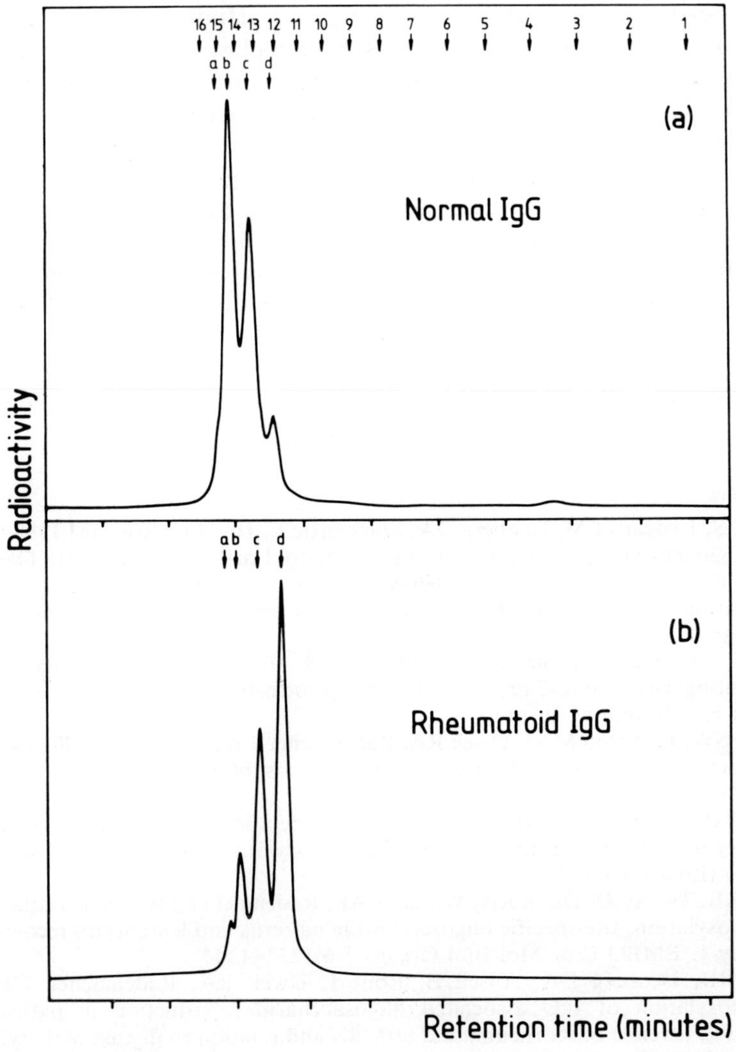

FIG. 10. Representative BioGel P-4 (~400 mesh) gel-filtration chromatograms of the asialo-oligosaccharides of total serum IgG from a healthy individual (a) and from a patient with rheumatoid arthritis (b) (from Parekh et al 1985).

Summary

In conclusion, these three examples illustrate that the analysis of dynamic changes in glycoform populations coupled with appropriate functional assays can give clues to the role of oligosaccharides in diversifying protein function.

The dispersion of *'peptide-fixed'* kinetic and affinity constants may be an important function of oligosaccharides. Such a plasticity may be necessary for proteins to operate in a variety of unexpected and non-optimal conditions (pH, temperature, etc.). Furthermore, carbohydrates attached to proteins can be used to impart seemingly subtle but important functional properties to the set of glycoprotein molecules in a *continuous* fashion (e.g. slow to fast activity tPA). Finally, future analysis of changes in glycoform populations in normal and disease states may provide us with the elusive role for oligosaccharides in biology. The shifts in the prevalence of glycoforms, such as occur in rheumatoid arthritis and may be central to the pathogenesis of this disease, clearly establish the importance of glycosylation in biology.

Acknowledgement

The Glycobiology Unit is supported by the Monsanto Co.

References

Axford JS, Lydyard PM, Isenberg DA, Mackenzie L, Hay FC, Roitt IM 1988 Reduced B cell galactosyltransferase activity in rheumatoid arthritis. Lancet II: 1486

Dube R, Rook GAW, Steele J et al 1989 Agalactosyl IgG in inflammatory bowel disease: correlation with C-reactive protein in Crohn's disease, but not in ulcerative colitis. Submitted.

Filley E, Andreali A, Steele S et al 1989 A transient rise in agalactosyl IgG correlating with free IL-2 receptors, during episodes of erythema nodosum leprosum. J Clin Exp Immunol, in press

Homans SW, Ferguson MAJ, Dwek RA, Rademacher TW, Anand R, Williams AF 1988 Complete structure of the glycosyl phosphatidylinositol membrane anchor of rat brain Thy-1 glycoprotein. Nature (Lond) 333:269–272

Parekh RB, Dwek RA, Sutton BJ et al 1985 Association of rheumatoid arthritis and primary osteoarthritis with changes in the glycosylation pattern of total serum IgG. Nature (Lond) 316:452

Parekh RB, Tse AGD, Dwek RA, Williams AF, Rademacher TW 1987a Tissue-specific N-glycosylation, site-specific oligosaccharide patterns and lentil lectin recognition of rat Thy-1. EMBO (Eur Mol Biol Organ) J 6:1233–1244

Parekh RB, Isenberg DA, Ansell B, Roitt I, Dwek RA, Rademacher TW 1987b Galactosylation of IgG associated oligosaccharides: reduction in patients with adult and juvenile onset rheumatoid arthritis and relation to disease activity. Lancet I:966–969

Parekh RB, Isenberg DA, Roitt IM, Dwek RA, Rademacher TW 1988a Age-related galactosylation of the N-linked oligosaccharides of human serum IgG. J Exp Med 167:1731–1736

Parekh RB, Dwek RA, Rademacher TW 1988b Rheumatoid arthritis as a glycosylation disorder. Br J Rheumatol 27:162–169

Parekh RB, Dwek RA, Thomas JR et al 1989a Cell type-specific and site-specific N-glycosylation of type I and type II human tissue plasminogen activator. Biochemistry, in press

Parekh RB, Isenberg DA, Rook GAW, Roitt IM, Dwek RA, Rademacher TW 1989b A comparative analysis of disease-associated changes in the galactosylation of serum IgG. J Autoimmunity 2:101–114

Rademacher TW, Homans SW, Parekh RB, Dwek RA 1985 Immunoglobulin G as a Glycoprotein. In: Kay J et al (eds) Genes and proteins in immunity. Biochem Soc Symp 51:131–148

Rademacher TW, Parekh RB, Dwek RA et al 1988a The role of IgG glycoforms in the pathogenesis of rheumatoid arthritis. Springer Seminars in Immunopathology 10:231–249

Rademacher TW, Parekh RB, Dwek RA 1988b Glycobiology. Annu Rev Biochem 57:785–838

Roitt IM, Dwek RA, Parekh RB et al 1988 The role of antigen in autoimmune responses with special reference to changes in carbohydrate structure of IgG in rheumatoid arthritis. J Autoimmunity 1:499–506

Wittwer AJ, Howard SC, Carr LS et al 1989 Effects of N-glycosylation on *in vivo* activity of Bowes melanoma and human colon fibroblast-derived tissue plasminogen activator. Biochemistry, in press

DISCUSSION

Stanley: What are the carbohydrate structures at site 184 in the slow-acting tPA? One structure is 'lying down' and one is 'standing up' in your model (Fig. 8).

Rademacher: In the slow-acting tPA there is a sugar lying over the surface of the fibrin-binding site. We assume that this sugar can move around, but we have no evidence for that. With the sugar present whole fibrin cannot bind, but small fibrin fragments may fit in under the sugar. We assume there is some form of equilibrium between the open and closed forms.

Stanley: If you compare fractions III and I of the tPA (Fig. 6), can you see structural differences in the carbohydrates that would explain your results?

Rademacher: We now need to do the entire analysis on each of the individual fractions from the lysine-Sepharose column. We have only looked at two of the main bands. To determine which glycoforms correlate with the slow- or fast-acting forms, we need to do a site analysis at 184 across the entire chromatogram.

Ruoslahti: If you deleted the glycosylation site, would all the resultant tPA be fast acting?

Rademacher: I would presume so.

Feizi: What is the structural basis of the fractionation you have done? Is it a matter of charge, size or other factors?

Rademacher: The fractionation is based on the ability of tPA to bind to lysine, because tPA binds to lysine residues on fibrin. This is well documented, and the fibrin-binding sites are known.

Feizi: On oligosaccharides?

Rademacher: No; on the actual tPA. We obtain a profile of the intact protein fractionated on lysine-Sepharose. The peptide is the same for each fraction. The molecules at one end of the profile (fast) will dissolve a clot. At the other end (slow), the molecules are very inefficient. The literature binding constant of tPA for fibrin is an average value.

Hughes: What should we be aiming for in engineered tPA? Do we need a slow acting, fast-acting or intermediate product? Can you get enough natural tPA to do fibrin chromatography, to see the fibrin-binding profile or even to do sugar analysis?

Rademacher: We obtained tPA from human colon fibroblasts which are non-transformed. In theory, they are immortalized but not tumorigenic. It is not yet possible to obtain natural tPA from human serum because it is present in extremely small quantities. We shall look at the enzymic patterns of tPA in human serum to see if they represent risk factors. In theory, a person with more slow tPA variant than fast variant will be susceptible to clots. Many publications have attempted to explain the conflicting actions of tPA; we think this may be how it happens.

Sharon: Dr Ruoslahti asked about engineered tPA in which the 184 site is not available for glycosylation. Has this been tested?

Rademacher: No, we haven't tested that engineered product.

Fukuda: Would you be confident enough to advise abolishing the glycosylation site at 184 to produce tPA for therapeutic purposes?

Rademacher: It depends what is required. In the initial bolus injection one wants to lyse every clot that is present and one does not care too much about new fibrin being deposited. The problem arises in terms of maintaining patency or in long-term therapy. In long-term therapy whether one needs a tPA that is slow or fast acting is open to debate.

Hindsgaul: In your work on IgG, have you compared the kinetics of galactosyltransferase before and after pregnancy?

Rademacher: We haven't been involved in the galactosyltransferase issue. We don't know if the amounts of IgG(0) reflect the galactosyltransferase activities in cells. This is interesting because, looking at the galactosylation of IgG with age, we find a parabolic curve, whereas Professor Kobata sees a flat curve of enzymic activity over the same age range. In the normal population, this may reflect clearance in young children and old people. Other people are looking at the role of heavy chain binding protein in this. Some of the effects we see in pregnancy may not be due to a change in galactosyltransferase activity. Heavy chain binding protein directs immunoglobulin intracellularly. The glycosylation changes are isotype specific; IgA and IgM do not reflect any of these changes, as far as we can tell.

Leukosialin, a major sialoglycoprotein defining leucocyte differentiation

Minoru Fukuda

La Jolla Cancer Research Foundation, Cancer Research Center, La Jolla, CA 92037, USA

Abstract. We have isolated a major sialoglycoprotein on leucocytes and found that this glycoprotein, termed leukosialin, is ubiquitously present on various human leucocytes (granulocytes, monocytes/macrophages and T lymphocytes). Our studies showed that leukosialin is significantly glycosylated by O-linked oligosaccharides (70 chains/molecule). The polypeptide portions of these molecules are, however, apparently the same, with a molecular mass of 38.5 kDa. The amino acid sequence derived from cDNA shows tandemly repeated O-glycan attachment sequences, and about 70% of the serine or threonine residues in the external domain are modified by O-glycans. The structures of those O-linked oligosaccharides are characteristic of each cell lineage and maturation stage. In particular, we have shown that O-glycans of leukosialin are converted from NeuAc(α2–3)Gal(β1–3)[NeuAc(α2–6)]-GalNAc to NeuAc(α2–3)Gal(β1–3)[NeuAc(α2–3)Gal(β1–4)GlcNAc(β1–6)]GalNAc during T cell activation.

1989 Carbohydrate recognition in cellular function. Wiley, Chichester (Ciba Foundation Symposium 145) p 257–276

The haemopoietic lymphoid tissue is one of the few that continuously renew themselves in the adult. Thus it is suitable for studying developmental changes. Haemopoietic stem cells and cell lineages for the various differentiation processes have been demonstrated. Pluripotent stem cells differentiate, probably by a multi-step process, into mature cells of various functions and morphology. The pathogenesis of many haemopoietic disorders can be related to abnormalities in stem cell differentiation and/or maturation processes, or in the cells that regulate these processes. Understanding the course of differentiation and maturation is therefore essential for elucidating the mechanisms of haemopoietic diseases. We have been particularly interested in the changes in saccharide structures during differentiation of haemopoietic cells. Our major concerns are to determine: (1) whether different types of mature cells can be distinguished by cell surface glycoconjugates; (2) whether carbohydrate markers can distinguish the pathways of divergence of pluripotent stem cells into terminally differentiated cells; and (3) whether carbohydrate markers can distinguish between differentiation processes in normal cells and their leukaemic counterparts.

This chapter summarizes the advances we have made since our last review (Fukuda & Carlsson 1986). I shall pay particular attention to leukosialin, which we have found to be ubiquitously present, but with significant differences in O-glycosylation, in various leucocytes (Carlsson et al 1986, Piller et al 1988a).

Isolation and characterization of leukosialin

Most nucleated blood cells derived from bone marrow possess only a limited number of heavily sialylated glycoproteins (Andersson & Gahmberg 1978). The major sialoglycoproteins on erythroid, myeloid and lymphoid cells appear to be of different sizes depending on the types of cells, but the apparent molecular masses are in the range of 95 000 to 150 000 Da, as estimated by SDS-polyacrylamide gel electrophoresis (SDS-PAGE).

We identified a major sialoglycoprotein, Gp 105, on the human erythroleukaemic cell line K562, which probably represents pro-erythroblasts (Fukuda 1980, Fukuda et al 1981). Gp 105 is a minor sialoglycoprotein on erythroblasts in culture, and is absent on mature erythrocytes (Fukuda et al 1980). Thus, we identified Gp 105 as an early-stage marker in erythroid differentiation. To characterize Gp 105 and to look for its expression on various cells, we decided to isolate it.

The glycoprotein was purified by affinity chromatography on wheat germ agglutinin-Agarose, preparative gel electrophoresis and Sephacryl S-300 gel filtration, and an antiserum was obtained against it. Cell surface glycoproteins of various human leukaemic cell lines (erythroid, myeloid, T and B lymphoid cells) were labelled by the periodate/NaB[$^3H_4$] method and immunoprecipitated by the antiserum. As shown in Fig. 1, the antiserum precipitated a sialoglycoprotein from all cell lines. For most cell lines, this precipitated sialoglycoprotein was the major sialoglycoprotein in those cells. However, the apparent molecular masses varied from 113 to 134 kDa. Because of its association with leucocytes we termed this glycoprotein 'leukosialin' (Carlsson & Fukuda 1986).

To determine whether the varied molecular mass of leukosialins from different cells arises from differences in the protein or carbohydrate part, the precursors obtained by pulse-labelling with [^{35}S]methionine of the cells were compared. The sizes of the proteins in each cell line are indistinguishable; the protein portions of these sialoglycoproteins are thus probably the same or very similar (Carlsson & Fukuda 1986).

Domain structure of leukosialin

The chemical nature of leukosialin was investigated by isolating cDNAs encoding it, using anti-leukosialin antibodies for the screening of λgt11 libraries (Pallant

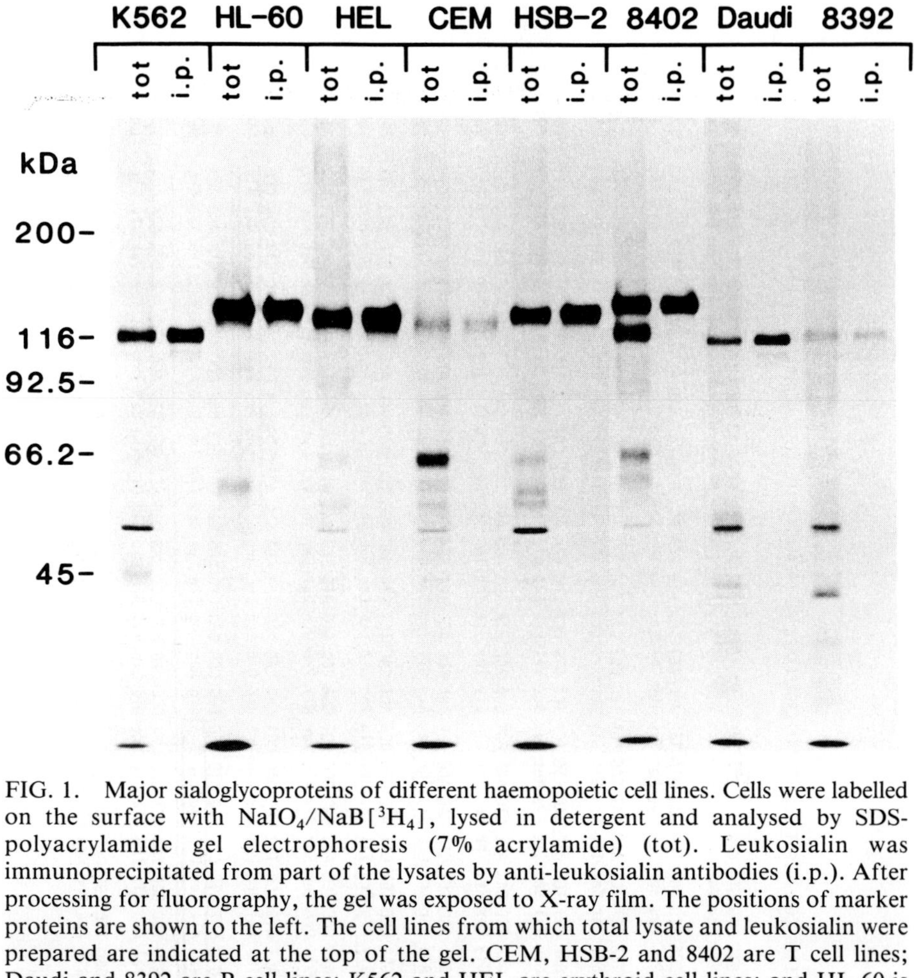

FIG. 1. Major sialoglycoproteins of different haemopoietic cell lines. Cells were labelled on the surface with $NaIO_4/NaB[^3H_4]$, lysed in detergent and analysed by SDS-polyacrylamide gel electrophoresis (7% acrylamide) (tot). Leukosialin was immunoprecipitated from part of the lysates by anti-leukosialin antibodies (i.p.). After processing for fluorography, the gel was exposed to X-ray film. The positions of marker proteins are shown to the left. The cell lines from which total lysate and leukosialin were prepared are indicated at the top of the gel. CEM, HSB-2 and 8402 are T cell lines; Daudi and 8392 are B cell lines; K562 and HEL are erythroid cell lines; and HL-60 is a myeloid cell line. (From Fukuda & Carlsson 1986.)

et al 1989). The amino acid sequences thus derived indicate that the polypeptide portion consists of 400 amino acids, of which 19 N-terminal residues form a putative cleavable leader peptide. Thus the mature protein has 381 amino acids. Residues 236 to 258 constitute a hydrophobic transmembrane portion which is flanked on the C-terminal side by a putative cytoplasmic segment starting with three arginines. The extracellular domain contains 234 amino acids, of which 93 residues are either serine or threonine. From the chemical composition and amino acid sequence it was calculated that 71% of these hydroxy amino acids were glycosylated by O-linked glycans. Some of the O-glycans are attached

```
GGAGCCTCGG GAGGTGGTGG AGTGACCTGG CCCCAGTGCT GCGTCCTTAT CAGCCGAGCC GGTCCCAGCT CTTGCTCCTG CCTGTTTGCC TGGAA    ATG GCC   101
                                                                                                        Met Ala   -18

ACG CTT CTC CTT CTC CTT GGG GTG CTG GTG GTA AGC CCA GAC GCT CTG GGG AGC ACA ACA GCA GTG CAG ACA CCC ACC TCC GGA GAG   188
Thr Leu Leu Leu Leu Leu Gly Val Leu Val Val Ser Pro Asp Ala Leu Gly Ser Thr Thr Ala Val Gln Thr Pro Thr Ser Gly Glu    12

CCT TTG GTC TCT ACT AGC GAG CCC CTG AGC TCA AAG ATG TAC ACC ACT TCA ATA ACA AGT GAC CCT AAG GCC GAC AGC ACT GGG GAC   275
Pro Leu Val Ser Thr Ser Glu Pro Leu Ser Ser Lys Met Tyr Thr Thr Ser Ile Thr Ser Asp Pro Lys Ala Asp Ser Thr Gly Asp    41

CAG ACC TCA GCC CTA CCT CCC TCA ACT TCC ATC AAT GAG GGA TCC CCT CTT TGG ACT TCC ATT GGT GCC AGC ACT GGT TCC CCT TTA   362
Gln Thr Ser Ala Leu Pro Pro Ser Thr Ser Ile Asn Glu Gly Ser Pro Leu Trp Thr Ser Ile Gly Ala Ser Thr Gly Ser Pro Leu    70

CCT GAG CCA ACA ACC TAC CAG GAA GTT TCC ATC AAG ATG TCA TCA GTG CCC CAG GAA ACC CCT CAT GCA ACC AGT CAT CCT GCT GTT   449
Pro Glu Pro Thr Thr Tyr Gln Glu Val Ser Ile Lys Met Ser Ser Val Pro Gln Glu Thr Pro His Ala Thr Ser His Pro Ala Val    99

CCC ATA ACA GCA AAC TCT CTA GGA TCC CAC ACC GTG ACA GGT GGA ACC ATA ACA ACG AAC TCT CCA GAA ACC TCC AGT AGG ACC AGT   536
Pro Ile Thr Ala Asn Ser Leu Gly Ser His Thr Val Thr Gly Gly Thr Ile Thr Thr Asn Ser Pro Glu Thr Ser Ser Arg Thr Ser   128

GGA GCC CCT GTT ACC ACG GCA GCT AGC TCT CTG GAG ACC TCC AGA GGC ACC TCT GGA CCC CCT CTT ACC ATG GCA ACT GTC TCT CTG   623
Gly Ala Pro Val Thr Thr Ala Ala Ser Ser Leu Glu Thr Ser Arg Gly Thr Ser Gly Pro Pro Leu Thr Met Ala Thr Val Ser Leu   157

GAG ACT TCC AAA GGC ACC TCT GGA CCC CCT GTT ACC ATG GCA ACT GAC TCT CTG GAG ACC TCC ACT GGG ACC ACT GGA CCC CCT GTT   710
Glu Thr Ser Lys Gly Thr Ser Gly Pro Pro Val Thr Met Ala Thr Asp Ser Leu Glu Thr Ser Thr Gly Thr Thr Gly Pro Pro Val   186

ACC ATG ACA ACT GGC TCT CTG GAG CCC TCC AGC GGG GCC AGT GGA CCC CAG GTC TCT AGC GTA AAA CTA TCT ACA ATG ATG TCT CCA   797
Thr Met Thr Thr Gly Ser Leu Glu Pro Ser Ser Gly Ala Ser Gly Pro Gln Val Ser Ser Val Lys Leu Ser Thr Met Met Ser Pro   215

ACG ACC TCC ACC AAC GCA AGC ACT GTG CCC TTC CGG AAC CCA GAT GAG AAC TCA CGA GGC ATG CTG CCA GTG GCT GTG CTT GTG GCC   884
Thr Thr Ser Thr Asn Ala Ser Thr Val Pro Phe Arg Asn Pro Asp Glu Asn Ser Arg Gly Met Leu Pro Val Ala Val Leu Val Ala   244
                *

CTG CTG GCG GTC ATA GTC CTC GTG GCT CTG CTC CTG CTG TGG CGG CGG CGG CAG AAG CGG CGG ACT GGG GCC CTC GTG CTG AGC AGA   971
Leu Leu Ala Val Ile Val Leu Val Ala Leu Leu Leu Leu Trp Arg Arg Arg Gln Lys Arg Arg Thr Gly Ala Leu Val Leu Ser Arg   273

GGC GGC AAG CGT AAC GGG GTG GTG GAC GCC TGG GCT GGG CCA GCC CAG GTC CCT GAG GAG GGG GCC GTG ACA GTG ACC GTG GGA GGG  1058
Gly Gly Lys Arg Asn Gly Val Val Asp Ala Trp Ala Gly Pro Ala Gln Val Pro Glu Glu Gly Ala Val Thr Val Thr Val Gly Gly   302

TCC GGG GGC GAC AAG GGC TCT GGG TTC CCC GAT GGG GAG GGG TCT AGC CGT CGG CCC ACG CTC ACC ACT TTC TTT GGC AGA CGG AAG  1145
Ser Gly Gly Asp Lys Gly Ser Gly Phe Pro Asp Gly Glu Gly Ser Ser Arg Arg Pro Thr Leu Thr Thr Phe Phe Gly Arg Arg Lys   331
```

```
TCT CGC CAG GGC TCC CTG GCG ATG GAG GAG CTG AAG TCT GGG TCA GGC CCC AGC CTC AAA GGG GAG GAG GAG CCA CTG GTG GCC AGT  1232
Ser Arg Gln Gly Ser Leu Ala Met Glu Glu Leu Lys Ser Gly Ser Gly Pro Ser Leu Lys Gly Glu Glu Glu Pro Leu Val Ala Ser   360

GAG GAT GGG GCT GTG GAC GCC CCA GCT CCT GAT GAG CCC GAA GGG GGA GAC GGG GCT GCC CCT TAA GTG TCG GTG AAT AGT GAG GCT  1319
Glu Asp Gly Ala Val Asp Ala Pro Ala Pro Asp Glu Pro Glu Gly Gly Asp Gly Ala Ala Pro end                                381

GGAGGCCGGA ATCTCAGCCA GCCTCCAGCA CCTTCCCTCT CACCATCCCA CTGCCCCCTC GCTCCCATGT TTCCACCCGG CACCCTGATC CTCACCCGAA TCTCC  1424

TTTTTTTTTT TCTTTTGAGA CAGAGTTTCG CTTTGTCGCC CAGGCTGGAG TGCAATGCAC GATCTCAGTT CACTGCAACC TCTGCCTCCT AAGTTCAGGC GATTC  1529

TCCTGCCTCA GCTTCCCGAG TAACTGAGAT TACAGGCACC CACCACCATG CCCAGCTGCT TTTTTGTATT TTTGGTAGAG ATGGGGTTTC ACCATGTTGG CTAGG  1634

CTGGTCTCAA ACTCCTGACC TCAGGTGATC TACCTGCCTC AGCCTCCCAA AGTGCTGAGA TTACAGACAT GAGCCTCCGC GCCTTGCCTC CTCACCCACC TCTTC  1739

ACTCTGAATC CTCATGAGGC TTCTCAGCCC TGGATTTCCT GCTGCCATCC TCACCCAGCA CCCACAACTA GCGCCTGGGC AGGGCAGGGC TGGCACCTCT CAACG  1844

TCTGTGGACT GAATGAATAA ACCCTCCTCT TACAAATGCC AAAATTCATT CAGCTTTGAT GATAAACACT GAGGCCCAAT GGCCTTTATC ATCTAGGGAG TATCA  1949

AGAATGAGCA AGAGGCTAAC TCAGCGTGAG TTACCCTGGG AAAGGAAAGA AGAAATGGTT CATATAGGAC ACACATAGAT ACCTTCAAGG GTGTTTGTAG AGTTG  2054

TTTCTTAAGT AGTTGGTTTC CTTCACAGAA AGTTCTTAAA CTCAGAATAT ACCCATCCAT GCACCCCACC AGCAATACAA AGACCCCAAA CAAAAAATTA CTATA  2159

TTCTTACCCT ACGCATGGCT TCCTCCTCTT CTTGACGCTT TTCATAATGT GCAAAGTCAT CAAAGATTGA GGTGGTATGC TTGAAAGTAG CAATTATTTT AAGCA  2264

CTTGCTTAGC TTTTTCAAGG GACG                                                                                          2288
```

FIG. 2. The nucleotide sequence and the deduced amino acid sequence of leukosialin. The predicted N-terminus of the mature protein is partly based on the identification of a limited number of residues from N-terminal sequence analysis (amino acids A, V, Q, E, P and L at residues 4, 5, 6, 12, 13 and 14, respectively). Sequences determined from two internal peptides of leukosialin are underlined. The transmembrane region (amino acids 236–258) is indicated by a box. The N-terminal amino acid is enclosed in a square and the potential N-glycosylation site is indicated by an asterisk. The tandemly repeated sequences (residues 127–198) are underlined by a dotted line and the boundaries of each repeat are indicated by a half-arrow (from Pallant et al 1989).

to tandemly repeated sequences, each sequence consisting of 18 amino acids (Fig. 2).

The cytoplasmic segment is relatively large and contains several potential sites for phosphorylation by protein kinase C. We have demonstrated that leukosialin is indeed phosphorylated by protein kinase C (Piller et al 1988a). Human and rat leukosialin (Killeen et al 1987) share extensively homologous structures in their cytoplasmic portions, which may indicate some functional significance for that segment.

The gene for human leukosialin has been localized to chromosome 16p11.2 band (Pallant et al 1989). This is also the location of the α-subunit of LFA-1, Mac-1 and p150,95, members of the leucocyte adhesion receptor family (Corbi et al 1988). It is possible that the cytoplasmic portion of leukosialin interacts with cytoskeletal proteins, permitting a signal transduction from the extracellular domain to the cytoplasmic segment and, eventually, to the cytoplasmic components of cells.

O-linked oligosaccharides on leukosialin vary with cell lineage

To study what causes the apparent differences in molecular mass of leukosialin, we analysed the structures of saccharides attached to leukosialin from K562, HL-60 and HSB cells (representing erythroid, myeloid (promyelocytic cells) and T lymphoid cells, respectively). Cells were labelled with [^{3}H]glucosamine and leukosialins were isolated by immunoprecipitation. Glycopeptides were isolated by pronase digestion of the glycoproteins and Sephadex G-50 gel filtration. The glycopeptides were subjected to alkaline-borohydride degradation and oligosaccharides were isolated by Sephadex G-50 gel filtration followed by QAE-Sephadex column chromatography. The structures of purified oligosaccharides were elucidated by specific exoglycosidase digestion, methylation analysis, and Smith degradation. HL-60 and HSB-2 leukosialins have a tetrasaccharide neutral core polymer, whereas the neutral core oligosaccharides from K562 are mostly disaccharide.

These results indicate that each erythroid or myeloid (and T lymphoid) cell line expresses a characteristic set of O-linked oligosaccharides which differ in core structure as well as in sialylation (Fig. 3). In HL-60 and HSB-2 cells, the Gal(β1–3)GalNAc core is further elongated by a Gal(β1–3)-GalNAc(GlcNAc to GalNAc) (β1–6)-*N*-acetylglucosaminyltransferase (Enzyme 1 in Fig. 3) to form GlcNAc(β1–6)[Gal(β1–3)] GalNAc, core 2 (see Brockhausen et al 1985). This branching is followed by galactosylation and sialylation to form NeuAc(α2–3)Gal(β1–4)GlcNAc(β1–6)[NeuAc(α2–3)Gal(β1–3)]GalNAc. In K562 cells, Enzyme 1 may be missing and the disaccharide core, Gal(β1–3)GalNAc, is, instead, sialylated by Enzyme 2, followed by Enzyme 3, to form NeuAc(α2–3) Gal(β1–3)[NeuAc(α2–6)]GalNAc, or it is sialylated by Enzyme 3 to form

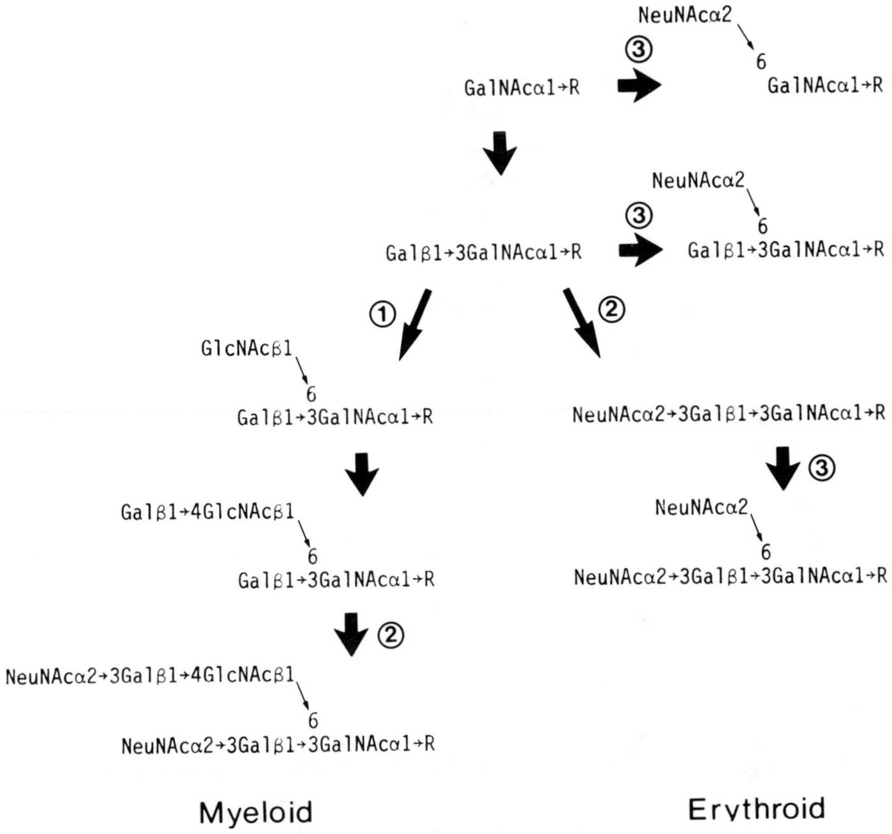

FIG. 3. Postulated scheme for biosynthesis of leukosialin-attached O-linked oligosaccharides in K562, HL-60 and HSB-2 cells. The biosynthetic steps were hypothesized with reference to Beyer et al (1979) and Brockhausen et al (1985). Enzyme 1, Gal(β1–3)GalNAc–R (GlcNAc to GalNAc)(β1–6)-*N*-acetylglucosaminyltransferase; Enzyme 2, Gal(β1–3)GalNAc–R (sialic acid to Gal)(α2–3)-sialyltransferase; Enzyme 3, GalNAc–R(α2–6)-sialyltransferase. K562, HL-60 and HSB-2 cells represent erythroid, myeloid and T lymphoid cells respectively. HL-60 and HSB-2 leukosialins contain almost the same sets of O-linked oligosaccharides (from Carlsson et al 1986).

Gal(β1–3) [NeuAc(α2–6)] GalNAc. Enzyme 3, however, may be missing in HL-60 cells, which represent myeloid cells. Thus it appears that each cell type has a characteristic set of glycosyltransferases which form O-linked oligosaccharides.

Structures of O-linked oligosaccharides depend on the maturation stages of myeloid cells

Analysis of O-linked oligosaccharides isolated from normal granulocytes, chronic myelogenous leukaemia (CML) cells and acute myelogenous leukaemia

(AML) cells showed that these different cell types contain significantly different sets of O-linked oligosaccharides (Fukuda et al 1986). In granulocytes, the major oligosaccharides are NeuAc(α2–3)Gal(β1–4)GlcNAc(β1–6)[NeuAc(α2–3)Gal(β1–3)]GalNAc and Gal(β1–4)GlcNAc(β1–6)[NeuAc(α2–3)Gal(β1–3)]GalNAc. In addition, the tetrasaccharide core is further elongated by the 'extension enzyme' (see Piller & Cartron 1983, van den Eijnden et al 1988), (β1–3)-*N*-acetylglucosaminyltransferase. In CML cells, the majority of the oligosaccharides are in a fully sialylated form—NeuAc(α2–3)Gal(β1–4)GlcNAc(β1–6)[NeuAc(α2–3)Gal(β1–3)]GalNAc. In AML cells, on the other hand, a large proportion of O-linked oligosaccharides remain as NeuAc(α2–6)Gal(β1–3)GalNAc, Gal(β1–3)[NeuAc(α2–6)]GalNAc or NeuAc(α2–3)Gal(β1–3)[NeuAc(α2–6)]GalNAc. Because CML and AML cells, respectively, represent less mature and least mature myeloid cells, these results show that structures of O-linked oligosaccharides vary in the different maturation stages along the same cell lineage.

Studies using stage-specific reactivity to a monoclonal antibody produced similar results for O-glycans attached to leukosialin in the erythroid cell lineage (Bettaieb et al 1988). This monoclonal antibody, GA3, apparently recognizes O-glycans as part of the epitope and the structure(s) recognized by the antibody are present only in certain stages of maturation.

Leukosialin as a T cell differentiation antigen

It is known that glycoproteins that resemble leukosialin are altered when T lymphocytes are activated. Kimura & Wigzell (1978) and Andersson et al (1978) found that murine or human activated T lymphocytes express a major cell surface sialoglycoprotein which is larger than its counterpart in non-stimulated T lymphocytes and also migrates more slowly when sialic acid is removed. We tested the possibility that these glycoproteins are actually leukosialin and found, as shown in Fig. 4, that activated T lymphocytes do indeed express leukosialin which is larger than that expressed in non-stimulated T lymphocytes.

The structures of O-linked oligosaccharides attached to leukosialin have been determined on both activated and resting T lymphocytes. Resting T lymphocytes almost exclusively express NeuAc(α2–3)Gal(β1–3)[NeuAc(α2–6)]GalNAc, whereas activated T lymphocytes mainly express NeuAc(α2–3)Gal(β1–3)[NeuAc(α2–3)Gal(β1–4)GlcNAc(β1–6)]GalNAc. The activities of six glycosyltransferases involved in the synthesis of these oligosaccharides were measured, and changes were found in the activities of two key enzymes: whereas resting T lymphocytes have negligible activity of (β1–6)GlcNAc transferase, activated T lymphocytes express a significant amount of the enzyme. At the same time, (α2–6)-sialyltransferase is reduced appreciably after T cell activation. Because both enzymes compete for the same precursor substrate, the coordinate changes in their activities are responsible for the dramatic change of O-glycans

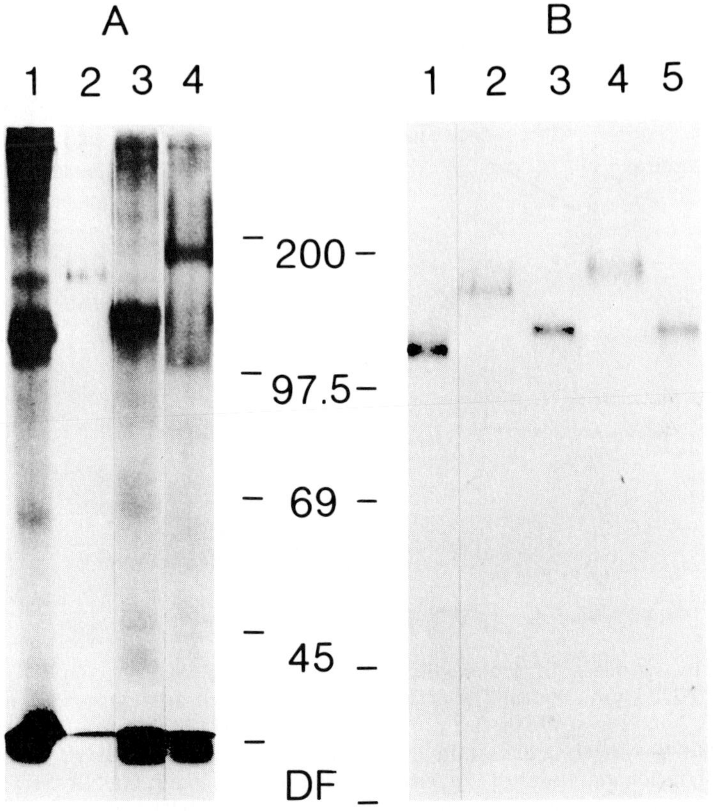

FIG. 4. Tritium-labelled glycoproteins from total cell lysates (A) and after immunoprecipitation by leukosialin-specific antiserum (B) analysed on 7% SDS-PAGE gels followed by fluorography. A, total lysates of cell surface-labelled resting (lanes 1 and 2) and activated (lanes 3 and 4) T lymphocytes from peripheral blood; B, immunoprecipitated leukosialin from cell surface-labelled resting thoracic duct lymphocytes (lanes 1 and 2) and activated T cells from peripheral blood (lanes 3 and 4) and synovial fluid (lane 5). The cells were labelled with $NaB[^3H]_4$ after oxidation with $NaIO_4$ (A, lanes 1 and 3; B, lanes 1 and 3) or after treatment with sialidase followed by galactose oxidase (A, lanes 2 and 4; B, lanes 2 and 4). In B, lane 5, the cells were labelled with $[^3H]$glucosamine. DF, dye front (from Piller et al 1988).

on leukosialin during the activation of human T lymphocytes (Piller et al 1988b) (Fig. 5).

Patients suffering from the rare immunodeficiency Wiskott-Aldrich syndrome (Parkman et al 1978) show decreased levels of leukosialin and/or alterations in its glycosylation (Remold-O'Donell et al 1984). This disease is associated with the X chromosome and in the patients the lymph nodes and spleen are gradually

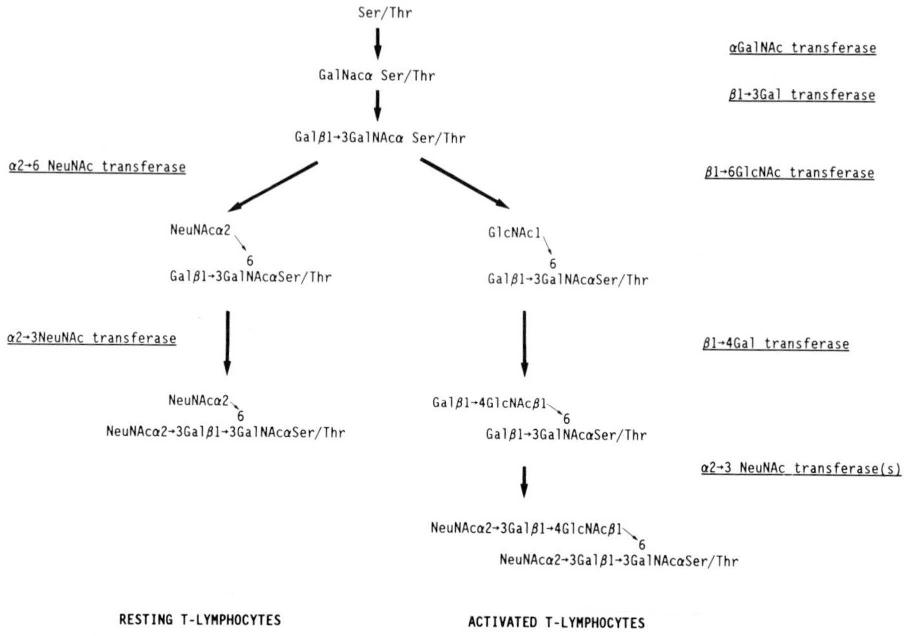

FIG. 5. Biosynthesis of leukosialin-attached O-linked oligosaccharide in resting and activated T lymphocytes. The activities of six glycosyltransferases were measured. Resting T lymphocytes express no detectable Gal(β1–3)GalNAc–R (GlcNAc to GalNAc)(β1–6)-*N*-acetylglucosaminyltransferase, whereas activated T lymphocytes express a significant amount of that enzyme. On the other hand, GalNAcSer(Thr)-(α2–6)-sialyltransferase is decreased after T lymphocytes have been activated (from Piller et al 1988).

depleted of T lymphocytes (Cooper et al 1968), suggesting that leukosialin or its carbohydrate could be involved in lymphocyte function. It has been shown that leukosialin is involved in activation by a pathway different from that associated with the T cell antigen receptor complex (Mentzer et al 1987). We have found that T lymphocytes of Wiskott-Aldrich patients show activated forms of O-linked oligosaccharides, despite the fact that these cells are not functionally active (Piller et al 1988c). As already mentioned, we have localized the leukosialin gene to chromosome 16 (Pallant et al 1989). The results suggest that the abnormality in Wiskott-Aldrich syndrome could be due to a defect in one of the glycosyltransferases, which may be coded by the X chromosome. It will be interesting to determine whether the changes in oligosaccharides during T cell activation adapt the T cells to their altered interactions with their environment or play a more direct role in the altered functions of activated T lymphocytes.

Acknowledgements

The work from our laboratory was supported by grant RO1 CA 33895 awarded by the National Cancer Institute. I would also like to thank Tami Clevenger for secretarial assistance.

References

Andersson LC, Gahmberg CG 1978 Surface glycoproteins of human white blood cells. Analysis by surface labeling. Blood 52:57–67

Andersson LC, Gahmberg CG, Kimura AK, Wigzell H 1978 Activated human T lymphocytes display new surface glycoproteins. Proc Natl Acad Sci USA 75:3455–3458

Bettaieb A, Farace F, Mitjavila M-T et al 1988 Use of a monoclonal antibody (GA3) to demonstrate lineage restricted O-glycosylation during terminal erythroid differentiation. Blood 71:1226–1233

Beyer T, Rearick JI, Paulson JC, Prieels J-P, Sadler E, Hill RL 1979 Biosynthesis of mammalian glycoproteins. Glycosylation pathways in the synthesis of non-reducing terminal sequences. J Biol Chem 254:12531–12541

Brockhausen I, Matta KL, Orr J, Schachter H 1985 Mucin synthesis: UPP-GlcNAc:GalNAc-R 3-N-acetylglycosaminyltransferase and UDP-GlcNAc:GlcNAcβ1→3GalNAc-R (GlcNAc to GalNAc) 6-N-acetylglucosaminyltransferase from pig and rat colon mucosa. Biochemistry 24:1866–1874

Carlsson SR, Fukuda M 1986 Isolation and characterization of leukosialin, a major sialoglycoprotein on human leukocytes. J Biol Chem 261:12779–12786

Carlsson SR, Sasaki H, Fukuda M 1986 Structural variations of O-linked oligosaccharides present in leukosialin isolated from erythroid, myeloid and T-lymphoid cell lines. J Biol Chem 261:12787–12795

Cooper MD, Chase HP, Lowman JT, Krivit W, Good RA 1968 Wiskott-Aldrich syndrome: an immunological deficiency involving the afferent limb of immunity. Am J Med 44:499–513

Corbi AL, Larson RS, Kishimoto TK, Springer TA, Morton CC 1988 Chromosomal location of the genes encoding the leucocyte adhesion receptors LFA-1, Mac-1 and p150,95. J Exp Med 167:1597–1607

Fukuda M 1980 K562 human leukemic cells express fetal type (i) antigen on different glycoproteins from circulating erythrocytes. Nature (Lond) 285:405–407

Fukuda M, Carlsson SR 1986 Leukosialin, a major sialoglycoprotein on human leukocytes as differentiation antigens. Med Biol (Helsinki) 64:335–343

Fukuda M, Fukuda MN, Papyannopoulou T, Hakomori S 1980 Membrane differentiation in human erythroid cells. Unique profiles of cell surface glycoproteins expressed in erythroblasts in vitro from three oncogenic stages. Proc Natl Acad Sci USA 77:3474–3478

Fukuda M, Koeffler HP, Minowada J 1981 Membrane differentiation in human myeloid cells: expression of unique profiles of cell surface glycoproteins in myeloid leukemic cell lines blocked at different stages of differentiation and maturation. Proc Natl Acad Sci USA 78:6299–6303

Fukuda M, Carlsson SR, Klock JC, Dell A 1986 Structures of O-linked oligosaccharides isolated from normal granulocytes, chronic myelogenous leukemia cells, and acute myelogenous leukemia cells. J Biol Chem 261:12796–12806

Killeen N, Barclay AN, Willis AC, Williams AF 1987 The sequence of rat leukosialin (W3/13 antigen) reveals a molecule with O-linked glycosylation of one-third of its extracellular amino acids. EMBO (Eur Mol Biol Organ) J 6:4029–4034

Kimura AK, Wigzell H 1978 Cell surface glycoproteins of murial cytotoxic T lymphocytes. I. T145, a new cell surface glycoprotein selectively expressed on Lyl-2+ cytotoxic T lymphocytes. J Exp Med 147:1418–1434

Mentzer SJ, Remold-O'Donell E, Crimmins MAV, Bierer BE, Rosen FS, Burakott SAJ 1987 Sialophorin, a surface sialoglycoprotein defective in the Wiskott-Aldrich syndrome, is involved in human T-lymphocyte proliferation. J Exp Med 165:1383–1392

Pallant A, Eskenazi A, Mattei M-G et al 1989 Identification of cDNAs encoding human leukosialin and localization of the leukosalin gene to chromosome 16. Proc Natl Acad Sci USA 86:1328–1332

Parkman R, Rapport JM, Geha R et al 1978 Complete correction of the Wiskott-Aldrich syndrome by allogeneic bone marrow transplantation. N Engl J Med 298:921–927

Piller F, Cartron J-P 1983 UDP-GlcNAc:Galβ1→4GlcNAcβ1→3 *N*-acetylglucosaminyltransferase. Isolation and characterization in human serum. J Biol Chem 258:12293–12299

Piller F, Piller V, Fox RI, Fukuda M 1988a Human T-lymphocyte activation is associated with changes in O-glycan biosynthesis. J Biol Chem 263:15146–15150

Piller V, Piller F, Fukuda M 1988b Leukosialin is phosphorylated in its endoplasmic domain. J Cell Biol 107:722a (abstr)

Piller F, Weinberg K, Parkman R, Fukuda M 1988c Resting T-lymphocytes in Wiskott-Aldrich syndrome carry O-glycans specific for activated normal T-cell. J Cell Biol 107:191a (abstr)

Remold-O'Donnell E, Kenney DM, Parkman R, Chairs L, Savage B, Rosen FS 1984 Characterization of human lymphocyte surface sialoglycoprotein that is defective in Wiskott-Aldrich syndrome. J Exp Med 159:1705–1723

van den Eijnden DH, Koenderman AHL, Schiphorst WECM 1988 Biosynthesis of blood group i-active polylactosaminoglycans. Partial purification and properties of an UDP-GlcNAc:N-acetyllactosamine β1→3 N-acetylglucosaminyltransferase from Novikoff tumor cell ascites fluid. J Biol Chem 263:12461–12471

DISCUSSION

Feizi: What do you believe the defect to be in Wiskott-Aldrich syndrome?

Fukuda: The defect could be in glycosylation. The Wiskott-Aldrich syndrome is linked to chromosome X, but the leukosialin gene is localized to chromosome 16. It is possible that one of the glycosyltransferases is linked in X.

Feizi: Could there be a depletion of lymphocytes for another reason, the cells with the activated T cell oligosaccharides arising from an attempt at replenishing? Could these simply be young lymphocytes that are analogous to stimulated lymphocytes?

Fukuda: If these are young lymphocytes, we don't expect to see the activated form. Essentially, this is a chronic activation.

Sharon: Many years ago we showed that immature thymocytes in both mouse and man will bind peanut agglutinin, whereas the mature thymocytes will not. More recently we found that the binding in the immature thymocytes is accounted for by two glycoproteins, with relative molecular masses of 110 and 180 (De Maio et al 1986a, b). The 110 kDa protein becomes sialylated and the

180 kDa one is absent in the mature cells. You have shown that leukosialin, molecular mass about 110 kDa, is present in thymocytes. Is it present in the whole cell population or is it only found in mature or immature cells?

Fukuda: We want to study that. That is why we looked at human thymocyte oligosaccharides. We didn't see clear-cut results in terms of asialylation. We see complex patterns in two almost distinct bands, but there are smeared bands between them. We are trying to separate thymocytes with different carbohydrates.

Sharon: That is easy to do with peanut agglutinin. We speculate that there may be a lectin in the thymic cortex which has a similar specificity to that of peanut agglutinin and is one of the β-galactoside-specific lectins that holds the thymocytes there as long as they are immature. With maturation, sialylation occurs and the cells are free to migrate from the cortex (Sharon 1983). Levi & Teichberg (1983) have some evidence for the occurrence of such a lectin in the thymus.

Kobata: We analysed the mucin-type sugar chains of normal and choriocarcinoma human chorionic gonadotropin (hCG) and found a marked change in the tumour hCG, from the normal Gal(β1–3)GalNAc core to a Gal(β1–4)GlcNAc(β1–6) [Gal(β1–3)] GalNAc core. Could it be a growth-related change?

Fukuda: That's a good question. This is a possibility.

Ruoslahti: Is this change in oligosaccharides found only in leukosialin or in every protein?

Fukuda: Two glycoproteins in lymphocytes, leukosialin and T200, as Dr Sharon mentioned, have O-linked glycans. We haven't looked at T200, but when a monoclonal antibody is raised against activated T cells it usually recognizes a carbohydrate epitope that T200 and leukosialin share. It is therefore likely that the proteins both show this change.

Ruoslahti: Aren't many T lymphocytes terminally differentiated rather than growing? That would contradict the suggestion that this change is associated with proliferation.

Gahmberg: In many cases there is association with growth, because this change is seen, for example, in the mixed lymphocyte reaction or after activation by lectins. In leukaemic cells you also see this higher band. On the other hand, chronic lymphocytic leukaemic cells are not activated—most cells look like resting cells and show the lower band.

Fukuda: Because chronic lymphocytic leukaemic cells show only the lower band, despite the fact that they are proliferating, I believe this phenomenon is not related to growth.

Hart: I would like to comment on Professor Sharon's suggestion about cortical lymphocytes. If you radiolabel thymocytes with purified sialyltransferases, either the one that adds sialic acid (α2–6) to Gal(β1–4)GlcNAc or the one that adds sialic acid (α2–3) to Gal(β1–3)GalNAc, without prior sialidase treatment, the only protein that is labelled is T200. The rest of the proteins, such as Thy-1, appear to be fully sialylated. If you do the same thing in mature lymphocytes or in medullary mature T cells or medullary thymocytes, you don't see any labelling in the absence of sialidase treatment. Thus, it appears that T200 may be selectively undersialylated, and its exposed galactosyl moieties may play a role in keeping immature lymphocytes in the thymus.

Sharon: We think gp180 may be related to T200 (De Maio et al 1986b).

Gahmberg: Using galactose oxidase/$NaB[^3H]_4$ labelling you see the same thing. T cells cannot be easily labelled, but thymocytes are. You see both bands.

Sharon: As I mentioned, there is also evidence for the presence in thymic stroma of a galactose-specific lectin (Levi & Teichberg 1983).

Feizi: We have some information about O-linked chains on the T200 glycoproteins of T and B lymphocytes (Childs et al 1983). On B cells, the O-linked chains chromatograph as long-chain, poly-*N*-acetyllactosamine type structures, and the glycoproteins react with anti-i. On T lymphocytes, O-linked chains with the same chromatographic profile are present, but they must be substituted in some way, because the glycoproteins have negligible i-antigen activities.

Fukuda: We see more long-chain structures in HL-60 cells than in B cells. In HL-60 cells the chains are very heterogeneous in molecular mass and the leukosialin band is smeared. The chains are probably of the poly-*N*-acetyllactosamine type.

Jessell: There have been reports that leukosialin is present in neural tissue. Have you looked?

Fukuda: We haven't done that yet, although we made a cDNA library from brain tissue. It is possible that leukosialin is expressed on both lymphoid and brain tissue, as is Thy-1.

Stanley: Does the phosphorylation state change in the activated T lymphocyte?

Fukuda: We have not studied that in detail yet. We have defined the phosphorylation sites. Phosphorylation can be inhibited by the inhibitor for protein kinase C.

Stanley: Are those *in vitro* studies?

Fukuda: No, we worked on intact cells. To define the phosphorylation site we made a synthetic peptide and incubated it in the phosphorylation system. The interesting question is whether activation can induce more phosphorylation.

Ruoslahti: Did you compare the digest of the whole protein with that peptide, to make sure that they were the same?

Fukuda: Yes, we did two-dimensional electrophoresis to ensure this.

Drickamer: At what stage during the erythroid branch does leukosialin go away? Are leukosialin and glycophorin mutually exclusive, or is there a cell type with both?

Fukuda: Leukosialin is completely absent from erythrocytes.

Gahmberg: K562 cells contain a lot of leukosialin.

Fukuda: That is a very early cell. After incubation with erythropoietin, nuclear erythroblasts contain a small amount of leukosialin. This is down-regulated during further maturation and the amount of glycophorin increases.

Ruoslahti: But are there cells that have both?

Fukuda: I think so, but I haven't tried to separate them.

Gahmberg: At least leukaemic cells have both.

Ruoslahti: But everything is always a mixed population, unless you clone or look at a single cell.

Drickamer: You could use cell sorting to separate them.

Gahmberg: Glycophorin from immature red cells has a different sugar composition from that in mature cells. It doesn't react with several monoclonal antibodies against glycophorin. When red cells mature, glycophorin A acquires more typical glycophorin sugar chains while leukosialin decreases. But you certainly find both at the same time in some cells, such as K562 ones. We have red cells that lack glycophorin A, but leukosialin is never found in mature red cells.

Sharon: Did you activate the cells using a mitogenic lectin?

Fukuda: Cells were incubated with anti-T3 antibodies and interleukin 2. Interestingly, phytohaemagglutinin induces the same effect, but concanavalin A doesn't. We also used cells obtained by mixed lymphocyte culture and they show the increase in molecular mass.

Ruoslahti: Did you look at phosphorylation when you did this?

Fukuda: The studies on phosphorylation came later—we have not finished that work.

Hughes: Do you have any ideas about function?

Fukuda: In the Wiskott-Aldrich syndrome, lymphoid cells are gradually depleted from the lymphoid system, both lymph nodes and thymus. Activated T lymphocytes do not return to the lymph nodes; they remain in the circulation. These facts suggest that if, by activation or chronic activation, the oligosaccharides change into the complex type, they are no longer recovered by the lectin in the lymphoid system, so the cells continue to circulate and eventually die.

Hughes: But why are there hundreds of carbohydrate chains per molecule? If it was a recognition function, one might expect just one or two chains to be used.

Fukuda: The interaction should be stronger because the glycoprotein is polyvalent. In addition, almost all blood cells need highly sialylated proteins to prevent them sticking together.

Gahmberg: Some human cells, such as Ena(−) red cells, do not have such proteins.

Fukuda: Those cells still have glycophorin B and C.

Gahmberg: But there are also mutants that lack both glycophorin B and A.

Fukuda: Yes, but I have never seen cells which lack sialoglycoprotein entirely.

Hughes: In the Ena(−) mutant, glycosylation of other membrane glycoproteins, particularly Band 3, goes up, which may in part compensate for the depletion of glycophorin carbohydrates.

I was struck by the similarity between glycophorin and leukosialin, in that both contain multiple O-glycosylations. I wonder if such glycoproteins are structural proteins for cells in the circulation, with some sort of protective function as they go through venules and so on. The carbohydrate moieties might not have a recognition function at all.

Fukuda: That is possible. Why then do we need a different sialoglycoprotein in leucocytes and red cells? I think it is something to do with cell shape. One way to answer this question is to continuously express leukosialin in erythroid cells. This may be possible by transfecting the leukosialin gene with an erythroid-specific promoter. Transgenic mice may provide a good experimental system.

Feizi: You mentioned the tandem repeats in the leukosialin sequence. Have you compared your sequences with sequences from epithelial mucins? Are there any homologies? Is a pattern emerging?

Fukuda: There are no homologies so far. I started work on leukosialin because I wanted to know what structures are acceptors for O-glycans. It's still not clear what is the critical signal for O-glycosylation by *N*-acetylgalactosaminyltransferase.

Stanley: Can activated human T lymphocytes from patients be further activated with lectins?

Fukuda: We tried to activate the function in several ways. In the disease state there is a discoordination between phenotypes and we can't get any expression of the T cell receptor.

Hart: One theme emerging from this meeting is the influence of protein structure on the processing of oligosaccharide chains. This selective undersialylation of specific molecules in the thymus is a good example. Identical oligosaccharide structures on molecules such as Thy-1 are fully sialylated, but on particular proteins, perhaps leukosialin and T200, they are undersialylated. Even sialyltransferases seem to be greatly affected by the protein–oligosaccharide structures.

Fukuda: There is a small difference here; Thy-1 has mainly N-linked oligosaccharides, whereas these are mainly O-linked.

Muramatsu: We are studying poly-*N*-acetyllactosamine in embryonal carcinoma (EC) cells and in early mouse embryos. These cells have a large number of poly-*N*-acetyllactosamine oligosaccharides and the major components are of extremely high molecular mass, around 10 000 kDa, which is larger than the poly-*N*-acetyllactosamine-containing proteins found in many other cells. We expect many ligands to react with poly-*N*-acetyllactosamine, which would be important in embryogenesis (Muramatsu 1988). Barry Shur proposes that galactosyltransferase is involved in the recognition, but whether galactosyltransferase is present on cell surfaces is a matter of debate. There is enzymological evidence, but it is much better to use histochemical methods.

As Ten Feizi pointed out, polyclonal antibodies previously used to demonstrate the localization of galactosyltransferase on cell surfaces were contaminated with anti-carbohydrate antibodies. Therefore we investigated the problem by preparing a monoclonal antibody to galactosyltransferase. This was done in collaboration with T. Suganuma in the Anatomy Department of our medical school. We purified the galactosyltransferase from EC cells, immunized a rat and produced monoclonal antibodies. We confirmed the specificity of the monoclonal antibody: it weakly, but consistently, inhibited galactosyltransferase activity; no anti-carbohydrate activity was detected. This antibody was used to stain a section of mouse small intestine. Electron microscopy showed that the Golgi complex was heavily stained, as expected. In EC cells, the Golgi area was stained but we also detected significant and reproducible staining on the cell surface.

Ruoslahti: Can you tell whether the galactosyltransferase is inside or outside the membrane, or can you only say it is membrane-associated?

Muramatsu: It is on the membrane, but it is hard to tell whether it is really external.

Barondes: One way to clarify this is to stain intact cells.

Muramatsu: Our cells were treated with detergent to improve accessibility. In intact cells, the staining is much weaker.

Ruoslahti: There is no reason to expect the enzyme to be inside the membrane. This galactosyltransferase has a signal sequence.

Feizi: Have you done surface immunofluorescence with viable cells?

Muramatsu: We could not obtain a clear-cut result.

Feizi: What about the surface of the intestinal cells? With rabbit antibodies to human milk galactosyltransferase the strong immunofluorescence of the stomach and small intestine epithelial cells was abolished by absorbing of the antibodies with blood group substances (Feizi et al 1987). This casts doubt on the occurrence of galactosyltransferase in concentrated form at the epithelial surface.

Muramatsu: We have looked at that, and the staining pattern is different. In particular, the villi in the distal part are not stained.

Hart: I am surprised by how this controversy lingers on. The evidence that Barry Shur (Lopez et al 1985, Shur & Neely 1988) has accumulated that galactosyltransferase is on the cell surface seems overwhelming. He has done α-lactalbumin- and antibody-binding studies. Some of Joel Shaper's antibodies were used to show that galactosyltransferase is on the surface. I am convinced there is no sugar nucleotide out there, and galactosyltransferase might not be acting as an enzyme, but I find it amazing that people still refuse to believe that the transferase is on the surface, given the evidence published in the past few years.

Muramatsu: Most of the work, including the α-lactalbumin experiment, needs independent confirmation. But our immunohistochemical data using EC cells supports the presence of galactosyltransferase on the cell surface.

Nagai: The monoclonal antibody specifically inhibited the catalytic activity of the enzyme, but polyclonal antibodies may sometimes be directed to different sites on the protein molecule.

Hart: Joel Shaper has monoclonal antibodies and has mapped the peptides that they are against.

Watkins: Jurgen Roth's work on the immunolocalization of sialyltransferase and *N*-acetylgalactosylaminyltransferase in intestinal cells has shown that the enzymes are present in mucin droplets within the cells as well as in the extracellular mucus layer (Roth et al 1986, 1988). In the colon the enzyme is definitely in surface mucin. Torken Ørntoft has demonstrated fucosyl- and *N*-acetylgalactosaminyltransferases in cell-free colonic mucin (Ørntoft et al 1987).

Drickamer: Has anybody tried to quantify the amount of protein? That seems to be the relevant question. It would surprise me if 1% of it didn't get to the cell surface: the sorting can't be as accurate as all that.

Watkins: The gold labelling in the Golgi itself and in the mucin droplets appears to be about equal in Jurgen Roth's electron micrographs, but I don't know how quantitative that is.

Feizi: Attempts have been made to fractionate plasma membranes from intestinal epithelia, in order to quantify glycosyltransferases (Boyle et al 1986). The bulk of galactosyltransferase activity was in the Golgi fraction, with less than 1% in the plasma membrane fraction.

Watkins: But for other enzymes there is not much difference between the activity of cell-free mucin and the corresponding tissue.

Fukuda: For me, it's more important that if we have an enzyme we should be able to test the function. As galactosyltransferase is a special enzyme, we should learn what it is for. We cannot do that by ascertaining whether it is on the cell surface or not.

Stanley: A system which looks quite interesting with respect to galactosyltransferase is the isoproterenol-stimulated rat parotid gland. Marchase et al (1988) have done cell fractionation and in stimulated animals they have observed a fortyfold increase in plasma membrane-associated galactosyltransferase. Their hypothesis is that this transferase is usually mostly in the Golgi with some on the cell surface, but under certain conditions the amount at the surface is increased.

Feizi: It would be nice to design methods of detecting enzyme activity *in situ*.

Ruoslahti: That's a good idea. Could you design a substrate that precipitates?

Feizi: We have resolved to do that.

Nagai: I would like to make a general point about differentiation. During differentiation, alterations of cell populations are very important. Why do particular cells survive in differentiation and developmental lineages? An unknown selective mechanism might be operating. Dr Fukuda has discussed using the carbohydrate characteristics as the marker for a particular cell population. I wonder whether the expression of a particular carbohydrate structure is associated with such a postulated cell selection, or with the accelerated or controlled growth of the cells. In the thymus, nearly 90% of the cells die. The question can be raised as to whether carbohydrate chains are associated with the patterns of proliferation and differentiation involved in the mechanism underlying cell death in the thymus.

References

Boyle F, Snape S, Duane P, Cook N, Peters T 1986 Galactosyltransferase activity is not localized to the brush border membrane of human small intestine. Biosci Reps 6:171–175

Childs RA, Dalchau R, Scudder P, Hounsell EF, Fabre JW, Feizi T 1983 Evidence for the occurrence of O-glycosidically linked oligosaccharides of poly-N-acetyllactosamine type on the human leucocyte common antigen. Biochem Biophys Res Commun 110:424–431

De Maio A, Lis H, Gershoni JM, Sharon N 1986a Identification of glycoproteins that are receptors for peanut agglutinin on immature (cortical) mouse thymocytes. FEBS (Fed Eur Biochem Soc) Lett 194:28–32

De Maio A, Lis H, Gershoni JM, Sharon N 1986b Identification of peanut agglutinin-binding glycoproteins on immature human thymocytes. Cell Immunol 99:345–353

Feizi T, Thorpe SJ, Childs RA 1987 Blood group genetic markers on human milk galactosyltransferase: relevance to the immunohistochemical approach to enzyme localization. Biochem Soc Trans 15:614–617

Levi G, Teichberg VI 1983 Selective interactions of electrolectins from eel electric organ and mouse thymus with mouse immature thymocytes. Immunol Lett 7:35–39

Lopez LC, Bayna EM, Litoff D, Shaper NL, Shaper JH, Shur BD 1985 Receptor function of mouse sperm surface galactosyltransferase during fertilization. J Cell Biol 101:1501–1510

Marchase RB, Kidd VJ, Rivera AA, Humphreys-Beher M 1988 Cell surface expression of 4β-galactosyltransferase accompanies rat parotid gland acinar cell transition to growth. J Cell Biochem 36:453–465

Muramatsu T 1988 Developmentally regulated expression of cell surface carbohydrates during mouse embryogenesis. J Cell Biochem 36:1–14

Ørntoft TF, Greenwell P, Watkins WM 1987 Activity of *H/Se*, *Le*, *X* and *A* gene-coded glycosyltransferases in normal and malignant distal colon. Proc IX Int Symp Glycoconjugates, Lille, France. Abstract E108

Roth JDJ, Taatjes J, Weinstein J, Paulson JC, Greenwell P, Watkins WM 1986 Differential subcompartmentation of terminal glycosylation in the Golgi apparatus of intestinal absorptive and goblet cells. J Biol Chem 261:14307–14312

Roth J, Greenwell P, Watkins WM 1988 Immunolocalisation of blood group A gene specified α-acetylgalactosaminyltransferase and blood group A substance in the trans-tubular network of the Golgi apparatus and mucus of intestinal goblet cells. Eur J Cell Biol 46:105–112
Sharon N 1983 Lectin receptors as lymphocyte surface markers. Adv Immunol 34:213–298
Shur BD, Neely CA 1988 Plasma membrane association, purification and partial characterization of mouse sperm β1,4-galactosyltransferase. J Biol Chem 263:17706–17714

Final general discussion

The diversity of N-linked oligosaccharides on human immunodeficiency virus

Feizi: The human immunodeficiency virus (HIV) is interesting to us because 50% of the envelope glycoprotein of this virus is carbohydrate, on some 20 N-linked chains. We are trying to establish whether these are attachment factors in the host.

We (Mizuochi et al 1988a,b) have investigated the N-linked chains of a recombinant form of the envelope glycoprotein gp120 (cloned from HTLV-III B isolate) generated in Chinese hamster ovary (CHO) cells at Genentech. The oligosaccharides were released by hydrazinolysis, fractionated by lectin affinity and Bio-Gel P4 column chromatographies, and subjected to Dr Kobata's microsequencing procedures (Mizuochi et al 1981, 1982, Yamashita et al 1982) and Dr Hakomori's methylation procedure (Hakomori 1964). By electrophoresis we found that 40% of the released oligosaccharides are neutral and the rest are acidic. The acidity is due to sialylation—sialidase treatment renders the oligosaccharides neutral.

The majority of the neutral structures are of oligomannose type (Fig. 1). The rest are an array of hybrid- and complex-type structures which we characterized after desialylation. There are hybrid structures of two types, f_3 and g_1, with one or two 'unprocessed' mannose residues, respectively. There are two types of monoantennary structures, f_1 and f_2, with and without a core region fucose residue. Among the biantennary structures g_2, h and i, those with two terminal galactose residues are the other major components (34%). There are at least 12 triantennary structures with the 6–2 and the 4–2 core region branching patterns in fractions j_1, j_2, k_2, k_3, l_2, and l_3, and at least four tetraantennary structures in fractions k_1 and l_1, with the extended *N*-acetyllactosamine sequence on some outer chains. Thus there are at least 29 structures after desialylation. Because the di- and the multiantennary structures are either fully or partially sialylated, we estimate there to be about 100 structures on this glycoprotein. This is an extraordinary diversity on a single glycoprotein species.

Microsequencing of natural gp120 derived from H9 lymphoblastoid cells infected with HIV indicates that the N-linked chains are similarly diverse. Geyer et al (1988) have reached the same conclusion from their metabolic labelling studies of natural gp120. We should bear this diversity in mind when we consider the role of the oligosaccharides in interactions with the host components at various stages of the life-cycle of the virus.

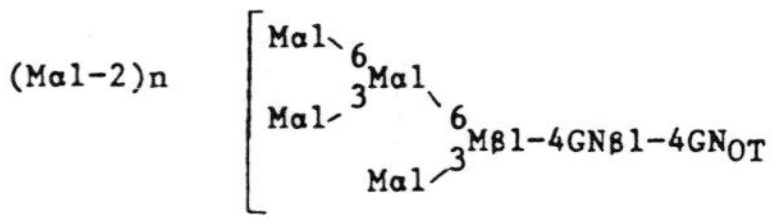

n=0 : a (3.9%)
n=1 : b (3.9%)
n=2 : c (6.9%)
n=3 : d (10.4%)
n=4 : e (7.5%)

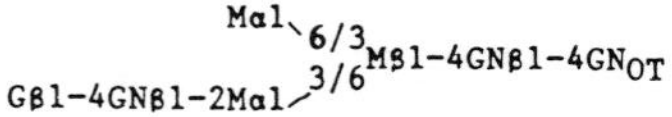

f_1 (1.1%)

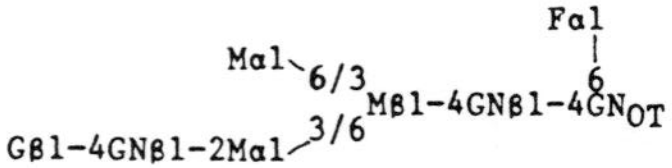

f_2 (1.3%)

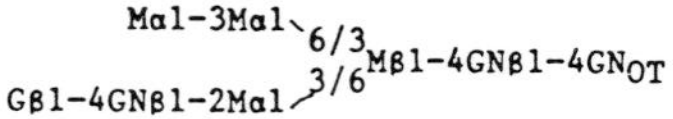

f_3 (1.3%)

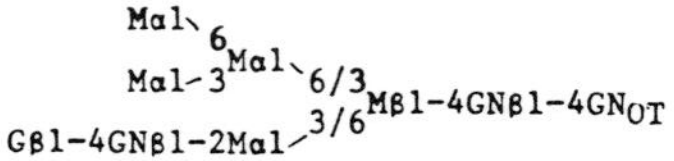

g_1 (3.1%)

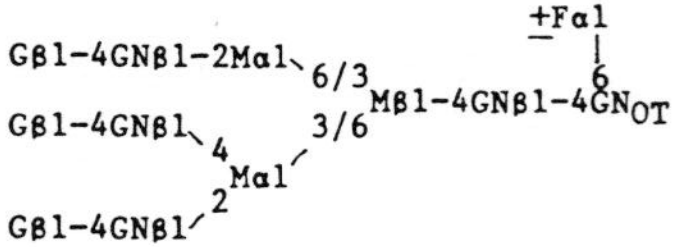

g_2 (3.1%)

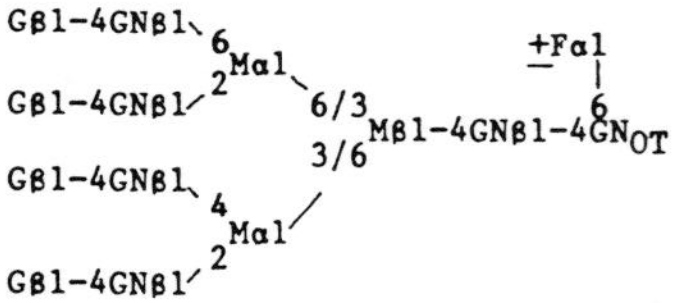

h_1 (9.3%)

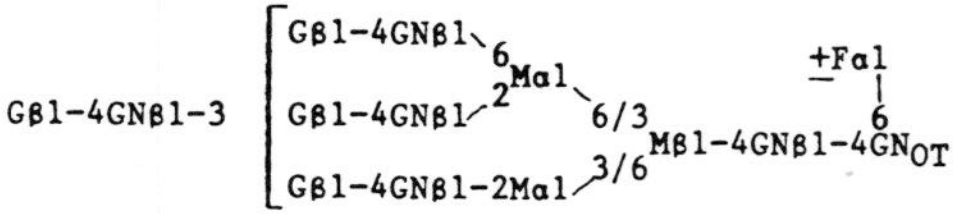

h_2 (0.8%)

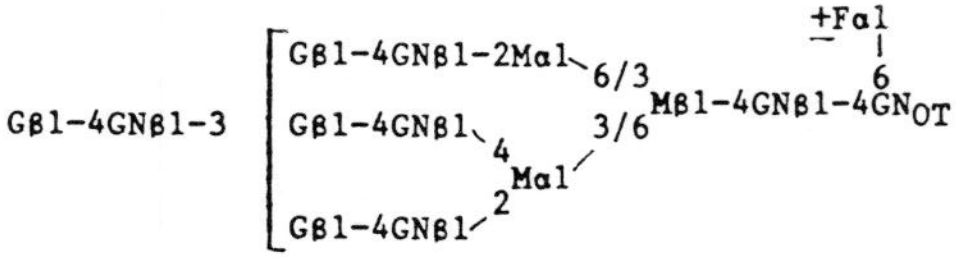

h_3 (0.2%)

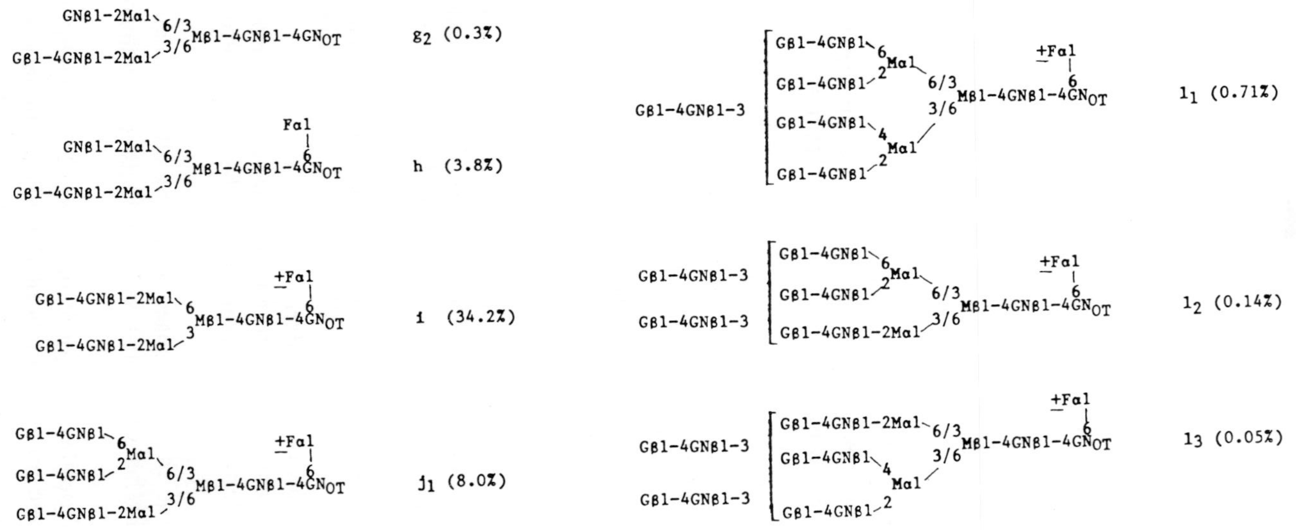

FIG. 1. (Feizi) Structural assignments for oligosaccharides isolated from the recombinant envelope glycoprotein gp120 of HIV. Designations a, b, c, f_1, f_2 etc. refer to oligosaccharides isolated by lectin affinity and gel permeation chromatographies. The molar percentages refer to the proportions of the total sialidase-treated oligosaccharides a–l. G, galactose; GN, *N*-acetylglucosamine; M, mannose; F, fucose. +F indicates oligosaccharides that are about 90% fucosylated. From Mizuochi et al (1988a).

Stanley: Could these be extended poly-*N*-acetyllactosamine chains?

Feizi: We only found evidence for two disaccharide units in series, as shown in Fig. 1. More extended structures are not present, despite their presence in other molecules on this cell line (Childs et al 1980, Li et al 1980).

Fukuda: CHO cells lack (α2–6)-sialyltransferase. If you have a tetra- or triantennary structure, the major part is always in the trisialylated form and not fully sialylated.

Feizi: These are, as far as we can tell from methylation analysis on total glycopeptides from gp120, only (α2–3)-sialylated structures.

Sharon: Dr Fukuda, I believe you recently examined erythropoietin from CHO cells and did not find much oligomannose.

Fukuda: The absence of oligomannose is not the result of lack of the enzyme. It rather reflects the accessibility for processing.

Sharon: In gp120 about half the carbohydrate chains are of the oligomannose type. This is very different from other proteins produced in the same cells.

Fukuda: That is a good example of how a protein determines the processing.

Ruoslahti: When you make a recombinant protein you push the synthesis. I wonder how much that affects the heterogeneity.

Feizi: If anything, the oligosaccharide profile of the recombinant gp120 is simpler than that of the natural glycoprotein, which is why it is taking us longer to complete our study of the latter.

Fukuda: Lysosomal membrane proteins have a similar array of oligomannose, typical complex-type N-glycans, and poly-*N*-acetyllactosamine structures.

Feizi: There are 20 potential sites and some of them must have alternative structures.

Hart: There are 23 potential sites for N-linked glycosylation and glycosylation tends to be very site-specific in terms of heterogeneity. So you would expect enormous ranges of structure. It's almost analogous to looking at 23 different glycoproteins.

Feizi: When I presented these results at an AIDS-Directed Programme meeting of the Medical Research Council, the difficulty of crystallizing such a heterogeneous macromolecule caused concern. It is possible that there are cell-to-cell variations in the oligosaccharide structures. It would not surprise me if there are many sub-species forms of the glycoprotein with different glycosylation patterns.

Carver: Only eight of the sites are conserved among all the isolates found so far. A vast amount of this heterogeneity may be irrelevant to binding of gp120 to CD4 (the cell surface target for the virus).

Feizi: Heterogeneity is altogether a major feature of gp120. Apparently, within a given individual there are many variants of the protein moiety which are changing all the time. The virus that is isolated from a patient is the one that is easy to culture in the system used.

Hart: We have data that bear on the influence of protein structure on glycosylation (Dahms & Hart 1986). Dimeric LFA-1 and Mac-1 receptors share the same polypeptide in a β-chain. We investigated how the association of a polypeptide chain would influence its glycosylation using the macrophage P388 cell line, which makes both receptors simultaneously. We found that four of the five glycosylation sites on β-chains associated with LFA-1 and Mac-1 were glycosylated totally differently, depending on the α-chain with which they were associated.

Ruoslahti: That is an interesting observation. Are you still cautious about the identity of the β-subunits?

Hart: The best evidence for the identity of the subunits are the mutations (Springer et al 1984) where all the integrins that share this β-chain are missing. We also did extensive peptide mapping and there were no detectable differences by HPLC.

Sharon: Is this reminiscent of the situation with glycoprotein hormones, where chain glycosylation affects assembly (Matzuk & Boime 1988)?

Hart: It is exactly analogous.

References

Childs RA, Kapadia A, Feizi T 1980 Expression of blood group I and i active carbohydrate sequences on cultured human and animal cell lines assessed by radioimmunoassays with monoclonal cold agglutinins. Eur J Immunol 10:379–384

Dahms NM, Hart GW 1986 Influence of quaternary structure on glycosylation. J Biol Chem 261:13186–13196

Geyer H, Holschbach C, Hunsmann G, Schneider J 1988 Carbohydrates of human immunodeficiency virus. Structures of oligosaccharides linked to the envelope glycoprotein 120. J Biol Chem 263:11760–11767

Hakomori S 1964 A rapid permethylation of glycolipid and polysaccharide as catalyzed by methylsulfinylcarbanion in dimethyl sulfoxide. J Biochem (Tokyo) 55:205–208

Li E, Gibson R, Kornfeld S 1980 Structure of an unusual complex-type oligosaccharide isolated from Chinese hamster ovary cells. Arch Biochem Biophys 199:393–399

Matzuk MM, Boime I 1988 The role of the asparagine-linked oligosaccharides of the α-subunit in the secretion and assembly of human chorionic gonadotropin. J Cell Biol 106:1049–1059

Mizuochi T, Fujii J, Kisiel W, Kobata A 1981 Studies on the structures of the carbohydrate moiety of human prothrombin. J Biochem (Tokyo) 90:1023–1031

Mizuochi T, Taniguchi T, Shimizu A, Kobata A 1982 Structural and numerical variations of the carbohydrate moiety of immunoglobulin G. J Immunol 129:2016–2020

Mizuochi T, Spellman MW, Larkin M, Soloman J, Basa LJ, Feizi T 1988a Carbohydrate structures of the human-immunodeficiency-virus (HIV) recombinant envelope glycoprotein gp120 produced in Chinese-hamster ovary cells. Biochem J 254:599–603

Mizuochi T, Spellman MW, Larkin M, Soloman J, Basa LJ, Feizi T 1988b Structural characterization by chromatographic profiling of the oligosaccharides of human immunodeficiency virus (HIV) recombinant envelope glycoprotein gp120 produced in Chinese hamster ovary cells. Biomedical Chromatography 2:260–270

Springer TA, Thompson WS, Miller LJ, Schmalsteig FC, Anderson DC 1984 Inherited deficiency of the Mac-1, LFA-1, p150,95 glycoprotein family and its molecular basis. J Exp Med 160:1901–1918

Yamashita K, Mizuochi T, Kobata A 1982 Analysis of oligosaccharides by gel filtration. Methods Enzymol 83:105–126

Chairman's summing-up

E. Ruoslahti

La Jolla Cancer Research Foundation, Cancer Research Center, 10901 North Torrey Pines Road, La Jolla, CA 92037, USA

I would like to make two general observations about the meeting. First, I have been struck by the idea that carbohydrates may increase the functional versatility of proteins. A number of speakers have emphasized this point. It makes sense to think that when a protein is transcribed from a gene it may be too perfect in its homogeneity. An unmodified protein may be good enough for the relatively constant intracellular conditions, but a protein that leaves the cell and meets various conditions, such as different temperatures, pHs and salt concentrations, may require some structural and functional flexibility. The enormous variation possible in carbohydrates might give a glycoprotein such flexibility that, depending on the conditions, the right variant would always be present. It would seem that such a relatively undefined role for carbohydrates will be difficult to prove or disprove, but the approaches that have been discussed here might enable us to begin to accomplish that.

As to the more defined functions for carbohydrates implied in the title of the meeting, I shall take the risk of offending some participants and say that I am not yet entirely convinced that carbohydrates perform such roles. Some of the systems described are promising with regard to establishing recognition functions but we need more examples elucidated at the molecular level. It seems clear that molecular genetics will be a powerful tool for generating such results. Transgenic mice and variant cells will permit observation of the consequences of, say, the deletion or addition of a carbohydrate structure at the level of the whole organism or intact cells. Analysis of carbohydrates at that level will require a great effort to clone transferases, lectins, and proteins that carry carbohydrates. That is a big task, but one that is likely to be very rewarding.

Index of contributors

Non-participating co-authors are indicated by asterisks. Entries in bold type indicate papers; other entries refer to discussion contributions

Indexes compiled by John Rivers

Subject index